SCIENCE AS A WAY OF KNOWING
The Foundations of Modern Biology

SCIENCE
AS A WAY
of KNOWING

The Foundations of
Modern Biology

JOHN A. MOORE

HARVARD UNIVERSITY PRESS
CAMBRIDGE, MASSACHUSETTS
LONDON, ENGLAND 1993

Library of Congress Cataloging-in-Publication Data

Moore, John Alexander, 1915–
Science as a way of knowing : the foundations of modern biology /
John A. Moore.
p.   cm.
Includes bibliographical references (p.   ) and index.
ISBN 0-674-79480-X (acid-free paper)
1. Biology—Philosophy.   2. Biology—History.   I. Title.
QH331.M59   1993
574'.09—dc20   92-46325
CIP

Designed by Gwen Frankfeldt

This volume is dedicated with the deepest thanks to my wife,
Betty C. Moore,
and to my friend and associate in science for nearly half a century,
Ingrith Deyrup-Olsen.
It would never have been produced without their
constant encouragement, support, and advice
based on their critical knowledge of the biological sciences.

Preface

We generally think of the Scientific Revolution of the sixteenth and seventeenth centuries as the movement that shaped the modern world. Nicolaus Copernicus (1473–1543), Andreas Vesalius (1514–1564), Francis Bacon (1561–1626), Galileo Galilei (1564–1642), Johann Kepler (1571–1630), William Harvey (1578–1657), Robert Hooke (1636–1703), Isaac Newton (1642–1727), and their like spoke for a new breed of men who rejected received truths and sought to understand the natural world in naturalistic terms. But it was a very slow revolution—two and a half centuries separated the birth of Copernicus and the death of Newton.

In marked contrast, we live in the time of a scientific revolution characterized by great speed and enormous accomplishment. During the nineteenth century, astronomy, physics, chemistry, geology, and biology began their modern development. They became more rigorous, more conceptual, and interdisciplinary. Today most major questions in science have general answers that, although not final, have been established as true beyond reasonable doubt. But science is never complete, and each new discovery produces new questions. In fact, we have now reached the point where ever more sophisticated questions can be asked and new techniques are available for seeking answers to many of them.

Science as a Way of Knowing tells the story of the development of concepts in the biological sciences—the sciences of life. The complexity of life itself has made this intellectual journey difficult. Nevertheless, in our time biology has become the most active, the most relevant, and the most personal science, one characterized by extraordinary rigor and predictive power. As the twentieth century comes to a close, we have a depth of understanding of that most distinctive and puzzling feature of life—its ability to self-replicate—that is satisfying to almost everyone, scientist and nonscientist alike. This understanding of self-replication made possible with the procedures of science has been an outstanding achievement in intellectual history. From an account of these advances in understanding in the pages that follow, we will see how long it has taken and how indirect and seemingly unrelated were the observations and experiments that eventually provided the answers.

Four major topics are developed in this volume: "Understanding Nature," "The Growth of Evolutionary Thought," "Classical Genetics," and "The Enigma of Development." Evolution, genetics, and developmental biology form the core of conceptual biology: all deal with the fundamental characteristic of life—its ability to replicate over time. Much of the material in these chapters first appeared in eight essays in the *American Zoologist* from 1984 to 1990, as part of the "Science as a Way of Knowing" project. The stimulus for that project, which was sponsored by twelve scientific and educational organizations, was the widespread feeling that human beings have become so numerous and are consuming resources so avidly that the earth cannot long continue to support our way of life. We are overwhelming the natural cycles that have made life on earth possible for well nigh 4 billion years. The "Science as a Way of Knowing" project sought to help remedy these problems by providing materials to assist in understanding. There can be no future for the human experiment unless a critical mass of involved people understands that the laws of nature constrain our activities and that our solutions to these problems must be based on knowledge and not blind adherence to fads.

The advice of many people was followed in developing the original versions of these essays. They have been acknowledged previously with sincere thanks, but I wish to note again that the advice of Betty C. Moore and Ingrith Deyrup-Olsen has been essential throughout. At Harvard University Press, Howard Boyer first suggested revising the essays for a general audience, Michael Fisher shepherded the manuscript through the review process, and Susan Wallace skillfully molded the essays into this single volume. Many thanks to all.

J.A.M.

Contents

SCIENCE AS A WAY OF KNOWING
The Foundations of Modern Biology

Introduction

N early 4 billion years ago, in the violent millennia of the Hadean world, some organic molecules achieved the ability to make more of themselves from the simpler chemical substances of the primeval seas in which they occurred. This ability to self-replicate at the expense of the environment was the beginning of life on earth. Life itself involves a conflict between two antagonistic phenomena: Self-replication can produce, in theory, an infinite number of products; yet, in reality, the world that supplies the materials for replication is finite. Life, then, is a tension between the infinite and the finite. It is inevitable that the finite prevails.

This struggle between the demands of life and the finite ambient world has resulted in the incredible variety of life with us today, as well as the wondrous fossils of life lived long ago found in sedimentary rocks. Each of the many kinds of life—each biological species—represents a different way of exploiting the living and nonliving world. Had the mechanisms of self-replication never changed, life on earth today, had it survived at all, would consist only of a uniform primeval ooze. But diversification ensued, and today there are untold and largely unknown millions of species, each representing a different way of making a living and leaving offspring. This diversification is a consequence not of a "desire" to diversify but of the imperfect nature of replication itself in a finite world.

But life has not been easy, and there have been many crises during the last 4 billion years. As recent as 10,000 years ago, North America was home to horses, camels, mammoths, mastodons, saber-toothed cats, and many other species that are no longer with us. They totally disappeared, leaving a much depleted variety of mammals. The cause is unknown. Approximately 230 million years ago another catastrophe exterminated approximately 90 percent of life in the sea. Again, the cause is unknown.

One of the greatest biocides may be occurring now; if not, it is a distinct possibility for the near future. In contrast with previous cases of mass extinction, the current one is taking place in decades, not in thousands of years. Its cause is that tension between the needs of organisms for resources and the limited quantity of resources in the environment. Until the time of man, there had always been an equilibrium between those needs and the ability of the environment to provide. Organisms do not destroy the chemical substances they use for life; they merely borrow them. For example, human beings take in food, water, and air, consisting mainly of compounds of carbon, hydrogen, oxygen, nitrogen, and sulfur, and during life eliminate these molecules in the form of carbon dioxide, water, urea, and feces. At death all the borrowed molecules are returned to the surrounding world. Green plants and microorganisms use the molecules eliminated by human beings and other animals for food and for the substances of their own bodies. Animals neither deplete the atmosphere of the oxygen required for their lives nor saturate the atmosphere with the carbon dioxide that is a waste product of their metabolism. This is because the photosynthetic reactions in the cells of green plants use that carbon dioxide and other molecules to synthesize organic compounds, leaving oxygen as a waste product. The activities of animals and green plants, then, are in such exquisite equilibrium that the concentrations of oxygen and carbon dioxide in the atmosphere are nearly constant. Both life and the environment are sustained because the environment is farmed, not mined.

Times have changed. Human beings are now making such extraordinary demands on the environment that the natural cycles can no longer provide a seemingly unlimited supply of resources. Our rate of borrowing has become too great. Tropical forests are being cut at a rapid rate for their timber or to make new agricultural land. The fertility of soils in many parts of the world is being depleted by poor agricultural practices. Species and their habitats are being destroyed by human activity at a rate never experienced before. There is overuse not only of the normally renewable resources—air, water, food, timber—but of nonrenewable resources as well. The minerals and fossil fuels that have become the basis of civilization are being consumed at a rate that will ensure their depletion before many more generations have passed.

And the waste products of civilization now exceed the ability of the environment to deal with them effectively. A few generations ago it was common practice for towns to use a local river for water and then to empty untreated sewage into it. Towns downstream would use that same river as a source of water and as a dumping ground for their sewage. This was possible because, as the saying goes, "A river purifies itself in 10 miles." The waste of the towns was food for the bacteria and other microorganisms that lived in the river and in the mud at the bottom. That bacterial purification system works so long as there are towns, not cities. If the human population becomes too large, the amount of sewage overwhelms the ability of the microorganisms to purify the water in a short distance.

Today there is another difficulty. Microorganisms can use human bodily wastes and garbage for food, but they cannot metabolize many of the industrial poisons that are now added to our old-fashioned wastes—a huge list of toxic substances emitted from mines, nuclear reactors, factories, automobiles, and homes. In modest amounts many can be handled by the environment, but when they are dumped in massive quantities, there may be unacceptable consequences.

What human beings seem often to lose sight of is that in the natural world different species form communities that enhance the life of the various species within the community. Plants provide places for birds to nest and food for many animals. Microorganisms cycle the substances from corpses, through themselves, and back to the plants and animals of the community. Natural communities are balances of interacting and interdependent species. Disruption of those balances may severely impact some or all of the members of the community.

One cannot overemphasize the importance of the interdependence and interrelations of organisms. Not only are they necessary for one another, but their activities mold the nonliving environment, making it a better place for life. For example, the composition of today's atmosphere is a product of life. The atmosphere of the primitive earth was devoid of oxygen, and organisms relied on anaerobic processes (those not requiring oxygen). When green plants appeared, the oxygen that was an end product of photosynthesis began to enter the atmosphere. Subsequently, organisms evolved that could use oxygen in their energy-producing metabolic reactions. The difference was dramatic. When glucose is used for energy, aerobic metabolism provides 18 times as much energy as does anaerobic metabolism. Today, nearly all complex animals and plants are aerobic.

Because of the size of modern cities and the amounts of waste products they produce, natural cycles can no longer maintain the purity of the atmosphere in most metropolitan areas. Waste gases are destroying the ozone layer, which protects living creatures from harmful radiations; others may result in a warming of the globe, the consequence being widespread disruptions of agriculture. The list of abuses of our life-support systems is long, disturbing, and growing.

These matters are discussed so prominently by the mass media that it is not necessary to elaborate further. The conclusion reached by many thoughtful scientists is that the demands human beings are now making on the environment simply cannot be continued.

Very difficult decisions will have to be made if we are to have a sustainable human society in a sustainable environment. Many of those decisions will require extensive knowledge of biology. We have reached the point in history, therefore, when biological knowledge is the *sine qua non* for a viable human future. Such knowledge will be especially necessary for the leaders of society—in government, industry, business, and education—but the tough decisions will have to be supported by an informed electorate. A critical subset of society will have to understand the nature of life, the interactions of living creatures with their environment, and the strengths and limitations of the data and procedures of science itself. The acquisition of biological knowledge, for so long a luxury except for those concerned with agriculture and the health sciences, has now become a necessity for all.

Material to help achieve that understanding is the subject of this book. First there will be an account of how human beings over the millennia have sought to understand nature. There will follow a survey of some of the basic areas of conceptual biology—evolution, genetics, and developmental biology.

This will be a long and detailed story so it is useful to provide a thumbnail sketch of the basic concepts of biology that will inform and provide perspective. Such an overview, even when simple, will allow one to fit the parts to the whole. Parts are best understood when they can be seen in relation to an entire structure—in this case the conceptual structure of biological science.

A Brief Conceptual Framework for Biology

At any moment of time life appears not as a continuum (as is the case with the air or the oceans) but sequestered in individuals. These individuals consist solely of the atoms that are common in the nonliving world, but those atoms are usually built into tremendously complex molecules, such as nucleic acids, proteins (including enzymes), carbohydrates, and fats. Life, then, is an expression of complex and dynamic chemical reactions. These reactions occur continuously, until ended at death, and involve chemical substances entering and leaving the body. Most of these reactions require energy, which in the last analysis comes almost entirely from the light of the sun. These complex reactions of life are programmed by the nucleic acids DNA and RNA.

The key characteristic of living organisms is their ability to make more of themselves—to reproduce—from chemical substances in the environment. The offspring resemble the parent(s) closely, as a consequence of their having received the hereditary substance DNA (and, in some microorganisms, the similar nucleic acid, RNA). Because the environmental resources available for making new individuals is finite, organisms have devised innumerable ways of maximizing the acquisition of resources. Green plants require only simple chemical substances—carbon dioxide, water, and salts plus light from the sun. Animals use for resources the bodies of green plants directly (they eat them) or indirectly (they eat animals that have eaten plants). Microorganisms use the bodies of dead plants and animals for resources. The interactions among these various sorts of organisms cycle renewable resources with such precision that they remain nearly constant.

The origin of species capable of exploiting different resources, or the same resources differently, is made possible by the nature of self-replication. The basic requirement for self-replication is precision. If an organism has the heredity makeup, the DNA, that enables it to survive in a particular environment, it is important to have that hereditary program transmitted to the offspring ("If it works, don't fix it"). But if the DNA were transmitted with complete accuracy, the organisms would be restricted to that particular environment and way of life for all time, or at least until some catastrophic change in the environment wiped out all of life. Fortunately for the history of life, replication of the genetic program is not completely accurate—there is some slight variation in the hereditary programs transmitted from parent to offspring. If in the long term these slight variations permit the exploitation of new resources—plants growing in a previously unexploited environment, marine animals able to live on land, normally free-living species becoming parasites, and so on—new places and ways of living become available. Over time, through a long series of variations, organisms with different structures, physiologies, and behaviors evolve. Every distinct species should be viewed as an experiment, successful for the moment, in coping with a specific habitat in a specific way.

Thus, built into the phenomenon of organic reproduction is a constant pressure to expand and to diversify life. The result is that, today, there are uncounted millions of species—each representing a slightly different way of obtaining resources and, hence, of living. Three major fields of biology are concerned with this basic phenomenon: genetics, developmental biology, and evolution.

Genetics deals with the ultimate physical basis for life—the genetic program that controls what an individual is and does, changes in that program, and the transmission of the program to the next generation. Throughout the world of life the hereditary program consists of four kinds of nucleotides of DNA (rarely RNA) arranged in a specific linear structure—in a chromosome—like letters and words of a very long book. Functional units of DNA are known as genes. During reproduction the DNA is normally replicated with precision, but, rarely, changes in the arrangement of nucleotides occur. These changes are mutations that may program an individual with a somewhat different anatomy or physiology.

A more common source of variation is a result of sexual reproduction. Reproduction, especially in the more complex plants and animals, usually involves the interaction of females and males. During sexual reproduction there is a shuffling of genes in the formation of ova in females and sperm in males. Further genetic variability arises when genetically different kinds of ova and sperm unite.

Cells are the basic structural and functional elements in organisms. They represent the simplest level of organization that can exist independently. Ova and sperm are highly specialized cells, and their union, in fertilization, can produce a new individual, with a genetic program similar to but not exactly like that of the parents.

Developmental biology deals with the events and processes that convert the single cell formed at fertilization into an adult individual that may consist of thousands or millions of cells. The basic problem of development is how is it possible for a single cell, with a single genetic program, to produce an adult with dozens or hundreds of different types of cells. The general answer is that, at different times in the course of development, different genes become active in different cell types.

Since many more offspring are produced than can secure the resources to survive, any individuals with a genetic program that better equips them to obtain resources will have an advantage over other individuals. Those better-equipped individuals are said to be "selected." **Evolution** consists, then, of the natural selection of genetic variations that give individuals a better chance of surviving and leaving offspring. Over the course of time, the individuals of a generation will come to differ greatly from their remote ancestors.

Every tree, shrub, insect, and bird has an ancestry that extends back 4 billion years to the time when life first emerged. Over the course of that long ancestry, the forms of life have diverged, and every species represents a unique way of living. The bird is not better than the bee or the oak tree. Each began from a tiny fertilized egg or a bud of some sort and underwent a complex development, guided by the hereditary program of the species, to reach the adult state. Every organism is part of an interacting and interdependent community of life. Life provides for life.

Such a thumbnail sketch of how we understand life is a relatively recent achievement. Although the antecedents of this understanding go back to the dawn of history, most of it is a product of twentieth-century science. Before turning to a detailed account of evolutionary biology, genetics, and developmental biology in the nineteenth and twentieth centuries, we will first take a look at how science became a way of knowing the natural world from prehistory to the age of Darwin.

part one

UNDERSTANDING NATURE

chapter 1

The Antecedents of Scientific Thought

The natural world on which all life depends has been an enigma over the long course of human history. In some respects the ancient Sumerians, at the dawn of Western civilization, probably knew more about it than does the average city dweller today. The rise of civilization has been paralleled by an ever-decreasing personal contact with nature, despite our species' dependence on living organisms for food, fiber, medicines, tools, timber, and other resources.

The recent breaking of that direct contact with nature for so many people has had enormous consequences—life takes on new directions and new meaning when human beings no longer have the possibility of obtaining their food and other resources directly. For many people living in the more developed nations there is little firsthand experience with wind, rain, planets and fixed stars, the phases of the moon, natural communities of plants and animals, the rise and fall of the tides, and the cycle of the seasons. Long ago nearly all human beings were familiar with these phenomena when their lives depended on their personally obtaining resources from nature. But beginning about 5,000 years ago, when the first cities came into existence, the percentage of those securing their food directly from hunting, gathering, or farming slowly decreased—and so did their experiences with nature.

This decrease was gradual even in the technologically advanced nations. The first census of the United States, in 1790, counted 202,000 urban individuals, compared with 3,728,000, or 95 percent, living in rural areas. In 1910 the rural population was still slightly greater, 54 percent of the total, than the urban. More recent statistics divide the rural population into farm and nonfarm. By 1960 the farm population had dropped to 8.7 percent, and by 1988 to 2.0 percent, of the total population.

Animism, Totemism, and Shamanism

The manner in which human beings attempt to understand living creatures depends upon how they answer the question "What is the real world?" Historically there have been two answers, essentially incompatible. One is that the real world consists of matter, energy, and obvious things; the other answer is that the real world resides beyond the obvious. As pointed out by Sir James Frazer (1926, p. 3), one of the founders of modern anthropology: "On the one view, the world is essentially material; on the other, it is essentially spiritual. Broadly speaking, science accepts the former view, at least as a working hypothesis; religion unhesitatingly embraces the latter." We cannot know how prehistoric human beings answered these questions, but we can speculate on the basis of surviving artifacts and on what today's anthropologists tell us about the beliefs of the peoples they study. The various sources of information do suggest some general conclusions, namely, that there was a point of view that united the real world of the scientist with the world of hope, imagination, emotion, religion, and the supernatural.

This conclusion is based on the fact that, among the preliterate societies studied during the past two centuries by anthropologists, similar patterns of belief are found worldwide: for each "thing" in the natural world of our senses, there is a supernatural force—spirit, soul, specific energy, life—that accounts for the nature and behavior of the "thing." In its most general form this belief postulates a specific spirit in every rock, mountain top, animal, plant, spring, breeze. Thus, a rabbit or a tree or a spring owes its specific nature to a rabbit-spirit, tree-spirit, or a spring-spirit. A rabbit behaves like a rabbit because it has a rabbit-spirit. This pattern of thought, which says no more than that a rabbit behaves like a rabbit because it is a rabbit, does seem to lack a certain logical economy. Nevertheless, this ancient mode of explanation is accepted by many people even today as a satisfying way to understand the phenomena of nature.

This system of belief is known as *animism*. The term is derived not from "animal" but from a root common to both—the Latin *anima*, meaning soul or life. Animism sees a duality in the phenomena of nature: there is the phenomenon and the spirit of the phenomenon. Animism avoids the problem of seeking natural *or* supernatural explanations; spirit and substance are aspects of the same thing.

This way of looking at the world once prevailed on all continents, and it is important to ask why. One hypothesis could be that these ideas arose in one place and slowly spread throughout the world over many millennia. A more likely hypothesis, I believe, is that a generalized animism is a natural way for preliterate people to deal with such otherwise inexplicable phenomena as death, dreams, sleep, apparitions, illness, unconsciousness—and even aspects of thought itself. This is the analysis of Edward B. Tylor, who is considered, with Frazer, to be one of the founders of anthropology.

> The idea of the soul which is held by uncultured races, and is the foundation of their religion, is not difficult to us to understand, if we can fancy ourselves in their place, ignorant of the very rudiments of science, and trying to get at the meaning of life by what the senses seem to tell. The great question that forces itself on their minds is one that we with all our knowledge cannot half answer, what the life is which is sometimes in us, but not always. A person who a few minutes ago was walking and talking, with all his senses active, goes off motionless and unconscious in a deep sleep, to awake after a while with renewed vigor. In other conditions the life ceases more entirely, when one is stunned or falls into a swoon or trance, where the beating of the heart and breathing seem to stop, and the body, lying deadly pale and insensible, cannot be awakened; this may last for minutes or hours, or even days, and yet after all the patient revives. Barbarians are apt to say that such a one died for a while, but his soul came back again. They have great difficulty in distinguishing real death from such trances. They will talk to a corpse, try to rouse it and even feed it, and only when it becomes noisome and must be got rid of from among the living, they are at last certain that the life has gone never to return. What, then, is this soul or life which thus goes and comes in sleep, trance, and death? To the rude philosopher, the question seems to be answered by the very evidence of his senses. When the sleeper awakens from a dream, he believes he has really somehow been away, or that other people have come to him. As it is well known by experience that men's bodies do not go on these excursions, the natural explanation is that every man's living self or soul is his phantom or image, which can go out of his body and see and be seen itself in dreams. Even waking men in broad daylight sometimes see these human phantoms, in what are called visions or hallucinations. They are further led to believe that the soul does not die with the body, but lives on after quitting it, for although a man may be dead and buried, his phantom-figure continues to appear to the survivors in dreams and visions . . . Here then in few words is the savage and barbaric theory of souls, where life, mind, breath, shadow, reflection, dream, vi-

sion, come together and account for one another in some such vague confused way as satisfies the untaught reasoner . . .

It may have occurred to some readers that the savage philosopher ought, on precisely the same grounds, to believe his horse or dog to have a soul, a phantom-likeness of its body. This is in fact what the lower races always have thought and still think, and they follow the reasoning out in a way that surprises the modern mind, though it is quite consistent from the barbarian's point of view. If a human soul seen in a dream is a real object, then the spear and shield it carries and the mantle over its shoulders are real objects too, and all lifeless things must have their thin flitting shadow-souls. (Tylor 1881, pp. 342–343, 346)

Today's reader, who is less likely than the Victorians to think in terms of "savage" and "barbarian," may take pride in the fact that human beings at a preliterate stage of culture could develop such a consistent view of self and nature when both philosophy and science were disciplines of the distant future.

We have in animism the beginnings of the way our remote ancestors sought to understand what for them could not be understood and to control that which could not be controlled. To them the world was dominated by unseen supernatural forces, the spirits or souls, that were part of all the phenomena of the world that could be seen, heard, and felt. The spirits were the true essences—they had a life of their own, perhaps an immortal one, that survived demise and decay. By and large spirits were to be feared; calamities were their doing. In some instances, however, they could be propitiated by offerings.

Totemism is a social system in which clans and the members of the same clan choose some living or, rarely, nonliving object as a focus of reverence and source of protection. Usually the totem is regarded as the ancestor from which all members of the clan are descended. The forebear of the bear clan was a bear. Totemism has been found among preliterate peoples throughout the world, being especially well developed in North America and Australia. The name itself is derived from the language of the Ojibwa of North America. The totems vary with the locality, some of those of Native Americans being turtle, bear, wolf, crayfish, carp, dog, crane, buffalo, snail, beaver, eagle, pigeon, snake, coyote, turkey, and raven. It is not unusual for clan members to dress for ceremonies in such a way as to resemble the totem and to imitate its behavior in dances.

Clans can have nonliving totems as well: thunder, lightning, ice, water, floods, wind, rain, bone, or rainbow, for example. Every tribe has at least two clans, as there is a strict prohibition of sexual relations or marriage among members of the same clan. The penalty for ignoring these prohibitions can be death.

Shamanism is still another belief system that invokes supernatural phenomena and suggests how preliterate peoples of recent times and, by extension, of prehistoric times thought and believed. Shamanism was best developed in Siberia, but elements occur throughout the world. The shaman ("he who knows") is a tribal medicine man who, through the ability to achieve ecstasy, can leave his body and undertake various assignments. In societies where shamanism is practiced, it is believed that the shaman's skill is inherited.

The shaman's main role is to cure sickness, which results from an ill person's soul leaving the body. The shaman is often able to pursue and capture the departed soul and return it to its rightful owner. However, if all procedures fail and the patient dies, the shaman accompanies the soul on its long and difficult journey to the other world. As part of this ceremony the shaman, during his ecstasy, recounts in graphic detail the events of the journey.

Because of the worldwide distribution of animism among preliterate people, as well as the fact that it was the dominant form of belief when historical records first become available, it is reasonable to assume that it had a long prehistory—possibly back to the time when human beings first sought to "explain." Less is known about totemism and shamanism, but they too must be very ancient.

The Paleolithic View

We now switch from modern anthropology to an earlier time and another sort of evidence—bones, tools, and other artifacts—that may throw some light on how prehistoric human beings thought about nature.

The earliest evidence linking animals and human beings consists of the broken bones associated with early human habitations such as caves or hearth sites (the use of fire dates to about 1.5 million years BP = "before present"). Seemingly the bones were broken deliberately, which suggests that the marrow was eaten. About all we learn from such data is that primitive human beings experienced hunger and liked marrow.

The earliest stone artifacts so far discovered date from about 2.5 million years BP (Gowlett 1984). Various species of *Australopithecus* lived at that time and probably *Homo* as well. The details are still open to question, based on interpretations of new discoveries of bones and artifacts. It is clear, however, that marked climatic changes in northern Europe and North America about 1.6 million years BP had tremendous effect on human history. Great ice sheets repeatedly advanced and retreated during this Pleistocene epoch (to use the geological term), ushering in the Old Stone Age, or Paleolithic Age. The Paleolithic ended about 10,000 years BP, when farming began.

Homo sapiens sapiens, the subspecies of humanity to which we belong, dates from about 30,000 BP in Europe, 40,000 BP in the Middle East, 50,000 BP in China, and possibly more than 100,000 BP in Africa. This Late Paleolithic period in Europe saw the rise of the Cro-Magnon people, who were highly skilled in making tools from stone, bone, and presumably wood and other plant material (most of which would have decayed). The discarded bones at the campsites of these hunter-gatherers give us evidence of their vertebrate food—mainly horse and reindeer.

Representational art began at this time—about 30,000 BP. The many objects so far discovered can tell us something about what these early human beings thought—and they thought a great deal about animals: those they ate and those they presumably feared.

Two important sites are the caves of Lascaux in France and of Altamira in Spain. The radiocarbon date for Altamira is 13,540 BP; Lascaux is thought to be about 2,000 years older. The Lascaux cave was rediscovered in 1940 when a dog slipped into a hole. The youths accompanying the dog went into the hole to rescue the animal and found themselves in a large cave. The main chamber was about 30 meters long, 10 meters wide, and 7 meters high. The walls were covered with more than a hundred paintings—in black, yellow, and red. Lamps, in the form of simple saucers to hold oil and a wick, were scattered about the floor of the cave. Similar lamps were used as late as the early twentieth century by the Eskimos and Aleuts of Alaska.

The paintings themselves have a primitive strength and economy of line that appeals to the modern eye. The fact that the painters of the Lascaux caves can be characterized as great artists—nearly 200 centuries after they lived—points up an important difference between the humanities and the sciences. Great art is eternal; great science tends to be replaced by greater science. Cro-Magnon artists at the beginning of Western art, and Homer at the beginning of Western literature, can hold their own with painters and writers of today. By contrast, today's physics is far better than that of Isaac Newton, and our evolutionary biology is far better than that of Charles Darwin—in spite of the magnificent achievements of those notable Englishmen. Ezra Pound (1910) was not speaking for the sciences when he said, "All ages are contemporaneous."

We generally accept that the work of artists reveals the thoughts and passions of their times, so what was it that the Cro-Magnon artists of Lascaux painted? Animals—wild horses, wild oxen, deer, ibex, bison, cave lions, plus a few bears, birds, rhinoceroses, wolves, and human beings. Nearly all are now extinct. Many of the animals have been transfixed by spears, darts, or arrows. The one human portrayed is

1 *Paleolithic cave art. A drawing of a stag from the Altamira Cave in northern Spain, estimated to have been made nearly 14,000 years ago. This and Figure 2 are from sketches by the renowned historian of ancient man, the Abbé Henri Breuil.*

either dead or dying. Towering over him is a huge bison, with a spear that has passed entirely through its body. The bison has been eviscerated, and loops of the intestine are exposed. This wound might have been made by human hunters or by the rhinoceros that walks off to the left, defecating.

There has been endless discussion of the "meaning" of these cave paintings. The fact that they are so excellent, and in a place so difficult for the painters to work, suggests that they must have some profound significance. They are not casual doodles. Are they no more than representations of the local fauna and food animals? Do they represent cult figures? Could it be that by portraying an animal transfixed by a spear, the hunter imagined that he would be more successful in the hunt? Does the paintings' location, deep in a dark cavern, suggest that they were involved in religious ceremonies? Or could it be that these large-brained hunters simply liked to paint? There are no sure answers, and it is hard to see how a definitive answer could ever be obtained.

In any event, an incredible step toward civilization was taken by Paleolithic human beings at Lascaux and elsewhere. Imagine what an intellectual achievement it was to show that the objects of nature could be represented symbolically by a few lines. Think also of what this meant for communication and preservation of information. Here is an early stage of symbolic representation that went on to pictographs and then to an alphabet. There are nouns at Lascaux—horses, bison, and deer. There are also verbs—to hunt, to die, to swim, and to walk.

Before Lascaux was discovered in 1940, the most famous Late Paleolithic cave paintings were those of the Altamira cave in northern Spain. As was the case at Lascaux, it was discovered by a dog, in 1868. The ceiling of the Altamira cave is very low and thus the cave lacks the grandeur of Lascaux. Nevertheless it has magnificent paintings. Nearly all are of large mammals: bison, horses, deer, ibex, and, less commonly, oxen, boars, elk, and wolf. The portrayals are generally incised outlines that are filled in with red, black, and other pigments.

When all Late Paleolithic art forms are considered—paintings, carvings, incised figures, and sculpture—the same themes dominate. There is an overwhelming concern with the larger mammals. Human beings are represented infrequently and, when they are, in a crude and often distorted manner. The human head, so important in later art, is sometimes so strangely portrayed in males that, had it not been attached to something vaguely resembling a body, proper placement in the scheme of classification would be in doubt. Females fared somewhat better in

Late Paleolithic art. One common style is the Venus figurine. This is a female typically portrayed with huge breasts, hips, and buttocks and often a small and nearly featureless head. Many are clearly pregnant. Not surprisingly, it is generally agreed that the Venus figurines are fertility symbols.

The very earliest artifacts so far discovered were strictly utilitarian. They consist of weapons for hunting and protection as well as tools for domestic purposes: darts, spears, bows, arrows, clubs, axes, hammers,

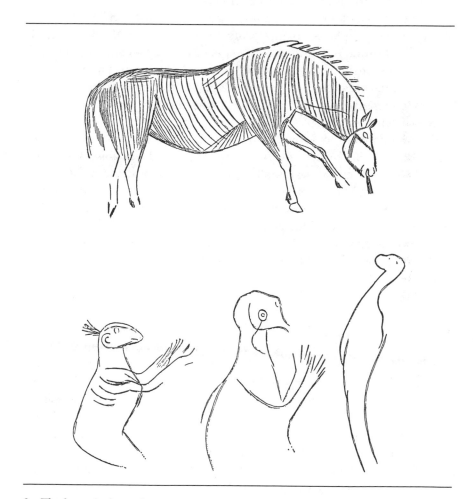

2 *The horse is from the cave at Marsoulas in France. The figures shown below it are from Altamira and are thought to be human beings.*

knives, scrapers, awls, and needles. The Late Paleolithic objects that we identify as "art" may reflect a major change in the way our ancestors looked upon nature. Thus, in Late Paleolithic times, when *Homo sapiens sapiens* arrived in Western Europe, something was added to culture that was not directly utilitarian. For whatever reason, these early people began to make objects that we call "art." A tool that has been engraved is no more effective than one that is not. And surely one cannot claim a direct utilitarian value for the painted walls of Lascaux or an ivory Venus figurine. Both may have had great symbolic value and, if so, that is an important step toward humanity.

Just as animals had occupied an important position in the thoughts of Paleolithic people, so plants came to dominate in Neolithic times. The domestication of plants was the spectacular achievement of human beings at the onset of the Neolithic period. This new and more reliable source of food was seen as a gift of the gods. In some instances the gift was indeed personal—the food plant was thought to have sprouted directly from the body of an immolated deity.

Plant foods were to make civilization possible. Slowly the improved techniques of the farmers made a surplus of food possible, and some plant foods, the cereals especially, could be grown in excess of immediate needs and stored for long periods. Thus there need be no continuous search for food as for the Paleolithic hunter-gatherers, and settled life was possible. Food for a clan could be raised on a much smaller plot of land than would be necessary for gathering wild plants and hunting animals. Thus communities could be established and, in time, cities; the root of civilization became feasible.

Mesopotamia

By approximately 3000 BC symbols that were used millennia before, such as those in cave paintings and engravings on stone and bone, had evolved into the cuneiform writing of the early Sumerians. This had been a slow evolution that began with crude drawings, for example of a horse, and ended as a group of wedge-shaped incisions on a tablet of wet clay, having lost all resemblance to a horse. Nevertheless, each group of incisions would have a constant relation to some thing and hence would serve the vital functions of communication and preservation of information. These cuneiform tablets were dried and became most durable—many of them have been discovered by archaeologists.

Another millennium and a half, to about 1400 BC, passed before a true alphabet appeared, that is, a small group of symbols to represent the individual sounds of a word. The economy of an alphabet is tremendous. The five letters of "horse" can be combined in innumerable ways to convey bits of very different information: she, he, hero, eros, roe, rose, ore, sore, shore, her, shoe, hoe, so, and so on. The vocabulary of preliterate people was probably very small, but, once writing became widespread, it would have become much richer. Literacy is autocatalytic. And at that point in history it became possible for future generations to know what their ancestors really thought—that is, if they wrote it down and if the tablets were preserved.

The Sumerians and other people of ancient Mesopotamia—present-day Iraq—have left us a rich literature that tells us how they looked upon nature. In this region, dominated by the Tigris and Euphrates Rivers, are the remains of the oldest known civilization, antedating that of Egypt, China, and the Indus Valley, and the one that, more than

3 Petroglyphs made by Native Americans. These are part of a large number on the rocks of Big Petroglyph Canyon, China Lake, California. They vary in age from a few hundred to a few thousand years old. Mountain sheep, which were hunted for food, were a favorite subject.

any of the others, has influenced Western civilization. What do we find in their writings about their view of the natural world?

The Sumerians believed that nature is divine and that human beings and other creatures were created by divine forces. This point of view, which remains robust to this day, is very different from what we accept in the sciences. The sciences make a sharp distinction between what is "natural" and what is "supernatural." The first refers to the objects and processes that obey the impersonal laws of nature. These laws involve constant relationships between causes and effects. They are invariant —given the same circumstances and conditions, a specific cause will be followed always by a specific effect. Neither human desires nor forces outside of nature can affect that outcome. We contrast this scientific view with explanations that involve supernatural forces that may be capricious and so do not obey invariant rules. These can never be studied by the basic approaches of science, which are observation and experiment. We are here in the realm of belief, not rational science. Belief is a pattern of thought that has characterized the human mind over all of its history, for it is comforting for many to accept that there are forces far more powerful than those available to human beings— forces that control the destinies of individuals and nations. This is a mind-set, however, that has been singularly unsuccessful in furthering an understanding of natural phenomena—the task of science.

For the Sumerians, nature and the divine were essentially the same. Like Christian fundamentalists today, they too had creation myths that accounted for the world as a whole. The goddess Nammu, representing the sea, produced two offspring asexually: An, the masculine god of the sky, and En-ki (or Ea), the female goddess of the earth. They reproduced in the standard Sumerian manner (sexually), and their off-spring was En-lil, the god of the atmosphere. Thus all of nature—sea, land, air, and sky—had a divine counterpart.

There were many other gods and goddesses. In fact, the Sumerians appear to hold the world's record—their pantheon numbered about 5,000 deities. Very important among them was Inanna, the goddess of the planet now known as Venus and the goddess of love. It is probable that she, as well as other deities, had an extremely ancient history— going back to Paleolithic times. The antecedents of Inanna are to be found in the Venus figurines of the Late Paleolithic, and she is wor-shiped in later centuries and by different peoples as Ishtar, Astarte, Aphrodite, and of course Venus. Thus that basic characteristic of life, reproduction, is guided by the divine.

The Mesopotamians did not believe that the world was eternal—instead, it had been created. Human beings were also created, and there were several suggestions for how this might have been accomplished: from the mixing of the blood of two gods; molded from clay by one god; from sprouted seeds; or formed directly by a goddess.

The most famous of their creation myths was *Enuma Elish*. It was written down by a Semitic people, the Akkadians, who conquered the Sumerians. In their belief system, the first divine couple was the male Aspu and the female Tiamat. She was often portrayed as a dragon. Later, numerous other gods appeared, including Marduk. Eventually there was trouble, and the gods resorted to conflict to settle their disputes. This was the first "Mother of All Battles" to engulf the area of Iraq. Two factions formed, one taking Marduk and the other Tiamat as their leader. Battle was imminent, and it was agreed that the outcome would be decided by single combat between the leaders. The conflict

4 *The Akkadian god Marduk pursuing his rival, the goddess Tiamat, as recorded in the Babylonian story of creation.*

that was to change all history began. Tiamat was about to swallow Marduk, but he caused the winds to blow into her mouth and distend her body. After thus incapacitating her, Marduk shot an arrow into Tiamat's heart and eviscerated her. He then cut her into halves—one half becoming the dome of the sky and the other the earth.

The Sumerians' religion has many counterparts with events described in the Old Testament. There was an evil serpent who tried to produce chaos in the world; there was a deluge that resulted in the extermination of life; En-Ki ate a forbidden fruit and lost immortality.

And the Sumerians thought about that mystery of all mysteries that has puzzled human beings over the centuries—disease. Since the causes of sickness were not apparent, and since that basic compulsion of human beings to understand could not be satisfied, by default the cause was placed in the hands of unknowable forces—the gods. Sickness, therefore, was a sign of divine displeasure. Sickness was to remain a great mystery until the discoveries of scientists such as Pasteur, Koch, and Semmelweis—more than 5,000 years into the future.

It is highly probable that these explanations of nature were basically the teachings of the priests—a group of individuals who assigned themselves the tasks of explaining the will of the gods, which they alone could understand, to ordinary folks. They exacted suitable payments for these services. One way to seek the return to health, therefore, was to propitiate the priests, who, if the price was right, could propitiate the gods.

But another way was practiced as well, and this eventually evolved into rational medical science. This involved the direct intervention by physicians. To a person bleeding from a deep wound it might seem more appropriate to try to stem the flow personally or to seek the help of another, a physician, rather than ask the gods to do it. Propitiating takes time.

Each of these two modes of dealing with medical problems had its professionals, especially in Akkadian times. The *asipu* physician examined the patient for possible omens and then decided how the gods were to be propitiated and the causative agents, evil spirits, exorcised. The *asu* physician, on the other hand, sought relief for his patient through the use of drugs and other medical procedures.

Sumerian texts list various drugs used in the treatment of common ailments. Kramer describes a cuneiform tablet dating to the end of the third millennium BC as "The First Pharmacopeia," and writes:

The Sumerian physician, we learn from this ancient document, went, as does his modern counterpart, to botanical, zoological, and mineralogical sources for his materia medica. His favorite minerals were sodium chloride (salt) and potassium nitrate (saltpeter). From the animal kingdom he utilized milk, snake skin, and turtle shell. But most of his medicinals came from the botanical world, from plants such as cassia, myrtle, asafoetida, and thyme, and from trees such as the willow, pear, fir, fig, and date. These simples [products of medicinal plants] were prepared from the seed, root, branch, bark, or gum, and must have been stored, as today, in either solid or powdered form. The remedies prescribed by [the] physician were both salves and filtrates to be applied externally, and liquids to be taken internally. (1959, p. 61)

5   In this Akkadian bas-relief, a lioness has her hindquarters paralyzed by an arrow that has cut her spinal cord. This reveals a surprising understanding on the part of the artist of the relationship of anatomy to clinical symptoms.

The gods were also responsible for the harvest. Their intercession was sought for ensuring desirable climatic conditions necessary for plant growth and for the control of pests. The Sumerians were talented agriculturalists and developed a complex technology for irrigation. The bounty of their farms made possible the support of an urban population and, hence, civilization. The earliest known cities are Sumerian: Eridu, Ur (the home of Abraham), Erech, and others date from about 3500 BC.

Over human history, one of the most interesting phenomena is the slow progress in those aspects of civilization that are science-based. In the case of medicine, for example, the Sumerian physicians, at the dawn of literate history, used techniques that were much the same as those of the Western world until a few centuries ago; and to this day medicine based on a useful knowledge of the causes of disease is still in an active stage of development. Until quite recently our pharmacopoeia was also derived mainly from plant products. The same remains true for many native peoples. And, seemingly, the supernatural approach continues in all cultures, with its appeals for divine intervention for the sick.

So much of what we are and do today appears to trace back to the very beginnings of civilization in Sumeria. The Sumerians were surely a talented and inventive people, but much of what they did must have been inherited from their preliterate ancestors. We base our opinions of the past mainly on the written record that has endured. No pre-Sumerian written record has been discovered, but we can assume that the Sumerians transmitted the accomplishments of their preliterate ancestors along with their own.

Egypt

Sumeria and Akkad came into history, had their remarkable, though brief, periods of eminence, and vanished into the desert sands. Egyptian civilization, on the other hand, began almost as early as that of Sumeria and was impressive from about 3500 BC to the time of the Roman conquest beginning in 58 BC, in the days of Julius Caesar, Cleopatra, Antony, and Pompey. Consequently, far more data exist that will allow us to reconstruct the Egyptian view of nature, and we find the familiar association of religious beliefs with animals.

> The relationships between gods and beasts were as intimate as they were ancient, and all kinds of creatures both wild and domestic were kept in temples for cult purposes. So, too, thousands upon thousands of carefully

embalmed bodies of all these cultic species from cats to crocodiles went into vast animal cemeteries. If one is looking for ways in which peoples differ from one another, zoomania can be recognized as an Egyptian peculiarity. (Hawkes 1973, p. 339)

According to Herodotus, the "father of history":

There are not a great many wild animals in Egypt, in spite of the fact that it borders on Libya. Such as there are—both wild and tame—are without exception held to be sacred . . . The various sorts have guardians appointed

6 *Egyptian deities. The goddesses Bastet* (left) *and Hathor* (right) *had many duties in common—protector of women and patroness of music and dance. Bastet is shown with a cat's head and Hathor with a cow's head.*

for them, sometimes men, sometimes women, who are responsible for feeding them; and the office of guardian is handed down from father to son. [The peoples'] manner, in the various cities, of performing vows is as follows: praying to the god to whom the particular creature . . . is sacred, they shave the heads of their children—sometimes completely, sometimes only a half or a third—and after weighing the hair in a pair of scales, give an equal weight of silver to the animals' keeper, who then cuts up fish (the animals' usual food) to an equivalent value and gives it to [the animals] to eat. Anyone who deliberately kills one of these animals, is punished with death; should one be killed accidentally, the penalty is whatever the priests choose to impose; but for killing an ibis or a hawk, whether deliberately or not, the penalty is inevitably death. (1954, pp. 127–128)

These seemingly severe penalties were related to what the animals symbolized. The ibis was regarded as the incarnation of the god Thoth. Thus ibicide was deicide. Thoth was the god of the moon and of great importance to scholars—being the god of learning, writing, and chronology. The falcon was sacred to the main deity, Horus, as well as the symbol of the pharaohs. Horus was god of the sky, whose outstretched wings protected the earth. At times his eyes were regarded as the sun and the moon. It is not surprising, therefore, that death was the penalty for killing a falcon.

Seemingly every common species of animal was associated in some way with the Egyptian gods. The cobra appears as an emblem, the uraeus, on the headdress of the pharaohs and was a symbol of sovereignty. The ram was the symbol of one of the most ancient of all the gods, Amun, who at various times was regarded as the god of water, the god of fertility (hence the ram, the "old goat"), the sun god, and, finally, as the soul present in everything. The jackal was associated with Anubis, the god of the dead. Seth began as a respectable deity but later became associated with evil. He controlled the desert and hence was an enemy of vegetation. In a battle with the god Horus, he lost his testicles. His main symbolic animal is imaginary, but other species such as the crocodile, fish, ass, pig, and hippopotamus are associated with him. Bees were the tears of Ra, the sun god. Bulls had various symbolic meanings, including being the soul of Ptah. Ptah did all sorts of things: invented crafts and created life. There were four god-animals associated with the canopic jars into which the internal organs were placed when a body was embalmed. A falcon jar held the intestines; an ape jar, the lungs; a jackal jar, the stomach; and a jar decorated with a human head held the liver. This fascinating list of the associations of animals with gods and human beings could go on and on.

It is interesting to speculate about what the Egyptians "really believed" to be the relation of animal to deity. Many, especially the more superstitious and illiterate people, may have assumed a near identity, especially if death was the sentence for killing the god's symbolic species. The priests may have had a more intricate opinion:

> Only rarely was the god regarded as the animal itself, except, for example, in times of religious decline. The individual animal was only an earthly image of the transcendent primeval image, the theriomorphic form of which expressed some particular aspect of a divine entity. Sacred animals were, therefore, the 'eternal soul' (as the ethnologist Frazer described it) or, as the Egyptians would say, the *ba* of the gods. The ram was the soul of Amun-Re, the Apis bull that of Ptah, and the crocodile was the *ba* of Suchos. (Lurker 1980, p. 26)

The deification of nature exemplified by the Sumerians and Egyptians was universal in ancient times. In some cultures, however, there was a trend toward reducing the pantheon from thousands to a few or even one main god—for example, among the ancient Hebrew people. There was also a trend toward reducing the close relation with animals and natural things and processes.

Animal symbolism remained strong until modern times. In churches throughout Italy, Spain, and France—those not scoured by the fury of the Reformation—animals abound. A lamb in sculpture or mosaic symbolizes Jesus. Three evangelists are represented by animals: John by an eagle, Mark by a lion, and Luke by an ox. The fourth, Matthew, is represented by a man. At times the mammals are shown with wings on creatures that also have arms, which does confuse the homology. The iconography of Western religious art of the Romanesque and Renaissance periods shows strong connections with the most ancient belief systems.

chapter 2

Aristotle and the Greek View of Nature

Western science traces its origins to Greece, not to the ancient civilizations of Mesopotamia and Egypt, for it was the Greeks who provided a new way of looking at nature. One of the most astonishing events in intellectual history is the sudden appearance, seemingly *de novo*, of naturalistic thought—so dominant in the science of Aristotle and the science of today. This is the procedure of basing explanations of natural phenomena on the things and processes of nature. For example, when ascertainable and specific meteorological conditions prevail, liquid water is precipitated from clouds as rain. This is in marked contrast to a supernatural or mythical explanation which assumes that rain is the tears of weeping gods. W. K. C. Guthrie credits Aristotle with contrasting these polar modes of thought:

> It is to Aristotle in the first place that we owe the distinction between those who described the world in terms of myth and the supernatural, and those who first attempted to account for it by natural causes. The former he called *theologi*, the latter *physici* or *physiologi*, and he ascribes the beginning of the new, "physical" outlook to Thales and his successors at Miletus, hailing Thales himself as "first founder of this kind of philosophy." (1962, p. 40)

"This kind of philosophy" has been fundamental for the advance of science.

The Greeks were not restricted to mainland Greece but had colonies throughout the Aegean and even along the coasts of the Black Sea, southern Italy, and elsewhere in the Mediterranean world. Miletus, a seaport on the coast of Ionia (now Turkey), was settled by Greeks about 1000 BC. It was the home of three philosophers who, in the absence of earlier evidence, are the first we know who systematically used naturalistic thought to explain natural phenomena. Thales (*ca.* 625–547 BC), the first, was followed by his pupil Anaximander (*ca.* 611–547 BC) and later by Anaximenes (*ca.* 585–528 BC). Among other problems, these Milesians were concerned with the basic materials of which all physical objects are composed. Thales thought the elemental substance was water, Anaximenes thought it was air, and Anaximander assumed some unknown and even more basic substance.

Thus these philosophers suspected that there was a common building block for all matter. That search was in principle ended when the English scientist John Dalton (1766–1844) provided acceptable evidence for atoms. In our century, the "indivisible" atoms have been resolved into a hierarchy of subatomic particles.

The specific hypotheses of the Milesians were of little value; as Wightman points out, it was their approach that was so novel and so important. Many things suggested to Thales

> that *if there is any one thing at the basis of all nature,* that thing must be water. If there is any one thing! This supposition, that is to say, the asking as it were of this question, constitutes Thales' claim to immortality. The fact that he made a guess at the answer, and a pretty good guess at that, is of minor importance. If he had championed the cause of treacle as the sole "element" he would still have been rightly honoured as the father of speculative science. True, others before him (such as Homer and Hesiod) had sketched the origin of the world from one substance, but they were not content to deal with *verae causae,* that is with things whose existence can be verified by observation. To attempt to explain the origin and process of the world by having recourse to gods and spirits endowed with special powers, is merely to beg the question, since the existence of such beings can never be proved (nor of course disproved) by the means wherewith we know that world. In a word, it was Thales who first attempted to explain the variety of nature as the modifications of something *in nature.* (1951, pp. 10–11)

Thus with the classical Greeks, particularly Aristotle, the analysis of nature becomes rational and the data empirical to a degree that suggests the metaphor of a mutation. Compared with what went before, mature Greek thought *is* different. Gone are the demons of creation and gone are the supernatural causes of natural phenomena. We find instead attempts, primitive and premature as they often were, to explain natural phenomena in terms of rules and regularities that derive from nature itself. Nature is interpreted in terms of nature, not by superstition and flights of fancy of the human mind. Science was under way.

We know, of course, that rationality was not forever the hallmark of human thought in the centuries following the classical Greeks. Other systems of thought remained in favor, and to this day there continue to be conflicts of the two dominant and antagonistic patterns of thought we have inherited—one based on the acceptance of supernatural forces, authority, or revelation to explain the phenomena of nature and another relying on observations, data, hypotheses, and verifiable conclusions.

The Science of Animal Biology

We start the science of animal biology with Aristotle (384–322 BC) because he was the first to ask the sorts of questions, seek appropriate kinds of data, and provide the type of answers that we employ in biology to this day. He provided a pattern for learning, which is among the most difficult of all steps in science. It consists of being able to ask a question in such a manner that data can be sought for the answer.

Aristotle, who spent much of his life in Athens, was a student of Plato and teacher of Alexander the Great. The extant biological works of Aristotle consist of *Historia Animalium*, which is a general biology of animals; *De Partibus Animalium*, a comparative physiology and anatomy of animals, the first animal physiology in any language; *De Motu Animalium*, dealing with movement and some aspects of psychology and metaphysics; *De Incessu Animalium*, also concerned with locomotion; *De Anima*, considering the vital principle of living things; *Parva Naturalia*, mainly about psychology; and *De Generatione Animalium*, Aristotle's treatment of developmental biology.

Scholars are reasonably sure that existing forms of these works are relatively accurate. During the Renaissance, when Aristotle's works became known in Western Europe from Arabic editions, it was suspected that the double translations from Greek to Arabic and from Arabic to Latin might have introduced errors. Subsequently manuscripts in Greek were discovered, and these are assumed to be closer to the originals. Some are suspected of containing not only errors made when the manuscripts were copied but also some creativity on the copyist's part or inclusions of material from later writers. When several different manuscripts of the same work are available, however, these errors and insertions can usually be detected and expunged. None of the extant manuscripts are very old. For example, the oldest of the nine most important Greek manuscripts of *Historia Animalium* dates from the twelfth or thirteenth century, and the rest are from the thirteenth to the sixteenth century. To keep this in perspective: the interval from Aristotle to the twelfth or thirteenth century is roughly the same as the interval from the end of the Roman Empire in the West (AD 476) to the present. There was ample opportunity for errors of transcription to creep in during the many centuries between Aristotle and Johannes Gutenberg (*ca* 1390–1468), when the printing of multiple copies made error less likely.

The major point is that someone in the Athens of the fourth century BC set the course for the science of biology. Let's call that someone "Aristotle," as it most probably was. Alfred North Whitehead said that all of Western philosophy is a footnote to Plato. It is not hyperbole for us to say that all of biology is a footnote to Aristotle. He defined the field, outlined the major problems, and accumulated data to provide answers—he set the course.

What did he do?

Historia Animalium is the oldest extant monograph of general zoology. In contrast with biological publications of today, it does seem rather loosely organized, since it covers everything he knew about all species—the structure, breeding habits, reproduction, behavior, ecology, distribution, and relationships of animals. It begins the analysis as though there is no science of holistic biology, which of course was true, and proceeds to develop one. Some of the things that were problems for him are no longer problems for us—they have become part of our conceptual framework and are taken for granted.

One's first experience of reading *Historia Animalium* is not unlike that of first reading Charles Darwin's *On the Origin of Species*. In each case one is constantly amazed at the information provided and the questions asked. "How could he possibly have thought of that?" Both works are instructive for showing how fine, disciplined minds work. The first few paragraphs of *Historia Animalium* begin the analysis of some very fundamental problems. Aristotle notes that the parts of animals are of two sorts. *Composite* parts are those that cannot be divided in such a manner that two basically identical parts result. The hand and the face are examples. All composite parts are composed of *simple* parts. These can be divided and the resulting halves will be essentially the same. Flesh, bone, and sinews would be examples.

This is an important concept: different complex structures are built from varying combinations of the same building blocks. This has been a powerful notion throughout the development of the science of biology but especially so in the last two centuries. In the early nineteenth century Schleiden and Schwann proposed that the bodies of all organisms are composed of the same building blocks—cells. Later the basis of Mendelian inheritance—genes situated on chromosomes—was found to be essentially the same in all organisms no matter how they might differ in form and function.

Next Aristotle notes that some individuals resemble one another in all their parts: "One man's nose or eye resembles another man's nose or eye." Such individuals are members of the same kind or species. There are also categories of individuals that resemble one another less closely, their parts being noticeably different. Some birds, for example, have long bills, others short; some have many feathers, others fewer. Such individuals can be said to belong to the same "genus," such as "birds" or "fishes." Aristotle is using "genus" not as we do today but for a larger category which we today would call a class.

Aristotle sought also to compare individuals that belong to different genera. Although they might be similar in some respects, their resemblances are merely analogous: "For what the feather is to a bird, the scale is to a fish" (486b21; see Suggested Reading for note on citations). Here we have the onset of a way of thinking that was to develop as one of the most important concepts relating to structure: homology versus analogy. Aristotle realized, of course, that there was variation among human noses, yet basically they were the same. We would say they are homologous structures. Although both feathers and scales have the same general function—covering the individual—Aristotle thought them different. They might have the same function, covering the body, but not the same structure. We call them analogous structures. This line of reasoning remains useful to this day. Consider this prophetic comparison in another work of the paired appendages of birds and fishes:

> Birds in a way resemble fishes. For birds have their wings in the upper [=anterior] part of their bodies and fishes have two fins in the front part of their bodies. Birds have feet on their under [=posterior] part and most fishes have a second pair of fins in their under-part and near their front fins. (*De Incessu Animalium*, 714b4)

Aristotle then goes on to make sense of the different ways animals live. Some animals live on land and others in water. Those that live in water show two basic patterns. Some, the fishes, take water in and pass it out over the gills. These are restricted to water and do not come out on land. Other species—crocodiles, otters, beavers—do not have gills through which water passes in and out. They may obtain their food in water, but they come out on land to breed: "Furthermore, some animals are stationary, and some move about. The stationary ones are found in water; no land-animal is stationary" (487b6). Just try to imagine what breadth of knowledge is necessary for Aristotle to make such a statement. A similar broad generalization is, "But no creature is able only to move by flying, as the fish is able only to swim, for the animals with leathern wings [as in bats] can walk" (487b23). The whole thrust of these opening sections of *Historia Animalium* is to reduce the great diversity of natural phenomena to a comprehensible conceptual scheme. Aristotle sought unifying themes, without which rational thought is difficult.

Aristotle notes that apes are intermediate between human beings and the other quadrupeds:

> Some animals share the properties of man and the quadrupeds, as the ape, the monkey, and the baboon. The monkey is a tailed ape. The baboon resembles the ape in form, only that it is bigger and stronger, more like a dog in face, and is more savage in its habits, and its teeth are more dog-like and more powerful.
>
> Apes are hairy on the back in keeping with their quadrupedal nature, and [have scant hair] on the belly in keeping with their human form . . . [The] face resembles that of man in many respects; in other words, it has similar nostrils and ears, and teeth like those of man, both front teeth and molars. Further, whereas quadrupeds in general are not furnished with lashes on one of the two eyelids, this creature has them on both . . .
>
> The ape has also in its chest two teats upon poorly developed breasts. It has also arms like man, only covered with hair, and it bends these legs like man, with the convexities of both limbs facing one another. In addition it has hands and fingers and nails like man, only that all these parts are somewhat more beast-like in appearance. Its feet are exceptional in kind. That is, they are large hands, and the toes are like fingers, with the middle one the longest of all . . . The creature uses its feet either as hands or feet, and doubles them up as one doubles a fist . . . It has neither hips, inasmuch as it is a quadruped, nor yet a tail, inasmuch as it is a biped . . . The genitals of the female resemble those of the female in the human species; those of the male are more like those of a dog than are those of a man (502a17–502b24).

There is no suggestion that Aristotle regards the apes and human beings as entirely different creatures—one like the gods, the others mere brutes. Yet one must not read between these lines and claim that Aristotle is a Darwinian evolutionist. Nevertheless he is emphasizing a natural phenomenon, the resemblance of human beings and the other primates, that was to puzzle naturalists until Darwin offered an explanatory hypothesis.

Historia Animalium deals mainly with the structure and habits (especially of reproduction) of an estimated 500 different species—both vertebrates and invertebrates. Much of what is said remains correct to this day. But Aristotle was not always right; for example, "The lion has its neck composed of a single bone instead of vertebrae" (497b16) and "Males have more teeth than females" (501b20). He also believed that the heart, not the brain, was the seat of consciousness and intelligence. Nevertheless, *Historia Animalium* is an extraordinary achievement. There was nothing like it before, and none known to me of a later date, that presents in a single small book such a broad sweep of observations and attempts to understand them in naturalistic terms.

The Parts of Animals

Many of the novel ways Aristotle looked upon nature have become so much a part of our thought patterns that they are taken for granted. One is that structures have functions. In fact, he seems to have separated structure and function less than is done today. When he reports on structure, it is nearly always as a prelude to speculations about modes of action and broader significances. Structure was regarded as basic to what is really important—function.

A companion work to *Historia Animalium* is *De Partibus Animalium*— "Of the Parts of Animals" or, as Aristotle would have preferred, "Of the Causes of the Parts of Animals." Here Aristotle tries to ascertain the functions of the structures of animals and speculates on their general significance. How should one start the investigation?

> It is clear that in the investigations of the natural sciences there must be agreed upon general rules by which the acceptability of the methods may be tested—independently of whether the statement is true or false. Should we, for example, discuss each species—man, lion, or ox—separately. Or should we first ascertain their common characteristics, there being many attributes which are identical that occur in many different groups of organisms—sleep, respiration, growth, decay, and death. I raise this, for at present there is not an agreed upon scheme. However, this much is plain, if we discuss them species by species we shall be repeating the same

descriptions for the many different animals since, for example, every one of the common characteristics mentioned above occurs in horses, dogs, and human beings alike. (639a, 639b)

Aristotle's answer is to emphasize the common attributes of different species—the similar ways that they solve the basic problems of life— which is the same approach that has proved so successful in all branches of biology. This approach emphasizes, also, the similarities among organisms in contrast with treating each one as entirely distinct from all other species—a point of view that must follow if each species represents a unique act of divine creation.

One of the great values of the physiological analyses in *De Partibus* is that they show vividly the enormous difficulties in asking questions when the techniques, basic information, and approaches are inadequate. The human body and the bodies of other organisms were black boxes to Aristotle: Food and air go in and solid and liquid wastes come out; what happens in between was a mystery. The opened body shows a bewildering collection of parts connected by tubes and strands. The living organism may be active in the extreme, but if its body is opened immediately after death, all is quiet. Nothing seems to be happening. Aristotle could see a huge liver, for example, but he could not see its function—that had to be inferred and, in the fourth century BC, many of the inferences were wrong or incomplete. Nevertheless, Aristotle made an enormous contribution: he asked the important questions and he sought naturalistic answers.

As examples of Aristotle's understanding of physiology, the following is what he believed the functions of the organs of the gut, blood, lungs, kidney, and bladder to be. For the most part he is discussing mammals, including human beings. He observed that an animal can neither exist nor grow without food. Therefore the two parts most necessary are those that take food in and those that eliminate the residues. The food is not used as such but is changed in the body by heat. The source of this heat is the heart. When the food needs to be broken up, this is done in the mouth by teeth. When the food is in small pieces it is easier for the heat to act upon it. The food then passes to the stomach and intestines. The food is transformed. In fact the final form of food is blood, which enters the blood vessels and is distributed to the parts of the body. The liver is a necessary organ for the use of the food in blood. "These and similar considerations make it clear that the purpose of the blood in living creatures is to provide them with nourishment . . . that is to say, nourishment for the parts of the body" (650b5, 15). I suspect that Aristotle's understanding of the requirements for food, and how the parts of the body receive it, was as good as that of the average person living today in the United States.

Aristotle was less successful in understanding the function of breathing. The lung is present for the sake of breathing. He suspected that the main function of breathing is to cool the body, which we know is indeed one of its functions. It must have been obvious to the ancients that the human body produces heat, especially in disease with fever. It was true, also, that one feels "hot" after strenuous exercise and one's rate of breathing increases. If Aristotle did not completely understand the function of breathing, he did understand the basis of inspiration and expiration: "When the lung rises up, the breath rushes in, and when it contracts the breath goes out again" (669a18).

Aristotle said that the purpose of the kidneys and bladder is to deal with residues. He notes an interesting relation between blood vessels and the kidney, and he has an astonishing insight into kidney function.

> The duct that extends from the great blood vessel to the kidneys does not terminate in the cavity of the kidneys but the blood is spent on the body of the kidneys. Thus no blood enters the cavity of the kidneys and none congeals there after death. Other channels come from the aorta to the kidneys; these are strong, continuous ones. This arrangement is for the purpose of enabling the residue to pass out of the blood-vessel into the kidneys, and so that, when the fluid percolates through the body of the kidneys, the excretion that results may collect in the middle of the kidneys, where the cavity is in most cases. From the cavity of the kidneys two sturdy channels [the ureters] lead to the bladder, one from each kidney. These contain no blood. (671b)

This account of kidney physiology just about takes us to the nineteenth century.

The Classification of Animals

As we have seen, Aristotle recognized clearly that there were biological similarities among animals. Some were so alike that they were the same kind, which he called species, as we do today. The different species seemed to fall into groups. The fewer the species in a group, the more resemblances there were among the group's members. For example, there were greater resemblances among one group of vertebrates, the birds, than among all vertebrates—fishes, amphibians, reptiles, birds, and mammals.

Aristotle did not attempt to provide a rigid hierarchal system of classification that would encompass all animals, but in his goal of seeking resemblances among organisms, instead of dealing with them one by one he found it useful to recognize some major groups. His two basic groups were those with blood and those without. The former group is nearly identical with vertebrates and the latter with invertebrates. Among those with blood, the vertebrates, the major subgroups were: "man; viviparous quadrupeds (mammals); birds; oviparous quadrupeds, the reptiles and amphibians (including some similar in most ways, such as the snakes); and fishes." Among the bloodless animals the main groups were: "soft and shelled animals (crustaceans); the 'softies' (cephalopods); insects; shelled animals (mollusks and echinoderms plus the sea anemones and sponges)." In addition to these main groups, Aristotle mentions that there are others that he does not name.

Of particular interest is Aristotle's extreme caution in classifying animals. He implies that others have used single character differences and that often this resulted in artificial groupings, which suggests that Aristotle had an intuitive grasp of what we mean today by a "natural" group, that is, one that is related through descent from a common ancestor. For example, he mentions that some naturalists have divided animals into those living on land and those living in water. That dichotomous scheme places some mammals (such as whales), some birds, and fishes in one group, and other mammals, birds, and the reptiles in the other group. He included snakes in the oviparous *quadrupeds*, despite their having not four legs but none at all. Nevertheless, he felt that they resembled lizards in so many other ways that they should be placed in the same class.

Aristotle's main contribution to classification was to maintain that there are natural groups—based on structure, physiology, mode of reproduction, and behavior. Of equal importance, I suspect, was his caution. In contrast with many taxonomists who came later, he did not try to fit the species to the system but realized that a system must deal with natural affinities.

The Aristotelian System

Aristotle's major contribution to biology was to seek understanding of natural phenomena in naturalistic terms, instead of believing them to be controlled by gods, demons, spirits, and forces of the supernatural world, as more ancient civilizations had done. For him, knowledge of natural phenomena comes from the application of disciplined thought to data acquired by observation. One believed what one saw, not saw what one believed.

Aristotle had far less success when dealing with astronomy and the physical sciences than with biology. With them he proceeded, as did Plato, to fit the facts of nature to a preconceived theory of the causes of natural phenomena. This is deductive science, which can be successful only if the basic theory is correct. Aristotle's theory was not correct, and his elaborate description of physical nature was eventually demolished, two millennia later, by Galileo and others. Aristotle's biology was largely inductive and, for that reason, proved far more enduring.

He said little about experimentation, which was to prove such a powerful tool for answering questions. In fact, the large-scale use of experimentation to test deductions was not to begin until the days of Galileo in the late sixteenth and early seventeenth centuries. Nevertheless, Aristotle may have employed it to some degree. How else could he have said that hair grows from the base, not the cut surface? The observation of sexual impotence in castrated animals is surely based on an experiment, even though the experiment was probably undertaken for nonscientific reasons. In addition, he takes advantage of some natural experiments: hair may begin to grow on a woman's chin once the menopause has passed.

It was this lack of experimentation that produced Aristotle's defective physics. It cannot be maintained that experiments in physics would have been impossible for him. He could have tested, for example, the commonsense notion that light objects, such as paper or feathers, fall more slowly than heavy ones, such as a rock. That notion was held, of course, because one observed, *in an experiment,* that a piece of paper floats gently to the ground in contrast with the rapid descent of a rock. Yet this result can be explained as due either to the relative masses of the two objects or to their shapes. Had Aristotle done further experiments, such as comparing the descent of one sheet of paper with another crumpled into a tight wad, the science of physics would have had a very different history. No doubt he saw no need to do such an experiment since he assumed he already knew the answer.

A basic component of Aristotle's view of nature was his notion of the existence of a phenomenon described by a word generally translated as "soul." The word has been translated more appropriately, I believe, as "living force," "life," or "vital principle." It has little, if any, relation to the Judeo-Christian meaning of soul, vague as that might be. No doubt the idea arose as an explanation of the difference between an animal alive and then killed. The form remains the same before and after, but there is some basic, though mysterious, difference. In fact, it is the soul that gives "form" to the "matter" of living creatures, and it exists only in relation to them. Whatever it is, Aristotle realized that it is restricted to what is alive and that its form varies with the complexity of the organism. Plants have the lowest level of "souls," theirs being concerned only with nutrition and reproduction. Animals differ from plants in having sense organs and a sentient soul along with the nutritive and reproductive soul as in plants. Human beings add a still higher level, the rational soul.

Aristotle believed that the soul is something apart from the basic elements—fire, earth, water, and air—possibly including a fifth element, *pneuma*. Aristotle was recognizing a phenomenon—the living state—and trying to provide ways for his rational soul to deal with it. But his hypothesis was not a very useful advance. Hypotheses in science are most productive when they associate different phenomena in testable ways. Aristotle's hierarchy of souls does no more than give names to some of the different things that living creatures do. The explanation that animals have sense organs that allow them to perceive the environment because they have a sentient soul really does not advance the argument.

This emphasizes a most important feature of scientific procedures: one cannot obtain answers unless one asks proper questions.

Basic Questions

In his biological investigations, Aristotle sought to describe in a systematic manner the structure of as many organisms as he could, ascertain the functions of those structures, discover the patterns of reproduction and development, and learn about behavior. But descriptive biology was only the beginning of the analysis for Aristotle. He appeared to have been far more interested in understanding the "hows" and "whys" of the observed phenomena. He considered what were to remain the most general questions that could be asked about the animal way of life. They were:

(1) What is the nature of the soul or *vital principle?* That is, what is responsible for giving matter the ability to exhibit the characteristics of living creatures. All creatures begin as a seed or egg and undergo a differentiation, grow, reproduce at maturity, and die. In contrast with the nonliving world, where objects do not seem to undergo continuous changes, life is characterized by constant change.

(2) How can this vital principle, so much the same in all living creatures—jellyfish, ferns, elephants, birds, trees, and human beings—be expressed in creatures so different from one another in form and function? *Is "being alive" basically the same for all creatures?*

(3) What is required for the *maintenance* of life?

(4) What is responsible for the *diversity* of living creatures—both the variation of the individuals of a species as well as the differences among species?

(5) What can account for the similarities among organisms and hence the possibility of recognizing *natural groups?* How can there be *hierarchies* of natural groups?

(6) What is responsible for *like producing like* or, as we would now say, genetic continuity? That is, what ensures that offspring resemble their parents?

It is important to remember that the answers to these questions are without practical consequences. No one in ancient Greece would have benefited economically or politically by knowing the answers. The questions were asked and answers sought by a very few individuals who could afford the luxury of speculation because of the intellectual joy it brought. Edith Hamilton caught this spirit in her paraphrase of a portion of *De Partibus:*

> The glory, doubtless, of the heavenly bodies fills us with more delight than the contemplation of these lowly things, but the heavens are high and far off, and the knowledge of celestial things that our senses give us, is scanty and dim. Living creatures, on the contrary, are at our door, and if we so desire we may gain full and certain knowledge of each and all. We take pleasure in a statue's beauty; should not then the living fill us with delight? And all the more if in the spirit of the love of knowledge we search for causes and bring to light evidences of meaning. Then will nature's purpose and her deep-seated laws be revealed in all things, all tending in her multitudinous work to one form or another of the beautiful. (1942)

chapter 3

Those Rational Greeks?

Aristotle's two-pronged scientific approach—collecting data and speculating about it—was not applied systematically by those who followed him. His successors engaged, for the most part, in what Thomas Kuhn, in *Structure of Scientific Revolutions* (1970), refers to as "normal science," that is, filling in the details of the explanatory paradigm accepted by the majority of scientists at a given time.

If the documents that have survived are any indication, the main interest among Aristotle's followers was the biology of human beings, and this meant studying the structure of the human body and trying to determine the function of its parts. Progress in biology was measured in centuries, with one notable exception.

Theophrastus and the Science of Botany

Although Theophrastus was younger than Aristotle, both were students of Plato and later were together in Aristotle's school in Athens—the Peripatetic. Theophrastus was born in 371 BC on the island of Lesbos, where Aristotle once lived, but most of his life was spent in Athens. After the death of Aristotle, Theophrastus became the master of the school. He died in 287 BC.

Theophrastus was a person of broad learning, but he is known to us almost exclusively as a botanist—his other works have not survived. His *Inquiry into Plants* attempts to do for botany what Aristotle did for animals in *Historia Animalium*. Nothing so important had been done before, so Theophrastus can be acclaimed as the Father of Botany.

Theophrastus began his *Inquiry* by noting both the special features of plants and the extent to which they resemble animals. He described the germination of many kinds of seeds, noting for example the origin of root and stem and the relation of their origin to the point of attachment of the seed to either the pod or stalk. In the cereals the root starts from the broader, lower part of the seed and the stem from the opposite end. Nevertheless, the root and stem are continuous. The legumes, on the other hand, produce root and stem from the same point on the seed, namely, where it had been attached to the pod (8.2.1–2).

Theophrastus had difficulty in dealing with the structures of plants, a consequence of the ephemeral nature of some of the parts. Animals do not lose their parts—arms, eyes, or stomachs. Plants do: leaves of many trees are lost in the autumn; fruits form and fall. He decided that the more important parts are roots, stems, branches, and twigs because these are more permanent. They correspond in their importance to the parts of animals. The roots enable the plant to draw nourishment from the soil.

He recognized sap, fibers, and veins. Not being aware of any names for those parts of plants, he decided to use the terms of structures that occur in animals. His reasoning for so doing is most interesting, with applicability far beyond this particular need. "It is by the help of the better known that we must pursue the unknown because the better known are things which are larger and plainer to our senses" (1.2.4).

He attempted to classify plants, and the categories he chose were trees, shrubs, undershrubs, and herbs. Theophrastus could not make these categories discrete and noted much overlap. Clearly he was not satisfied with his classification. He did have a good understanding of the differences that centuries later were to form the basis of a more acceptable classification: angiosperms and gymnosperms; monocots and dicots; the parts of flowers; and hypogynous, epigynous, and perigynous arrangements of flower parts.

Inquiry into Plants has descriptions of more than 500 species, and there is much information about domesticated varieties. As was the case with Aristotle's *Historia Animalium*, much of the information is based on what Theophrastus read or was told by others. He recorded that different parts of the then-known world may have different species and that climate is important in plant growth. He knew the importance of dung as a fertilizer and recognized many species of plant pests. He provided considerable information about timber trees and which kinds are best suited for building houses, ships, furniture or for making charcoal.

Aristotle and Theophrastus approached nature inductively and built a body of biological information on the basis of observations. The data so obtained were used to develop explanatory hypotheses. Their approach was to prove highly productive in later centuries with the addition of another procedure—the experimental testing of hypotheses. Their accomplishments were so noteworthy, and recognized as such at the time, that one might have hoped that their rational approach would be accepted by all.

Not so. Two millennia were to pass before biology regained the operational level provided by Aristotle and Theophrastus. Seemingly the Second Law of Thermodynamics works for intellectual pursuits as it does for physical processes. After Aristotle and Theophrastus there was a notable increase in entropy in the biological sciences. One example of conceptual entropy will be noted, Pliny.

The Roman Pliny

Gaius Plinius Secundus, a noble Roman, was born in AD 23 and died in the eruption of Vesuvius on August 24, 79, which buried a wide area of southern Italy near Naples, including Pompeii and Herculaneum.

Pliny was a man of great accomplishment. He served in the army, and at the time of his death was commander of the Roman fleet based in the Bay of Naples. He also had held various governmental posts. He is important for us because of his *Historia Naturalis,* an encyclopedic work covering the physical universe, geography, anthropology, biology, mineralogy, medicinal plants, and the fine arts. He believed that no other person had attempted such a broad survey. The work was dedicated to the Roman Emperor Titus.

Books 7 through 19 treat human beings, other animals, and plants. Nearly all of Pliny's information was derived from literature. He claims to have checked 2,000 volumes by 100 authors and extracted 20,000 facts. He was either overly modest or poor at addition, for Eichholz (1975) counts 473 authors and 34,707 facts. His *Historia Naturalis* is a valuable compilation of information from a wide variety of Greek and Roman works, some of which have not survived. Later authors have often criticized Pliny for his gullibility and lack of originality. The second criticism is strange indeed—one does not expect new discoveries to be first announced in an encyclopedia.

For Pliny, the world is to be regarded as a deity and eternal, with no evidence of a beginning or an end. Matter is composed of four elements: fire, air, earth, and water. There are seven planets among the many stars. The sun rules the seasons and the stars and is to be regarded as the supreme ruling principle and divinity of nature, a point of view that was held by Homer. The information Pliny reported about the movements of the planets, stars, and sun is astonishing. He knew that the sun is directly over the equator at the equinoxes and is directly over the Tropics of Cancer and Capricorn at the solstices. The sky belongs to the main god as the earth belongs to human beings. The earth is kind, gentle, and indulgent and lavishes us with her bounty—yet we abuse her in so many ways. We still do.

He discusses the differing opinions of scholars and the general public about the shape of the earth. The scholars maintain it is a sphere and that there are human beings everywhere. Thus the feet of those on opposite sides point to each other (hence, the Antipodes). Most people would doubt that the earth could be a sphere with people on the other side because they would fall off. Pliny points out that those on the other side would have the same worries about us, yet we seem to have no problem holding on. That might not be a generally accepted argument, but Pliny believed that all things are attracted to the center of the earth. He suggests that there are many reasons for believing the earth to be a sphere. For example, when a ship puts out to sea the top of the mast is the last part to disappear. The sea surrounds the earth and it does not fall off because the pressure for convergence toward the center prevents that happening.

There are five main climatic zones: two regions crushed by cruel frost and everlasting cold at the poles; two temperate zones, and a torrid zone. The torrid zone is so hot that the temperate zones are effectively isolated from one another. This is an astonishingly correct belief since, in Pliny's time, little was known of the world beyond the boundaries of the Roman Empire. No one, for example, had ventured across that terrible torrid zone.

The tides are related to the position of the moon and are at their highest and lowest points when the sun and moon are opposite one another (full moon) and when they are closest (new moon). The east-west portion of the earth known to Pliny, India to the Pillars of Hercules (Straits of Gibraltar), was estimated to be a distance of either 8,568 or

9,818 miles. The earth's circumference had been determined by Eratosthenes to be 252,000 stadia. (A Greek stadium was equal to 606.75 feet or 0.11 miles; thus the circumference of the earth would be 27,720 miles—not bad, considering that Eratosthenes' odometer for the small arc of the earth he measured was counting the paces of a camel.)

The most remote place to the northwest known to Pliny was Thule (possibly northwest Norway), where at midsummer there is no night and at midwinter no day. One day's sail away from Thule there is a frozen sea. As Pliny discussed areas ever more remote from the Mediterranean, the information tended to become fantastic.

In Book 7, Pliny began his zoology with a discussion of human beings, which for him are the highest species of animal; the others have been created by nature for him. He alone is born naked—all other creatures have feathers, fur, spines, or bark. Only man experiences grief, weeps, or knows luxury. He alone has ambition, avarice, undue appetite for life, superstition, or worries about what happens when he dies. Other species stand together, whereas man alone is most evil to his own kind.

In those far-off places, there are some truly strange human beings: some eat human flesh, others may have a single eye in the middle of the forehead, fight with griffins, have their feet turned backward, see better at night than during the day, cure snake bites by mere touch, have poisons in their bodies that kill snakes, serve as both males and females (the Androgyni), bewitch at a glance, have double pupils, stand looking at the sun all day, have eight toes on each foot, have only one leg and move by jumping, have no necks but eyes on the shoulders, and so on.

And there were giants. An earthquake in Crete exposed the bones of a creature estimated to be 69 feet high. They were assumed to be the remains of a god. (Such reports were surely based on observations. The fossils of very large marine and terrestrial species had been noticed since ancient times, but their significance was not known then.)

Pliny gave a superficial and none-too-accurate account of human reproduction, remarking that if the mother has too much salt in her diet the children are born without nails and that a sneeze following copulation is likely to cause an abortion. Male babies who are born feet first almost never have successful careers. Marcus Agrippa was so born, and although he reached high station he paid the penalty for feet-firstness when his daughters bore sons who were to grow up to become those most terrible of emperors, Caligula and Nero.

His remarks on heredity show the confusion that was to last to the twentieth century. Sound parents could have normal or deformed children, and the same was true for abnormal parents. Specific traits may be inherited for several generations. For example, a woman with white skin bore the daughter of an Ethiopian. The daughter was white but bore a son who was as black as his grandfather.

Some human beings are exceptionally strong, fleet of foot, have acute eyesight or noble character. Julius Caesar could dictate four letters at the same time; he could write and read or dictate or listen simultaneously. He fought 50 battles and was responsible for the death of 1,192,000 human beings.

Near the end of his natural history of human beings, Pliny explored the question of an afterlife. The soul and its various attributes

> are fictions of childish absurdity, and belong to an existence greedy for everlasting life. So also is the vanity of preserving men's bodies . . . What is this mad idea that life is renewed by death? What rest will the generations have if the soul retains permanent sensation in the upper world and the ghost in the lower? Most surely this sweet but improbable fancy destroys nature's chief blessing—death—and doubles the sorrow of one about to die by the thought of sorrow to come hereafter also. For if it is sweet to live, who can find it sweet to have done with life? (7.55.189, 190)

Pliny then went on to discuss animals living on land, beginning with the largest and, next to human beings, the most intelligent—elephants. Their virtues include honesty, wisdom, justice, and reverence for the sun and moon. They are humiliated if they lose their place of honor, as in a parade, and are shamed if conquered. Being modest, they mate only in secret, and adultery is unknown. They may fall in love with human beings. They are naturally gentle, and when captured are quickly tamed with barley juice. They are terrified by the squeal of pigs and are liable to flatulence and diarrhea. They eat with the mouth but drink and breathe with an organ that can be called a hand. They hate mice.

The last portion on zoology, in Book 11, considers more general matters. Pliny accepted spontaneous generation: the larvae of flies are generated out of dirt by the rays of the sun. There is some attempt at correlation: animals with blood have a head; creatures that possess only the sense of touch (sponges, and some shelled animals) have no head.

There are descriptions of the internal organs of human beings and some of other vertebrates, with the clear implication that the organs of different species correspond to one another. Much of the data concern structure and little is said about function. This is not surprising. Basic functions are nearly always the result of the interactions of molecules —unknown entities to the ancients.

The heart is believed to be the first organ formed when the embryo is in the womb. It has a definite movement, almost as though it were an organism within an organism. It is protected by the ribs and is the source of the vital principle and of the mind. Two large veins extend from it—carrying blood to all parts of the body. When the heart is wounded, the animal dies. The heart and other organs could be inspected by augurs, and the findings allowed them to predict the future. Julius Caesar had a person sacrificed on the first day he became dictator and that person was found to be without a heart. This was obviously a bad omen for the sacrificed person but also for Caesar—as events were soon to show.

Pliny summarized the literature reports on many mammals, other vertebrates, and some invertebrates, and seemingly accepted everything. As a consequence, much of what he recorded is pure fiction. Nevertheless, *Historia Naturalis* is tremendously important because it is a compilation of what was generally believed in the first century AD. It is a window to the science of long ago.

Hippocrates, the Father of Medicine

Although the purely biological problems that interested Aristotle were of little obvious usefulness, another branch of biology has always been cherished for its benefit to human beings. This is the biology of the human body, knowledge of which is essential for maintaining health and curing disease. The recorded history of Greek medicine extends from Hippocrates (*ca* 460 to *ca* 375 BC) to Galen (*ca* AD 130 to *ca* 200)—a span of nearly six centuries. But there was medicine before that; the skill of Egyptian physicians was well known in the Greek world. Greek medicine remained viable long after AD 200 because Galen was to remain *the* authority on human anatomy and physiology until the seventeenth century.

Hippocrates was born on the Aegean island of Cos to a family of physicians. His name is used to refer to a large body of medical writings, the "Hippocratic Corpus," which includes not only some works by Hippocrates but also those of other physicians of the late fifth and early fourth centuries BC. For the purposes of this discussion all will be credited to Hippocrates. He died about a half-century before Aristotle was born.

Hippocrates revolutionized medical science, and he did so in the typical Ionian Greek mode—explaining what he could in a rational, scientific manner, suggesting naturalistic hypotheses for what could not be accounted for, and avoiding supernatural forces in his explanations.

Central to Hippocrates' treatment of patients was the notion of the healing power of nature—nature could restore the body's physiology to its normal equilibrium state. We have here the core of a concept that was to prove of great importance in physiology. Illness was regarded by Hippocrates, and others long before his time, as due to an imbalance of the four humors: blood, which was made in the liver; phlegm, associated with lungs; yellow bile, associated with the gall bladder; and black bile, associated with the spleen. The four main diseases characterized by an excess of one of the humors were sanguine (an excess of blood), phlegmatic (excess of phlegm), choleric (excess of yellow bile), and melancholic (excess of black bile). Thus health was achieved when the four humors returned to their normal equilibrium concentrations. This notion was to find more formal expression in Claude Bernard's concept of the *milieu intérieur* of the nineteenth century and Walter Cannon's *homeostasis* of the twentieth century.

An important technique of Hippocrates was to record as carefully as possible the symptoms and courses of diseases. This technique, not to be systematically employed again until the sixteenth century, is the basis of modern medical diagnosis and prognosis. Physicians in those ancient times had few drugs or other means of treating diseases effectively. The more successful of them relied on diagnosing the ailment and then, on the basis of experience with previous cases, predicting what would transpire. This understanding was used in moral support of the patient; and, then as now, a patient's confidence in the physician was an important element in recovery. This benign approach to illness was especially effective at a time when many of the recognized treatments were dangerous and, as always, most patients get well anyway.

But Hippocrates and other effective physicians did have a program for maintaining health—a lifestyle that included proper rest, diet, and exercise. Moderation in all things was the key to good health. The Hippocratic physician was far less likely to do harm than were many who came after them, for example, those who bled George Washington to death or those who throughout the nineteenth century poisoned their patients with toxic chemicals or made them drug addicts.

Erasistratus

A notable physician who followed Hippocrates was Erasistratus, born about 304 BC on the Aegean island of Chios. Erasistratus appears to have been an excellent anatomist and even to have experimented. For example, wondering about possible emanations from the body, he placed a bird in a closed container with no food or water. The bird was weighed before incarceration and afterwards, together with its feces. The final weight of bird plus feces was less than the initial weight. Therefore something, possible emanations, had been lost. He dissected the brain of a hare, stag, and human being and correlated the extent of the convolutions with the animal's intelligence.

Erasistratus thought that all parts of the body were served by a vein, artery, and nerve. The vein brought food in the form of blood, the artery brought *pneuma*, and the nerve *psychic pneuma*. Blood was thought to be formed in the liver and distributed by the veins to all parts of the body. During the expansion of the heart, some blood enters the ventricle from the vena cava. This blood is pumped from the right ventricle through the pulmonary artery to the lungs. The tricuspid valve prevents the blood in the ventricle from returning to the vena cava, and the semilunar valves prevent blood in the pulmonary artery from returning to the right ventricle. Blood reaches the other parts of the body through the veins.

There is an obvious problem here of distinguishing "artery" and "vein." The classical physiologists realized that the vessels connecting the heart and lungs were different from those in other parts of the body—just as today we say that the pulmonary artery carries "venous blood" and the pulmonary vein carries "arterial blood." The ancient name for the pulmonary artery was "arterial vein," and for the pulmonary vein it was "venous artery."

Pneuma was thought to come from the atmosphere and pass through the nose to the lungs. It then goes in the pulmonary vein to the left ventricle. The left ventricle then contracts and sends *pneuma* through the arteries to all parts of the body. The bicuspid valve prevents the return of *pneuma* to the lungs. The lungs have no independent motion of their own. The chest expands and air rushes in because of *horror vacui*—that vacuum that nature was thought to abhor.

The *pneuma* carried to the brain was converted to *psychic pneuma*. This passes along the nerves to muscles, causing them to contract.

One might think at first that ignorance is to be preferred to Erasistratus' understanding of physiology. This is not the case. Although seemingly hopelessly wrong in many details, he was suggesting physiological principles that would prepare other investigators to think along productive lines. He realized that arteries, veins, and nerves serve all parts; that blood carries materials to those parts; that something in the air, *pneuma*, is required by the body's parts; that the brain, via the nerves, controls muscular contraction; that the heart pumps materials around the body; that the valves of the heart control the direction of the blood's movement; and that physiological functions are to be explained in naturalistic terms.

Erasistratus' beliefs, like all statements of science, are approximations to that elusive goal of "truth." Science is an accretive and self-correcting discipline and, generation after generation, its concepts become more precise and accurate. Many concepts have reached a stage where we say they are true beyond all reasonable doubt. Erasistratus was important in these developments because he recognized basic problems and suggested explanatory hypotheses that others could consider and test.

In a similar manner Aristotle's famous statement that "Nature never makes anything that is superfluous" (*De Partibus* 681b4) molded thought patterns in such a way that the concepts of adaptation and, finally, natural selection could be arrived at more readily. Most ideas that become important concepts probably start as rather fuzzy notions, and it may take great effort to remove the fuzz and uncover the notion.

Another noteworthy point concerning Erasistratus' knowledge of the function of arteries tells us much about the pitfalls of scientific research. He knew that a cut artery of a living animal, which theory held to contain *pneuma*, squirted blood. Rather than accept the evidence before him—that arteries as well as veins contain blood—he suggested a way out that would save the theory: when the artery was injured by the cut, it lost *pneuma* and, since nature abhors a vacuum, blood rushed in to fill up the space where the *pneuma* had been. Scientists then and now see through the eyes of theory—the operational paradigm of the moment.

Galen of Pergamum

Galen represented another important stage in the excruciatingly slow progress to better knowledge of human form and function. He held hegemony over the minds of physicians, anatomists, and physiologists for more than a millennium and was the last of the renowned Greek physicians of antiquity. In fact, he was the last notable human biologist until the sixteenth century.

Galen was born about AD 129 and died about 200. Pergamum was at this time a great center of Hellenistic civilization. It too was part of Ionia—home to nearly all who established Greek science, and hence the science of today. Galen began the study of medicine in his home city and then continued it in Smyrna and Alexandria. In Alexandria he was able to study human skeletons, but the dissection of human cadavers was no longer permitted as it had been when Erasistratus was there. Apart from the scant information Galen would have acquired from treating wounds in his patients, his knowledge of human anatomy would have come from predecessors like Erasistratus. He realized, however, that the anatomy of other primates was almost identical with that of *Homo sapiens*. In fact much of his description of "human anatomy" is really that of the rhesus monkey and Barbary ape.

After his studies, Galen returned to Pergamum and became a physician to the gladiators. This should have given him considerable experience with subcutaneous anatomy. When he was about 32 years old he went to Rome and began a most successful medical practice. He moved in high society and was a physician to many important people —the Emperor Marcus Aurelius was a patient and friend. After practicing in Rome he went home to Pergamum, then to Aquileia, and then back to Rome for a second time.

His mentors were Hippocrates for medicine and Plato for philosophy. He refers frequently to Erasistratus of Alexandria, though not always in a fair and balanced manner. In many respects Galen's knowledge was not much advanced over that of his famous predecessors, despite the fact that Hippocrates was born about seven centuries and Erasistratus more than four centuries before him. The equivalent antecedents for a biologist today would place one at about AD 1298 and the other at about 1554. A check of the biology of 1298 would have revealed a remarkable thing—the best source still would have been Galen! By 1554 the biological works of Aristotle were becoming better known; hence he joined Galen in being an unchallengeable authority. This was a sad fate for Aristotle and Galen, who had looked upon science as a way of knowing and not as a corpus to be accepted on faith.

Galen's importance lies not so much in what he personally discovered as in his summarization of Greek medicine from Hippocrates to his own times. And we must remember that much of what Hippocrates knew had come from much more ancient sources. Galen, then, summarized the profession as it had developed, painfully slowly, over a thousand years.

His basic clinical approach was that espoused by Hippocrates: keep the person well by proper diet, exercise, rest, and a pleasing environment. He emphasized the importance of making clinical observations in order to be able to identify the disease and to predict its course. Medical practice was to be based on the fact that the patient almost always recovers from an illness, except the last. Thus "nature" is the healer. The primary role of the physician is to be sympathetic and supportive of the patient, not aggressive in fighting the disease. This was a most reasonable approach in those periods of medical history when both the nature of disease and effective cures were almost entirely unknown—at least the Hippocratic physician was unlikely to do harm.

Galen summarized and augmented knowledge of form and especially of function. His *On the Natural Faculties* is a general summary of his understanding of human biology. One of his most important principles was that the human organism is an integrated whole—hence his attempt to treat the entire body as the functioning unit, not its constituent parts. By extension, this notion of an integrated whole could be held to apply to all organisms.

Galen held that there are three "faculties" of nature: embryonic development, growth, and nutrition. Nutrition results in the assimilation of food, but before this can occur the food must be changed. Otherwise how could beans, meat, bread, and other foodstuffs be changed to blood? Food in the mouth is chewed and mixed with saliva. It is changed, as can be shown by the fact that particles of food that stick between the teeth overnight are different from the original food. Food remains in the stomach until it is properly digested, then it moves through the pylorus. Its ability to pass into the pylorus is not a result solely of becoming liquid: if a pig is fed a mixture of flour and water and cut open three or four hours later, the food is still in the stomach. Thus the processing of the food requires digestion and not merely breaking it down into small particles. It is only when the food is digested that the normally closed pylorus opens and the food enters the small intestine. When the stomach is irritated by acidity, food leaves the stomach prematurely, even if it is not completely digested. The stomach undergoes peristaltic movements similar to those of the intestine. These are caused by the different coats in the wall. One coat causes traction and the other one peristalsis.

Galen is remembered most frequently for his attempts to understand the functions of heart, lungs, arteries, and veins. The problems relating to what we now call the circulatory and respiratory systems were those that ancient physiologists found most interesting and approachable. Galen's views were much the same as those of Erasistratus, his predecessor of five centuries. That is, blood is carried in the veins, *pneuma* in the arteries; the heart pumps both to all parts of the body. Galen did add some correct notions but, unfortunately, added some errors as well. Net progress in understanding was, therefore, about zero.

One of the correct notions was that arteries and veins are connected. "If you kill an animal by cutting a number of its large arteries, the veins as well as the arteries become empty of blood. This could never occur if there were not anastomoses between them" (3.15).

Another important notion was that there is a circulation of sorts, at least in relation to the heart and lungs. Historians of science have often been confused about what Galen did believe, but Fleming (1955) after a very careful analysis offers this version (I have substituted modern terms for the blood vessels and other structures):

> Blood on entering the right ventricle must pass by the one-way bicuspid valve inward, so that only an insignificant portion can relapse into the vena cava whence it came. Some of the blood passes directly from right to left through the interventricular septum. But much, and apparently most, of the blood moves into the pulmonary artery past the one-way semilunar valves from the ventricle. On contraction of the thorax, the blood in the pulmonary artery, its retreat cut off from behind, can only go forward into the venous system of the lungs. Whether the pulmonary veins then carry blood to the left ventricle is in question. [Galen] almost certainly thinks of the pulmonary vein as conveying the inspired air, in some form or another,—or at least some quality derived from the air,—from the lungs to the left ventricle. In the opposite direction smoky wastes are undoubtedly borne from the left ventricle to the lungs by way of the pulmonary vein. This process is made possible in Galen's view by the comparative insufficiency of the mitral valve opening into the heart. The blood in the left ventricle passes into the aorta through an aperture guarded by a one-way bicuspid or mitral valve.

Why those smoky wastes? It was believed that the heart produces heat which, in turn, suggests a combustion of some sort. The most likely way to rid the body of these wastes was via the lungs. This is an error, to be sure, but might it not have helped prepare the mind of others to accept that the body's heat is produced by processes not unlike those of ordinary burning? At least the heat was postulated as coming from natural processes and not from some sort of divine spark.

Galen is mainly remembered today for that seemingly egregious goof of believing that blood passes from the right ventricle to the left through pores in the interventricular septum. This is what he said:

> In the heart itself the thinnest portion of the blood is drawn from the right ventricle into the left through perforations in the septum between them. These perforations begin as pits with wide mouths and then become progressively narrower. It is not possible, however, actually to observe their extreme terminations, owing both to the smallness of these and to the fact that when the animal is dead all the parts are chilled and shrunken. (3.15)

It is fascinating to note how we evaluate individuals for proposing hypotheses that later prove to be correct compared with being incorrect. We may praise Galen for the hypothesis that somehow blood passes from pulmonary artery to pulmonary vein in the lungs, whereas we look unfavorably upon him for the hypothesis about the movement of blood through those postulated pores in the septum between the ventricles. He had no evidence for either. His hypotheses were invented to account for problems he saw in the movement of blood. He was recognizing important problems, and by suggesting possible hypotheses he was paving the way for others to determine whether the hypotheses are probably correct or probably incorrect. That is the way science works.

If Galen became an obstacle to scientific progress, it was the fault of those who came after him. His synthesis of human biology and medicine was the most magnificent available to the Western world from the second to the seventeenth century. Those who came after him did not imitate his inquisitive mind but only revered his words. Along with Aristotle the two were accepted as the final arbiters of "truth." If nature appeared to be at variance with their statements, that showed that either the observer or nature must be at fault. That was the fate of the two great biological scientists who began and ended the Greek miracle.

Galen had very definite ideas about the correlation of the parts of the body in different organisms—a thesis to be developed centuries later by Cuvier and others and to become one of the cardinal principles of comparative anatomy.

Those apes with arms and legs most resembling those in man lack both long canine teeth and long faces. Such apes have an upright gait, speed in running, a thumb on the hand, a temporal muscle, hair variously hard and soft and long and short. If you observe one of these characters, you can be sure of the others, for they always go together. Thus if you see an ape running swiftly upright, you may assume without close inspection that it is like a man. You can predict also that it has the other characteristics, namely, a round face, small canine teeth, and a moderately developed thumb . . . On the other hand, if any of these characteristics is different, all of the others will differ. (*On Anatomical Procedures* 6.1)

We observed the germ of this idea in Aristotle as well.

In *On the Natural Faculties* Galen not only tries to give an accurate account of human physiology but he criticizes those of his contemporaries who seem ignorant of facts or who espouse unproven notions. Apparently the medical profession was at a low point in the second century AD, which accounted for the modest hopes he held for his books.

I am not unaware that I shall achieve nothing at all or very little. For I find that a great many things which have been conclusively demonstrated by ancient authorities are unintelligible to most individuals today because of their ignorance—and because they are too lazy to learn. Even when they do understand they may not give an impartial account.

The fact is that he who wishes to know more than the multitude must far surpass all others as regards his nature and early training. When he reaches early adolescence, he must become possessed with an ardent love for truth—like one inspired. He must spend night and day learning thoroughly all that has been said by the most illustrious ancient authorities. Having learned all this, he must spend a prolonged period testing—observing what agrees with the ancient authorities and what does not—accepting one and rejecting the other. To such individuals my hope is that my treatise will prove of the very greatest assistance.

Yet such people may be few in number. For the others this book will be as superfluous as a tale told to an ass. (3.10)

Among "those other asses" were, in Galen's opinion, many of the Roman physicians of his time.

The Greek Miracle

The Greeks we have discussed were truly different from other ancient people. They discarded supernatural explanations and based their view of nature on what their senses told them, together with a rational analysis of those observations. They were inquisitive and had a burning desire to understand.

There is no doubt that the Greeks Aristotle and Galen were better biologists than the Roman Pliny. However, it is doubtful that the Greek and Roman masses differed in any appreciable manner from each other. The Greek miracle in biological science was based on a handful of individuals who probably had very little influence on their fellow citizens.

But progress in the arts, sciences, politics, and technology is always the product of a few individuals working in an environment that makes that progress possible. Thus, for our purposes, it is important to emphasize the accomplishments of the outstanding intellectuals of an age rather than attempt an evaluation of general opinion. Progress is a matter of chance and necessity: the chance appearance of a talented mind plus the absolute necessity of a favorable environment in which that mind may flourish. Such an environment was available for those interested in the biological sciences for a few centuries or so in ancient Greece and later in Alexandria. Then it ended.

Seemingly such was not the case in Rome, where the genius of its most talented citizens was turned to the problems of government, world conquest, empire, architecture, and public works instead of the basic sciences.

By the time of Galen's death the slow decline of the Roman Empire was under way, and in a few centuries it was to be replaced by a society in which the intellectuals held totally different views of nature and the place of human beings in it. Naturalistic patterns of thought were to be replaced by the supernatural. The Greek vision was to lie dormant for centuries as the Judeo-Christian worldview prevailed among the finest minds of those times and, as well, was pressed upon the multitudes. Progress in those aspects of civilization that require an unfettered mind—pure and applied science—came nearly to a halt.

chapter 4

The Judeo-Christian Worldview

All belief systems tend to close the mind. If one is wholly committed to democracy, for example, it is less easy to appreciate other forms of government. If one truly believes in the dogma of one religion, one cannot accept fully the dogma of another. True belief requires the acceptance of some things and the exclusion of others. Throughout human history the dominant belief systems of the majority have included a moral code and some force or deity that can, at will, abrogate the laws of nature. It is with this supernatural element that we now will be concerned.

We know of no belief system that challenged the supernatural mode of thought until the Ionian Greeks proposed to explain the phenomena of nature in objective terms. This mode of naturalistic thought appealed to some of the greatest thinkers of antiquity, but it was to be succeeded in the Western world by a period during which intellectuals returned to the supernatural mode, in the form of organized religion. There is little evidence that people in other parts of the world ever abandoned the supernatural mode of thought.

A fundamental difference between religious and scientific thought is that the received beliefs in religion are ultimately based on revelations or pronouncements, usually by some long-dead prophet or priest. These revelations and pronouncements become the dogma of the faith. Dogma is interpreted by a caste of priests and is accepted by the multitude on faith or under duress. Acceptance of a common dogma is one of the most cohesive forces in society so, not surprisingly, it tends to be forcefully promoted by priests and rulers, who may be greatly rewarded by so doing. In contrast, the statements in science are derived ultimately from the data of observation and experiment, and from the manipulation of these data according to logical and often mathematical procedures.

Since religion is not based on confirmable data, one can expect a great range of opinions about what it is and what it stands for. The following description of a fundamentalist type of Judeo-Christian thought has been widely accepted to this day and probably would not have been objected to by the Fathers of the Church: There is a supernatural person or force, God, who looks out for the welfare of individuals, especially those who worship him; there is a life after death for human beings, and this may involve a close association with God; individuals can request the assistance of God by prayer, and it may be granted; for the most part God leaves the working of the world to the natural order, but He has the power to override the laws of nature and do miraculous things—miracles being important evidence for God's existence.

As explained by the theologian John L. McKenzie (1965, p. 578), "Modern theology defines miracle as a phenomenon in nature which transcends the capacity of natural causes to such a degree that it must be attributed to the direct intervention of God." There are numerous miracles in the Old Testament. For example: a bush spontaneously catches fire yet is not consumed—a double miracle (Exodus 3:2–6); snakes can talk to human beings in a language that human beings can understand—at least Eve could (Genesis 3); sticks can turn into snakes and then back again (Exodus 3:2–4, 7:10–12); or sticks can turn into maggots (Exodus 8:16–19); water can be changed to blood (Exodus 3:8, 7:17–22); skin diseases can appear and disappear in an instant (Exodus 3:6–7); the motion of the sun can be reversed (Isaiah 38:8) or even stopped entirely (Joshua 10:12–14); loud noises can make the battlements of a city collapse (Joshua 6:20); and, of course, the greatest miracle of all was the creation of the earth and its living inhabitants (Genesis 1–2).

Between the times the Old and New Testaments were written, the Mediterranean world experienced the extraordinary phenomenon of Greek science and philosophy. This was a new way of thinking that influenced all intellectual activity. The conquests of Alexander carried Greek culture throughout the classical world, where it became known as Hellenistic civilization. Israel, as well as all other regions, was deeply affected, and secular Jews played an important role in Hellenistic times. Nevertheless, Greek naturalistic thought seems to have had no influence on those who wrote the New Testament. The New Testament is as full of miracles as the Old Testament. There are reports of walking on water, raising the dead, casting out demons, instant curing of long-standing diseases, calming storms, and vastly increasing the quantity of loaves and fishes.

The Old and New Testaments were available to the Christians of the early centuries AD. They were accepted as divinely inspired and became the spiritual basis of Christianity. There were, however, varying inter-pretations of what the Bible was saying, and a series of brilliant scholars attempted to discover its messages. Among these scholars St. Augustine was the most important. He, more than anyone else, set the pattern for Christian belief because of the sheer magnitude of his published works and the intellectual quality of his arguments—a pattern that was to dominate Western thought until the Renaissance.

The Bishop of Hippo

As the centuries passed, Christian dogma was to become more and more based on interpretations of scripture rather than on scripture itself, and St. Augustine was central to this development. It was felt that only the most learned could interpret some of the more obscure points. He attempted to explain such a difficult theological question as the nature of the Trinity. It is not immediately apparent how three entities—God the Father, Christ the Son, and the Holy Ghost—are really one, or why it is important to claim that they are. Augustine discussed this and many other matters and in so doing established the field of theology.

Augustine was born in AD 354 in North Africa, which at that time was part of the Roman world. He became bishop of the African city of Hippo, where he died in 430. He was active, therefore, about two centuries after Galen. This was a critical period for Christian religion and for the Roman Empire. The former was increasing in importance and power and the latter was declining and falling. When Augustine died, Hippo itself was under siege and was about to be vandalized by the Germanic tribe, the Vandals.

Although St. Augustine's interests were overwhelmingly theological, he did express opinions about natural events. For example, he said that natural phenomena are to be explained by theological methods. More-over, nothing is to be accepted save on the authority of scripture, since that authority is greater than all the powers of the human mind. In fact, all wisdom and everything worth knowing is to be found in the Bible. Thus scholarship is basically Bible study.

Augustine held that, at the beginning of creation, matter was made from nothing. When asked what the Creator was doing before he created the world, he replied that he would not quip, as others had done, that God "was preparing hell for people who pry into mysteries." It is better, he said, just to admit that you do not know.

The problem of the two accounts of creation in Genesis was "solved" by assuming that creation took place both in six days and in an instant. He thought that there was a good reason for six days: six is the first perfect number, but we must not say that it is a perfect number because God finished all his works in six days but that God did so in six days because six was a perfect number. But the second account of creation gave no time limit so it had been decided that it had occurred in an instant. For the most part, Augustine read the accounts of creation in Genesis as literally as possible. Nevertheless, some statements were not readily understandable and, hence, he and the greatest minds of the age attempted interpretations.

Augustine held open the possibility that not all creatures appeared fully formed during the first six days of creation. Some might have remained as dormant "seeds" and become activated later:

> Just as the seed contains all that is necessary in the course of time to grow into a tree, so the universe must be conceived as having had at the same time all things that were to be made in it since God created all things at the same time. [Included here are] also those things which earth and water produced potentially and causally prior to the time they came into being in the form they are now known to us. (*De Genesi ad litteram* 5.23)

Not only was God the Creator but He is necessary for the continued existence of all creatures:

> Indeed the power . . . and strength of the Creator . . . is for each and every creature the cause of its continued existence . . . If this strength were at any time to cease directing the things that have been created, they would cease to be. (6.22)

Augustine believed in the fixity of species:

> A bean does not grow from a grain of wheat, nor wheat from a bean. A beast does not give birth to a man, or a man to a beast. (9.32)

Some animals, such as frogs, mice, worms, and flies, were thought to be quite superfluous since they seemed to have no importance to human beings.

St. Augustine forcefully insisted on the authority of the Church, basing his view on apostolic succession. This was the belief that, beginning with the Apostles, there had been a God-directed succession of bishops of the Church whose opinions were reflections of God's will. The Pope, as Bishop of Rome, was held to be preeminent because he was in direct succession from the Apostle St. Peter, the "rock" upon which Christ said the Church would be built.

All this may seem of little importance to the development of science but quite the contrary: If Christianity had been just another one of the innumerable religions springing up in the dying days of the Roman Empire, Augustine's claims for complete authority and by implication for the infallibility of the Church might have been regarded merely as arrogant, amusing, or irrelevant. But the Christian Church was to become the dominant religious, political, and intellectual force after the passing of Roman civilization. As its power increased, so did its ability to demand obedience to what it claimed to be "truth." So Augustine's views became important because the Church became powerful.

It is hard to think of any attitude more devastating to science or to any intellectual discipline—even theology. Knowledge comes only if there is the freedom to seek it. Had the Church restricted its authority to matters of religion and morals, it might not have inhibited the growth of science. However, it claimed authority for everything on the basis of what it interpreted the Bible to be saying. A literal interpretation of scripture had the sun rising and setting, so any proposal that the apparent motion of the sun was a consequence of the earth's rotation was heresy and was to be dealt with as the gravest of sins.

Augustine accepted, of course, that nature was the work of an omnipotent God who, in the final analysis, was incomprehensible. It was clear to him, however, that the significant thing was to try to know God, so any interest in living creatures could be only for that purpose. But there was a more practical reason for not wasting time on nature study. The Bible was interpreted as indicating that the end of the earth was nigh. In the little time left there were more important things to do than to study the products of creation.

Augustine believed that the Creator had infused filth and carrion with the power to generate insects, worms, and many smaller creatures. This hypothesis made Noah's task of assembling *all* creatures in the Ark far simpler—and Adam would not have the huge task of giving names to all of them—a task unfinished to this day. Augustine had great difficulty understanding how wild animals living in faraway places could reach the Ark. He suggested that the transfer may have been accomplished by angels.

Although most theologians of the early Church read the scriptures as indicating a flat earth, Augustine was not so sure. But he was vehement in denouncing the view that human beings lived at the antipodes, that is, on the opposite side of the earth. If they did, they would not be able to see Christ descending from the heavens at the Second Coming.

He reported seeing a huge tooth—surely a fossil, but he would not have known that—and assumed that it had come from a giant of olden days. There were fierce theological debates about the age of the earth —some claiming 6,000 years, others 4,000. There was a strong argument for 6,000: since Adam had been created on the sixth day, it was reasonable to assume that the second Adam, Christ, would come after 6,000 years. In any event, Augustine believed one of the greatest heretical opinions would be to accept that Creation had taken place more than 6,000 years ago.

Augustine believed in devils, miracles, and magic, recounting that a drug placed in cheese could cause people to change into animals. He also taught that bones of the saints could heal the sick and that sickness was the result of demons.

One of Augustine's arguments was to have serious consequences for personal and public health for a millennium and a half. He denounced bathing because the Roman public baths were places where activities apart from cleansing the flesh were commonplace. In the West it was not until the beginning of the twentieth century that frequent bathing became common once again.

Augustine and most other theologians denounced the dissection of the human body. This prohibition had some strange consequences— among others the cessation of removing the flesh from the bones of fallen Crusaders. This procedure had made it more convenient to carry their bones home for burial.

Augustine taught that all of nature was good and all was the work of the Creator. If this was so, then life had a history—it had neither existed forever nor had it undergone a series of cycles as some Eastern religions maintained. It would have had a beginning, a period of increase as it populated the earth, and presumably it was to have an end. This was not evolutionary thinking, but it did prepare the way for the hypothesis of noncyclical changes over time—a point of view that was a step along the road of making biology a more conceptual science.

Again and again, in reading Augustine, one notes that much of what he believed about nature survives in the minds of many people of this day. For example, to Augustine it was necessary to remember the final goal—trying to know God—and not to be distracted by an undue interest in worldly phenomena. Augustine read St. John's first letter (2:15–16) as saying that a search for knowledge of earthly things is dangerous. To this day many people seem to accept that point of view. There is no question that St. Augustine discouraged free inquiry about earthly matters.

Free inquiry about religious matters became a hazardous enterprise as well—if you happened to hold a point of view that was considered wrong by more powerful individuals in the Church. There were innumerable arguments in the early Church about original sin, the sacraments, the Trinity, matrimony, the Mass, invocation of the Saints, authority of the popes, relics, demons and devils, the nature of sickness, inerrancy of the Bible, the proper date for Easter, baptism, and so on. The point of view that lost in the debates became heresy, and those who continued to uphold such a view were heretics. Later, a special institution, the Inquisition, was established to deal with these wrong-thinking people.

Critics of the early Christian Church often blame it not only for the destruction of Roman civilization but for the destruction of science as well. So far as Roman civilization was concerned, it can be argued that Christianity was only one of many factors that hastened Rome's fall. Decline and periods of anarchy were occurring as early as the third century AD. Even then the empire was about to be divided, since a single center at Rome could not rule the provinces extending from Britain to Mesopotamia and from Germany to the Sahara and defend their frontiers. Eventually the empire was divided into an eastern part based on Constantinople, now Istanbul, and a western part based on Rome.

By the end of the fourth century, Rome had lost most of its political importance. The city was sacked several times by Germanic invaders, and the last Roman Emperor in the West was deposed in 476. Anarchy ensued, and the Catholic Church was the sole institution of stability and power. It was inevitable, therefore, that its authoritarian belief system would replace all others. In 313 the Emperor Constantine granted religious freedom to all, which lessened the persecutions of the Christians—for a time—and began a spectacular increase in their numbers and importance.

As for the charge that Christianity destroyed science, there was hardly any to destroy. Rome had made little contribution to science, and even the imported Greek science was passing into oblivion. If we take the date of the fall of Rome to be AD 476, we are talking about *eight centuries* after Aristotle. Even Aristotle did not usher in a period of sustained achievement in science—in fact, it peaked with him and then largely declined. There was a brilliant, though brief, period of Hellenistic science centered in Alexandria in the two centuries before its absorption into the Roman Empire in 30 BC. Nevertheless, it is true that the attitudes of the Church prevented the development of science for more than a thousand years and inhibited it for centuries thereafter—and does so to some extent to this day.

Possibly those practical Romans saw no utility in science for science's sake. One could argue that the biological problems that were of interest to Aristotle would be useless in furthering commerce, industry, agriculture, or even human health. This was largely true of medical research as well. Knowledge of human anatomy and physiology, as synthesized by Galen, was of little importance in those days when diseases could not be recognized with certainty and, if they were, no effective medicines were available to cure them. Getting well depended to a large degree on "letting nature take its course." The outcome of illness would be the same whether or not the arteries contained *pneuma* or whether or not blood passed through those invisible pores in the interventricular septum.

Scientific knowledge did not become obviously useful until centuries later, nor did it expand to any appreciable degree. We draw the curtains, therefore, for a thousand years. During that long period the writings of St. Augustine were of basic importance, but not for science. During the early medieval period the intellectual's challenge was to be found in the problems of theology. This had its flowering in the eleventh century in the form of scholasticism.

Scholastic Thought

Scholasticism was a unification of theology and philosophy with the central goal of proving the existence of God. It would have been of little importance for science had it not been, for centuries, the dominant mode of thought of intellectuals—the group from which those with an interest in science would have been expected to emerge.

The scholastic method for arriving at truth, so widely used in the Middle Ages, has been much maligned by later scholars. A debased variation of it remains an important pattern of thought for many people to this day. The method accepts the opinions of others rather than data

7 *Mythological creatures from the twelfth-century Abbey Church of S. Antimo at Montalcino, near Siena, Italy.*

personally obtained by observation and experiment. Since the opinions of others might differ, a formal way of seeking "truth" became common: proposition, opposition, and resolution. That is, the question was raised and the supporting answers of accepted authorities were listed. Then the opposing answers were listed and, finally, an attempt was made to adjudicate the differences and reach some acceptable conclusion. This scholastic method was eminently suited for those whose disputations were on theological subjects. In fact, it is hard to think of any other way to decide such questions, short of bribery or violence.

Truth existed in the mind of God, and it was the task of mortals to fathom what that truth might be. The procedure was logical reasoning based on scripture, church dogma, and the opinions of revered philosophers. Thus, in the last analysis, all data were derived from revelation and right-thinking people. Faith came first, understanding later.

A notable exponent of scholasticism was Peter Abelard (1079–1142), who, however, exposed the fundamental weakness of the approach. In his famous treatise *Sic et Non* he lined up the "yes" and "no" opinions about the same question and showed that equally respected sources could hold diametrically opposed views. The Church did not find his point of view amusing. This was the Abelard who had an affair with the beautiful and loving Heloise. Her father felt strongly about that and had Abelard castrated to cool his ardor. It did.

Scholasticism precluded science. Even those who were interested in science looked to Aristotle and Galen for the answers, not to nature itself. This was a far cry from those naturalistic Greeks who probed nature with mind to obtain understanding in contrast to those who sought understanding by probing mind with mind. But slowly scholasticism revealed its inadequacy as a method of understanding man or nature, and the inquisitive mind turned elsewhere.

Islamic Science

During those long centuries when the Western view of nature was dominated by supernatural explanations, Chinese and Islamic civilizations were flourishing. China remained isolated from Europe until the thirteenth century, but learning in Islam was based on earlier Greek works. For example, in his famous *Canon of Medicine* Avicenna, or Ibn Sina (980–1037), combined the biology of Aristotle and Greek medical lore with what the Islamic physicians had discovered. When it became apparent that Islamic medicine was superior to that of Western Europe,

Avicenna's *Canon* became an authority in the West and remained so until the sixteenth century.

The Islamic scholars contributed greatly to mathematics and astronomy—most of the names we use today for the brighter stars are Arabic. Though their science was in the descriptive, data-gathering stage, with little conceptual advance, the Arabs are given much of the credit for keeping Greek science and philosophy alive during the early medieval period.

The Arab outlook on medicine, as contrasted with that in the Christian West, is shown by the following event that occurred in the twelfth century and was reported by an Arab physician:

> They brought me a knight with an abscess in his leg, and a woman troubled with fever. I applied to the knight a little [poultice]; his abscess opened and took a favorable turn. As for the woman I forbade her to eat certain foods, and I lowered her temperature. I was there when a Frankish [*i.e.* western European] doctor arrived, who said, "This man [meaning the Islamic physician] cannot cure them." Then, addressing the knight, he asked, "which do you prefer, to live with a single leg, or to die with both legs?" "I prefer," replied the knight, "to live with a single leg." "Then bring," said the doctor, "a strong knight with a sharp ax." The doctor stretched the leg of the patient on a block of wood, and then said, "cut off the leg with the ax, detach it with a single blow." Under my eyes the knight gave a violent blow. He gave the unfortunate man a second blow, which caused the marrow to flow from the bone, and the patient died immediately. As for the woman, the doctor examined her and said, "She is a woman with a devil in her head. "Shave her hair." They did so; she began to eat again—like her compatriots—garlic and mustard. Her fever grew worse. The doctor said, "the devil has gone into her head." Seizing the razor he cut into her head in the form of a cross. Then he rubbed her head with salt. The woman expired immediately. After asking them if my services were still needed, and after receiving a negative answer, I returned, having learned from them medical matters of which I had previously been ignorant. (Munro 1903)

Books on Beasts

The *Physiologus* began as a Greek work of the third century AD. The title means "One who knows nature," and rarely has there been a more inappropriate title, since so much was imaginary. The purpose was to explain the allegorical significance of animals, according to the Christian view of the animal kingdom. The *Physiologus* was widely copied and modified, and these copies were also known as *Bestiaries*. They were collections of stories about animals, real and imagined, modified to illuminate principles of morality and allegorical interpretations of the Bible and Church doctrine.

The following excerpts from a discussion of lions in a twelfth-century *Bestiary* give the flavor:

> Scientists say that Leo has three principal characteristics. The first is that he loves to saunter on the tops of mountains. Then, if he should happen to be pursued by hunters, their smell reaches up to him, and he then disguises his spoor with his tail. Thus the hunters cannot track him.
>
> It was in this way that our Saviour, the Spiritual Lion of the Tribe of Judah, once hid the spoor of his love in the high places. When sent by his Father he came down and into the womb of the Virgin Mary and saved the human race which had perished . . .
>
> The Lion's second feature is that he seems to sleep with his eyes open.
>
> In this very way, Our Lord while sleeping in the body was buried after being crucified. Yet his Godhead was awake. As it is said in the *Song of Songs,* "I am asleep and my heart is awake," or in the Psalm, "Behold, he that keepeth Israel shall neither slumber nor sleep."
>
> The third feature is this. When the lioness gives birth to her cubs, they are dead and are lifeless for three days. Their father comes on the third day and breathes on their faces and makes them alive.
>
> Just so did the Father Omnipotent raise Our Lord Jesus Christ from the dead on the third day. Quoth Jacob: "He shall sleep like a lion, and the lion's whelp shall be raised." (modified from White 1954)

Thus the animal kingdom was viewed as a reflection of Christian doctrine, and all intellectual interest in nature was organized on that basis. Animals and plants were thought of as symbols for various traits, and as such they played a most important part in the art of the Renaissance. Thus the reason that paintings of the Nativity usually include an ox and an ass can be traced to Isaiah 1:3, "The ox knoweth his owner, and the ass his master's crib." The basilisk, a mosaic of rooster and snake, was regarded as a symbol of the devil, the bee of industry, the bull of strength, the camel of temperance (it can go without drinking for many days), the cat of laziness, the rooster of watchfulness, the dog of fidelity, the dove of peace, the fly of sin, the fox of cunning, the locust of plagues, the hog of gluttony, the lamb of Christ (John 1:29— "Behold the Lamb of God"), the lion of courage, the peacock of immortality (it was believed that its flesh would not decay), and so on.

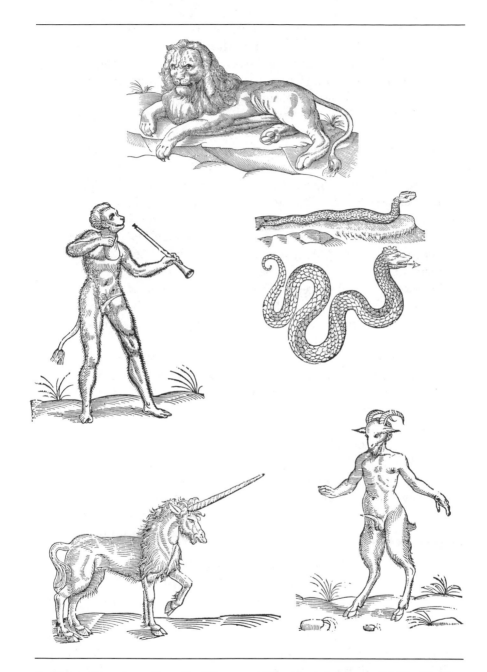

8 *Illustrations from Topsell's* History of Four-Footed Beasts *(1607). The lion (above) is a reasonably accurate representation of nature, but the same cannot be said for the fantastic figures below: a satyr, a basilisk (the king of the serpents), a unicorn, and an aegopithecus (a combination of ape and goat).*

Even to this day we attribute specific virtues to animals—fidelity of dogs, for example—that trace back to the medieval *Bestiaries.*

By the sixteenth century the *Bestiaries* were losing their influence because books about animals more in the style of Pliny—that is, folklore but without Christian symbolism—began to appear. The most famous of these was the *Historiae Animalium* published in five volumes, from 1551 to 1587, by the Swiss naturalist Conrad Gesner (1516–1565). This was translated into English and expanded by Edward Topsell (*ca.* 1572–1638) as *The History of Four-footed Beasts* (1607 and many later editions).

We can compare Topsell's approach to the lion with that of the twelfth-century *Bestiary* just described. He provides 25 large pages of information, required because of the importance and dignity of the lion. He notes the names used for the lion in various languages and the references to it in the Bible. Other animals may beget lions. For example, a concubine of the first king of Sardis bore him a lion. The soothsayers told the king that all places in Sardis where this lion walked would be safe forever. So the lion was taken round all the towers and battlements except for one—and years later that is where the soldiers of Darius breached the walls and took the city of Sardis.

Topsell notes that lions are found nowhere in Europe except in Greece. Essentially all that he has to say about the behavior of lions comes from literature, not from nature. The story about the lion's fear of white roosters is included—just as it was in the twelfth-century *Bestiary* and in Pliny as well.

Topsell describes the uses, mainly medical, of the species he discusses. Blood of the lion will cure cankers and, if smeared over the body, will protect human beings from all wild beasts; lion grease is equally protective; eating the flesh will prevent bad dreams; shoes made of lion skin will prevent gout; by smearing the body with fat from the lion's kidney, one will be protected from wolves; lion fat or feces can be mixed with ointments and used to cure acne; lion grease plus oil of roses as well as dried heart powder cure malaria; lion brains, when taken in drink, drive one mad; lion gall is useful in jaundice and eye diseases; fat from the lion's private parts can prevent conception. This is only a portion of the lion pharmacopoeia.

Topsell pays much attention to the lesser creatures as well. For example, he summarizes the available literature relating to "the vulgar little mouse" and then describes its medicines. A mouse can be skinned, cut in two, and placed over an arrow wound to help the healing process; if a mouse is beaten into pieces and mixed with old wine, the concoction will cause hair to grow on the eyelids; if skinned, steeped in oil, and rubbed with salt, the mouse will cure pains in the lungs; sodden mice can prevent children from urinating too much; mice that are burned and converted to powder are fine for cleaning the teeth; mouse dung, prepared in various manners, is useful for treating sciatica, headache, migraine, the tetters, scabs, red bunches on the head, gout, wounds, spitting of blood, colick, constipation, stones, producing abortions, putting on weight, and increasing lactation in women.

All this is a very far cry indeed from Hippocrates and his benign, rational medicine. How could it be that, generation after generation, people would put their faith in what Hippocrates of long ago and we today would regard as nonsense? Part of the answer is that when people are sick they want *something* to be done. In the absence of effective medicines, that something is likely to be anything—especially if the taste is terrible and the source is repulsive. That point of view goes back to the demons-are-the-cause-of-disease period when the effectiveness of the medicine depended largely on its being so unpleasant to the demons that they would depart the patient forthwith. A more potent reason is that these medicines from "the vulgar little mouse" seemed to be effective cures—after all, sick people nearly always got well. If they had used mouse dung in order to get well, and had gotten well, surely the remedy was effective. And we must not forget the well-documented power of placebos. In short, mouse dung *can be* effective —if one has faith in its efficacy. There are millions of individuals in the world today whose level of knowledge about animals and diseases shows no advance over that of Topsell.

In the *Bestiaries* there is no emphasis on personal observations, the need to check data, or naturalistic descriptions. Nevertheless, Topsell was able to expunge most theology from his book. Knowledge of the living world was improving, and we can recognize Topsell as closing out the biology of the Middle Ages. His *History* is a link between the fanciful, symbolic, theological descriptions of animals of the *Physiologus* and *Bestiaries* and more modern approaches to understanding.

Antecedents of a Revolution

By the thirteenth century essentially all of Greek philosophy and science, with their fresh and open-ended procedures, had become available to Western scholars in Latin translations from Arabic sources (which were themselves translations from Greek). Once the scholastics' awe of the Greek accomplishments was overcome and the bondage to accepted authority broken, scholars could imitate what the Greeks did, not parrot what they said. Science became possible once again.

Among the many important innovations that helped make the revival of science in the sixteenth and seventeenth centuries possible, of exceptional importance was the invention of movable type of Johann Gutenberg (*ca.* 1397–1468), which permitted the production of *identical* multiple copies of books. That invention was basic for reliable communication in science. No longer would the flow of information among individuals and between generations be dependent on the accuracy of a scribe, who might not only introduce errors of transcription but, even worse, attempt to be creative. Printed books could also be illustrated, which was a great boon for biology—with words alone it is hard to describe a clam shell, for example, to say nothing of a lobster.

The appearance and growth of universities in Western Europe was also important for progress in science. Science requires interactions among individuals, and traditionally universities have made that possible. Higher education was not readily available in classical times, although Plato and Aristotle had schools and the Library-Museum in Alexandria was noted for its scholarship. But such institutions were few and ephemeral. Universities appeared in the late Middle Ages, and by the sixteenth century there were many in Europe. They prepared students for careers in theology, law, and medicine. A vital service was the university library, which could maintain collections of books well beyond the means of individual scholars.

And finally, museums, which started anew in the sixteenth century, helped set the stage for a revival of biology. Conrad Gesner started a museum, and soon others appeared in many European cities. But not many sorts of animals could be kept in museum collections. Soft-bodied ones would rot quickly, since there was no known way then to preserve them. Some species with hard parts could be dried—insects, for example. Zoological collections, therefore, were restricted to materials such as dried arthropods, the bones of vertebrates, shells of mollusks, hard parts of coelenterates (corals, sea fans, and so on), and skins of animals and birds, until Robert Boyle (1627–1691)—who first distinguished elements from compounds, recognized the nature of chemical reactions, and the interrelations of pressure, volume, and temperature in confined gases—hit upon a way to preserve the soft structures of animals. He explained his method at a meeting of the Royal Society of London on May 7, 1666:

The time of the year invites me to intimate to you, that among the other Uses of Experiment, I long since presented the *Society,* of preserving Whelps taken out of the Dams womb, and other *Faetus's,* or parts of them, in *Spirit of Wine* [probably about 85 percent alcohol]; I remember, I did, when I was sollicitous to observe the Processe of Nature in the Formation of a Chick, open Hens Eggs, some at such a day, and some at other daies after the beginning of the Incubation, and carefully taking out the *Embryos,* embalmed each of them in a distinct Glass (which is to be carefully stopt) in *Spirit of Wine:* Which I did that so I might have them in readinesse, to make on them, at any time, the Observations, I thought them capable of

9 *The Imperato Museum in Naples, Italy, founded by Ferrante Imperato (1550–1625). Ferrante and his son explain the collection of curiosities, mainly animals, to visitors. Museums such as this stimulated interest in natural history.*

affording; and to let my Friends at other seasons of the year, see, *both* the differing appearances of the Chick at the third, fourth, seventh, fourteenth, or other days . . . When the *Faetus's*, I took out, were so perfectly formed as they were wont to be about the seventh day, and after, they so well retain'd their shape and bulk, as to make me not regret of my curiosity: And some of those, which I did very early this Spring, I can yet shew you. (Boyle 1666, pp. 199–200)

Scientific societies would play a key role in the advance of science after the sixteenth century. The first was the Royal Society of London, formed in 1662 for "the improvement of natural knowledge." Like its later sister institutions, the Royal Society fostered communication through lectures, informal discussion among its members, and publications—periodicals for shorter communications and, later, monographic series for more extensive scientific contributions. In its early years the Royal Society of London served as a center of scientific communication for all of Western Europe. Its long and honorable history continues right up to the present day.

chapter 5

The Revival of Science

By the sixteenth century the umbilicus to Aristotle was being severed, the world of nature was being accepted as fit for inquiry, and science based on observation and experimentation was becoming respectable. But there was no sudden springing to the barricades at the onset of the period that we now know as the Scientific Revolution. In fact, it had no obvious beginning—only a slow spread of a new way of defining the methods of obtaining knowledge of natural phenomena. Some historians date the onset of the Scientific Revolution at about 1660, near the time of the founding of the Royal Society of London and Sir Isaac Newton's studies of gravitation. Such a date, however, excludes most of the intellectual giants who truly gave us science as a way of knowing—Vesalius, Harvey, Bacon, Copernicus, Galileo, Kepler, and Brahe. For reasons about to be mentioned, my preference is 1543. No matter which birth date is selected, we can speak of a "revolution" because some exceptional contributions to natural knowledge were made in a relatively brief time.

The year 1543 saw three key events in the history of science. One was the recovery, translation, and publication of the works of the Greek physicist and mathematician Archimedes (287–212 BC). He had made astonishing contributions to mathematics and mechanics, and he was a notable inventor. He viewed the universe itself as a gigantic machine, operating on mechanical principles. This was a liberating notion in the sixteenth century, when the forces of nature were thought mystical and probably unknowable. The mechanics of Archimedes were basic to the work of Galileo and then to Newton's.

A second accomplishment in 1543 was the publication of *De Revolutionibus Orbium Coelestrium* by the Polish physician, clergyman, and astronomer Nicolaus Copernicus (1473–1543). Here was the beginning of modern astronomy, which held that, contrary to the opinion of the Greek astronomer Ptolemy, the sun, not the earth, is central. This hypothesis of heliocentrism lacked adequate proof. Only later was the Copernican theory verified beyond all reasonable doubt through data carefully collected by the Danish astronomer Tycho Brahe (1546–1601), refinements in the theory by the German astronomer Johannes Kepler (1571–1630), and observations by the Italian physicist and astronomer Galileo Galilei (1564–1642). (Science by this time was becoming truly international.)

Doing science in the sixteenth and seventeenth centuries called for great courage as well as a disciplined and open mind. As we have seen, the Church, never a supporter of open minds, had a fixed position on many scientific questions. Generations of Catholic scholars had studied the Bible, the available works of Greek philosophers and scientists, and the writings of the Fathers of the Church and on these had built a system of thought that became Church dogma—the official and sole system of permitted thought. One challenged that authority at great risk. The Church had declared the Ptolemaic astronomical theory correct because, among other things, it had placed the earth at the very center of the universe—the proper position for one of God's main creations. Copernicus knew the risk he was taking by demoting the earth to the position of a minor planet, and, being cautious, he dedicated *De Revolutionibus* to the Pope, indicating that his view was "just a theory." He probably saved himself much trouble by dying of natural causes very shortly after the book was published. Others were not so lucky. Giordano Bruno, a Dominican monk who opposed all dogmatism and in the main accepted the Copernican theory, paid for his intellectual independence by being burned at the stake in 1600. Luther had this to say about Copernicus: "This fool wishes to reverse the entire science for astronomy; but sacred Scripture tells us that Joshua commanded the sun to stand still, and not the earth." But Brahe, Kepler, and Galileo had done their work, and there was no escaping the conclusion that the earth rotates on its axis each day and circles the sun each year. Galileo had two trials by the Inquisition, and at the second in 1663 he was forced to make a public recantation of his belief in the heliocentric theory.

A third event of significance in 1543 was the publication of *De Humani*

Corporis Fabrica by Andreas Vesalius (1514–1564), a Belgian who became Professor of Anatomy at the University of Padua. Prior to this time Galen's anatomy was the authority. Vesalius was able to dissect human bodies and found that Galen was inaccurate in some instances. *De Humani Corporis Fabrica* has not only a complete description of gross anatomy but also beautiful illustrations by the Belgian artist Jan van Calcar and others (and not by Albrecht Dürer, as some have assumed). This was the beginning of modern anatomy and is a straight path to Grey.

Initially Vesalius had much opposition, since even suggesting that such an ancient and respected authority as Galen might have erred was not in the best of taste. One brave, free spirit who suffered because he thought otherwise was Michael Servetus (1511–1553), a scholar of broad interests, mainly theological, but also a serious student of Galen. In the course of his studies, he came to the conclusion that Galen was not correct in all matters. Servetus hypothesized, for example, that blood does not pass directly through those Galenic pores from right ventricle to the left but instead goes from the right ventricle to the lungs, where it picks up air, and then back to the left ventricle. Mainly because he questioned theological dogma but partly because he questioned Galen, whom the Church had named as the authority on anatomy and physiology, Servetus was captured while at prayer and, after a brief trial, was sent up in flames on October 27, 1553. Lest the reason be in doubt, one of his offending books was hung from his neck so it too was consumed on the pyre.

The Church was imprisoned by its own dogma. It had selected the best available scientific data before Copernicus and Vesalius and in that sense it was up to date and as correct as it could be. However, concepts that become dogma are not easily changed by better data that suggest new concepts.

Andreas Vesalius and the Study of Structure

The contributions of Copernicus and Vesalius differed fundamentally —a reflection both of the conceptual level of astronomy and biology of the sixteenth century as well as of the relative degree of complexity of the two fields. Copernicus was responsible for a true paradigm shift. By putting the sun in the center of the solar system, with the planets revolving around it, Copernicus laid the groundwork for modern astronomy.

In contrast with Copernicus, Vesalius was not responsible for a paradigm shift, but his contributions to anatomy were epoch-making. Vesalius was born in Brussels on New Year's Eve of 1514. He studied Galenic anatomy at Paris and then returned to Louvain, where he reintroduced the practice of human dissections. Later he went to the University of Padua, spent two days taking examinations, and was awarded a medical degree *magna cum laude*. The following day he was appointed to lecture on surgery and anatomy at this most famous of all medical schools at that time. This must be a record for academic advancement. Vesalius departed from tradition by conducting the dissections himself instead of following the custom of reading aloud from his lecture notes, which were based on Galen's writings, while an assistant pointed out the anatomical parts being mentioned.

Vesalius was a careful student of Galen but, as his experience with human dissections grew, he became convinced that there were many errors in Galen's descriptions—and he suspected the basic reason for these errors. Galen was probably never allowed to make a systematic and detailed dissection of the human body he was describing in such detail. That would have been forbidden when he lived, and what he had to say about human anatomy was based in part on the earlier work of the physician Herophilus, who did dissect human beings, plus his own dissections of apes and other mammals. Vesalius was able to work extensively on human bodies and came to realize that some of the things reported for human anatomy by Galen were really the anatomy of apes.

Because of Vesalius, knowledge of the human body became much more accurate. *De Humani Corporis Fabrica* is a huge, beautiful book. Most noteworthy are the splendid woodcuts, which convey information about structure in ways the written word never can. The artists, including Jan van Calcar, appear to have been students working in Titian's studio.

It is often believed that after Vesalius there was little to be discovered about human anatomy. This is most certainly not the case—discoveries continue to this day. But it must be admitted that no major organs have turned up recently. Vesalius was able to start with what was known of human anatomy, correct some of the errors, and add new material of his own. One of his greatest contributions was that his conclusions were based on what he saw and not on the received authority of Galen. He also maintained that medical students should learn anatomy by doing their own dissections.

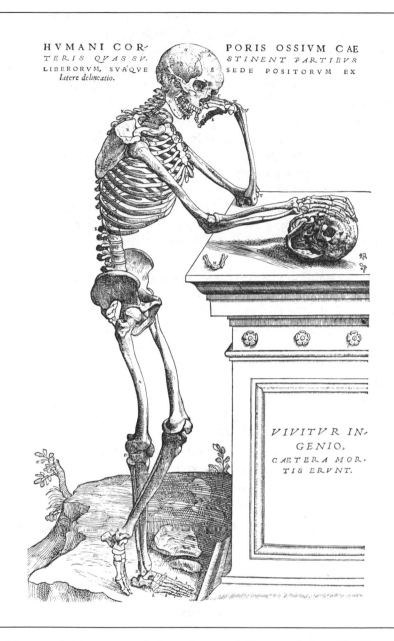

HVMANI COR-   PORIS OSSIVM CAE
TERIS QVAS SV-   *STINENT PARTIBVS*
LIBERORVM, SVAQVE   SEDE POSITORVM EX
latere delineatio.

VIVITVR IN-
GENIO,
CÆTERA MOR-
TIS ERVNT.

10   *A human skeleton contemplating a human skull, from Andreas Vesalius's* De
Humani Corporis Fabrica *(1543). Consider how very different this way of illustrating a skeleton is from current practice. The same philosophy is to be seen in
figures 11–14.*

In many instances Vesalius continued to rely on Galen for questions that could not be answered by careful dissections. He followed Galen in believing that blood was made in the liver and that nerves were hollow. Failing to find any pores in the interventricular septum, he marveled at the Creator for making it possible for blood to sweat from one side to the other. Vesalius's illustrations of "human" arteries seem to be based in part on Galen's apes.

Vesalius was severely criticized by many anatomists who continued to accept the Galenic tradition, but he soon had followers, especially among the younger anatomists. After the publication of his masterpiece, his anatomical studies were few and he became a practicing physician. For him the effective medical tools were diets, drugs, and surgery. Knowledge of anatomy was basic for the last. Apparently he was a gifted physician and for many years served Charles V, emperor of the Holy Roman Empire.

Vesalius's painstaking descriptions of human anatomy were far more important to surgeons than was Aristotle's concept that the structure of all mammals are but variations on a fundamental plan. Yet those who collect reliable data and those who formulate concepts form a symbiosis that is essential for scientific advances. Data are sought to answer questions; the answers may become important concepts; the concepts suggest new questions, and so on. Thus science can advance only if there are individuals who can collect data as well as those who can interrelate data to establish broad concepts. The metaphor of generals and soldiers may be applicable—neither alone wins battles.

Vesalius was not the first to obtain a near-modern understanding of human anatomy. In 1517, when Vesalius was three years old, Leonardo da Vinci was living in southern France. Cardinal Luis of Aragon and his party

> went to see Messer Lunardo Vinci the Florentine . . . This gentleman has written of anatomy with such detail, showing by illustrations the limbs, muscles, nerves, veins, ligaments, intestines and whatever else there is to discuss in the bodies of men and women, in a way that has never been done by anyone else. All this we have seen with our own eyes; and he said that he had dissected more than thirty bodies, both of men and women, of all ages. (MacCurdy 1938, p. 13)

Leonardo's observations were recorded in his notebooks only—he published neither descriptions nor illustrations. It is only within our century that his observations have become available and have been

11　In a series of plates Vesalius shows a flayed body with successive layers removed. Here the muscular system is entire and the specimen is enjoying a stroll through the countryside.

analyzed. The surviving illustrations are far better than those of Vesalius, in part because we see them in modern reproductions and not as the comparatively crude woodblocks available in Vesalius's time.

McMurrich (1930) expresses the opinion that "Vesalius was undoubtedly the founder of modern anatomy—Leonardo was his forerunner, a St. John crying in the wilderness." The moral of this for science is clear: Leonardo kept his information to himself and hence made no contribution to the science of his day. Science is a community effort, and it advances only when information is shared and the conclusions of one scientist are confirmed by others.

William Harvey and the Study of Function

William Harvey (1578–1657) was born 14 years after the death of Vesalius. He also was a physician to kings and a student of the human body—both the structure and function of its parts. He is remembered today mainly for his studies of the heart, circulation, and of embryonic development, which will be discussed in Part Four.

Although the importance of heart, blood, and lungs was accepted, there was only a vague understanding of their functions. Nevertheless, Vesalius, Harvey, and their fellow scientists proposed explanatory hypotheses that, at the time, could not be tested adequately. The Galenic hypothesis was that all parts of the body required "food" and "*pneuma*" for their living processes and that the blood vessels carried them. What was not known was how these substances were moved. As noted before, it was held that blood was made in the liver from materials received from the alimentary canal. This blood, which served as food for the parts of the body, was distributed by the veins—except to the lungs, which received blood from the pulmonary artery. The *pneuma* required by the parts of the body was carried to them by the arteries. The Galenic view was essentially correct: the parts needed food (=blood) and *pneuma*, and they received them from the veins and arteries.

Harvey, using the time-honored method of studying the same phenomenon in a variety of organisms, each giving a partial answer, observed the heart and blood vessels in many sorts of living vertebrates —the answer could never come from dead specimens. His experiments led to the conclusion that blood is circulating constantly throughout the body in a system of tubes: veins, heart, and arteries. He knew there

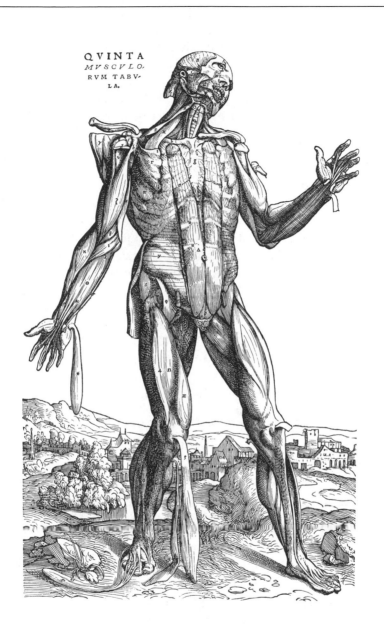

12 *Some of the outer muscles have now been removed and there are obvious signs of fatigue.*

must be connections between arteries and veins, now known as capillaries, but it remained for others to discover them.

This might seem like a trivial advance—just describing how blood moves—but it represents the first important discovery in physiology that was substantiated by observations. Harvey tested his hypothesis by extremely simple observations and analysis that were equally available to the Ionian Greeks. It may be pointless to ask why the Ionians never did what Harvey did. The answer may be that the level of maturity of biology was such that the questions and methods to solve them would not have occurred to any Ionian.

Once again we observe how difficult it has been to understand natural phenomena and how slowly come the answers. This may seem surprising considering the widespread notion that there are set procedures in science—the scientific method—that, if dutifully followed, will lead inexorably to new discoveries and deeper understanding. These methods, which have proved so effective, were poorly formulated before the seventeenth century.

In that century Sir Francis Bacon (1561–1626), Lord Chancellor of England, analyzed the ways scientists could gain understanding of natural phenomena. As de Solla Price (1975) put it, with deliberate hyperbole, "Francis Bacon plotted the [scientific] revolution and codified the scientific method."

Sir Francis Bacon's Great Instauration

Bacon provided a philosophical system for investigating natural phenomena and emphasized the importance of experimentation, which previously had not been commonly used for testing ideas. He was a dominant figure in exploring the nature of science and in what came to be known as "the scientific method" for gaining knowledge of the natural world. The essence of the scientific method was its rejection of the classical and medieval theological habit of starting the inquiry with a point of view that was accepted as true and then deducing the consequences (Bacon 1857–1874, 1937, 1960).

A classical example of this deductive reasoning was the acceptance of the Judeo-Christian God as the creator of the universe and all its inhabitants and then deducing what was thought to be some necessary consequences: creation occurred only a few thousand years ago; all species created remained the same; no species ever became extinct; pairs of all living creatures could be accommodated in a single boat, the

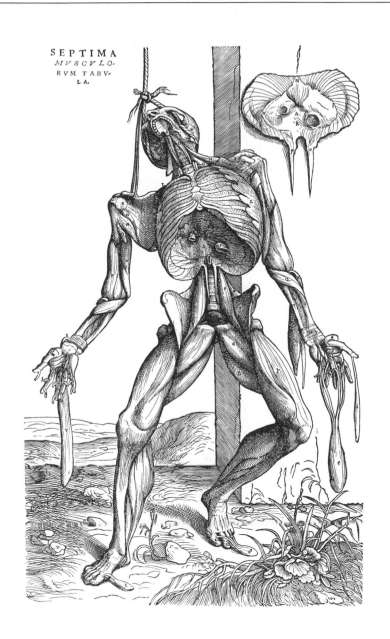

SEPTIMA
MVSCVLO-
RVM TABV-
LA.

13 *Most of the muscles and viscera have been removed. Clearly the body needs support.*

Ark. The Judeo-Christian worldview had been accepted as adequate for centuries—and remains so for many individuals today—but it leads to a very different view of nature than the one provided by modern science.

A diametrically opposed point of view began to develop as the Scientific Revolution emerged during the sixteenth and seventeenth centuries. Bacon's suggestion was to begin with data, not faith. That is, one should consider all known facts related to some natural phenomenon and then try to formulate hypotheses to explain those facts. This logical method of reasoning from the particulars to the general is known as induction—a procedure that was to give us the modern worldview.

Bacon's philosophy was presented in his *Instauratio Magna* of 1620. This was planned as a multivolume work, but only a small portion of it was ever published, the most notable being the *Novum Organum or True Suggestions for the Interpretation of Nature*. Even this was a preliminary abstract and consists of 129 aphorisms in Book 1 and 52 in Book 2. The old "Organon" consisted of the logical treatises of Aristotle, the procedures of which Bacon wished to replace.

His argument begins by pointing out how ineffective traditional attempts are for understanding nature. Bacon notes that, unless great care is taken, the things that the human mind imbibes tend to be "false, confused, and overhastily abstracted from the facts." This is often due to our interpreting our observations in terms of what we have *already assumed to be true*. We see what we believe rather than believe what we see. The consequence of this *a priori* approach is that "philosophy and the other intellectual sciences . . . stand like statues, worshipped and celebrated, but not moved or advanced." It is no wonder that our understanding of nature is "badly built up, and like some magnificent structure, without any foundation."

Bacon argued that reliable knowledge of the natural world comes from observing nature itself and not from probing the human mind. Nature was to be the arbiter "to commence the total reconstruction of sciences, arts, and all human knowledge"—his "Great Instauration" (renovation).

Every scientific investigation should begin by assembling all the data from observation and experiment that related to some natural phenomenon, taking great care lest erroneous information be included. That, of course, would lead to erroneous conclusions. Not only must the observations be made as accurately as possible, but often "neither the

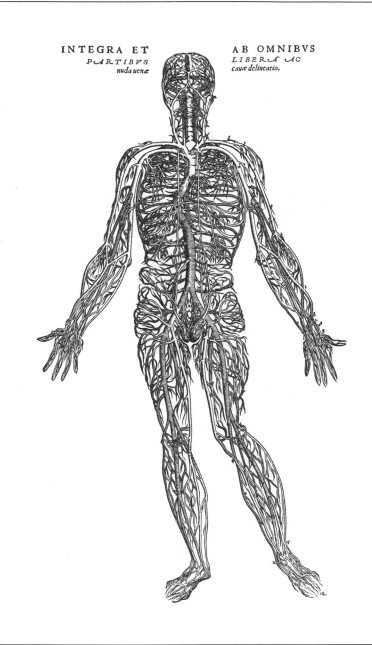

INTEGRA ET   AB OMNIBVS
PARTIBVS   *LIBERÆ AC*
nuda uenæ   *cauæ delineatio,*

14   *Vesalius's illustration of human arteries and veins.*

naked hand nor the understanding left to itself can effect much. It is by instruments and helps [for the mind] that the work is done."

The mind must guard against preconceived ideas if observations are to be accurately interpreted. This is extraordinarily difficult to achieve, since what we are, think, and do depends so greatly on our acceptance of the belief systems of the society in which we live and of the science that we profess. These belief systems become the idols to which we may submit and, to the extent we do, may lead to erroneous conclusions. Bacon lists four: the Idols of the Tribe, Cave, Marketplace, and Theatre. (Bertrand Russell recognizes still another, the Idols of the Schools, 1945, p. 544.)

The Idols of the Tribe consist of the erroneous preconceived ideas and fuzzy thinking common to one's tribe, or community.

The Idols of the Cave are the erroneous beliefs of each individual's mind—the mind being like an isolated cave. He notes especially how individuals tend to favor their own opinions and discoveries—a serious problem for us to this day.

> Men become attached to certain particular sciences and speculations, either because they fancy themselves the authors and inventors thereof, or because they have bestowed the greatest pains upon them and become most habituated to them. But men of this kind, if they betake themselves to philosophy and contemplations of a general character, distort and colour them in obedience to their former fancies . . . (Book 1, Aphorism 54)
>
> And generally let every student of nature take this as a rule,—that whatever his mind seizes and dwells upon with peculiar satisfaction is to be held in suspicion . . . (Book 1, Aphorism 58)

The Idols of the Marketplace are the semantic problems that arise when people try to communicate and use words differently. The words of our language were developed for everyday use; not infrequently, they are unsuitable or insufficiently specific for use in science.

The Idols of the Theatre consist mainly of adhering to philosophical and theological modes of thought where "truth" is deduced from *a priori* premises. He notes, for example that some have attempted to found a system of natural philosophy (that is, natural science) on the first book of Genesis. He advises, however, that "We be sober-minded, and give to faith that only which is faith's."

And there are more general problems, for it is not "to be forgotten that in every age Natural Philosophy has had a troublesome adversary and hard to deal with; namely, superstition, and the blind and immoderate zeal of religion." Or, most discouragingly, in schools and universities "similar bodies destined for the abode of learned men and the cultivation of learning, everything is found adverse to the progress of science . . . But by far the greatest obstacle to the progress of science . . . is found in this—that men despair and think things impossible."

After these lengthy discussions of what he regarded as the procedural and philosophical errors of the past that made progress in science difficult or impossible, Bacon introduces his new approach with this charming metaphor.

> Those who have handled sciences have been either men of experiment or men of dogmas. The men of experiment are like the ant; they only collect and use: the reasoners resemble spiders, who make cobwebs out of their own substance. But the bee takes a middle course; it gathers its material from the flowers of the garden and the field, but transforms and digests it by a power of its own. Not unlike this is the true business of philosophy; for it neither relies solely or chiefly on the powers of the mind, nor does it take the matter which it gathers from natural history and mechanical experiments and lay it up in the memory whole, as it finds it; but it lays it up in the understanding altered and digested. Therefore from a closer and purer league between these two faculties, the experimental and the rational (such as has never yet been made) much may be hoped. (Book 1, Aphorism 95)

Thus Bacon believes that his is the first suggestion for the systematic use of experimental methods to understand natural phenomena. It is not at all obvious, however, how one should try to be a bee. In Book 2 of *Novum Organum* he provides an example of what he has in mind by an analysis of what could be the true nature of heat. How is one to understand this phenomenon that is a constant feature of our ambient environment?

First, one should assemble all the readily available information about heat. Bacon gives three tables of relevant data. The "Table of Existence and Presence" enumerates many phenomena associated with heat: rays of the sun, meteors, thunderbolts, volcanic eruptions, flames, sparks, burning solids, quicklime sprinkled with water, horse-dung when fresh, strong wines, some spices, acid poured on the skin, and even intense cold.

A second "Table of Deviation and Absences" lists phenomena where we might expect heat but do not find it. For example he notes that although light from the sun's rays is hot, light from the moon and stars is cold. Furthermore, there have been instances where a person's hair is surrounded by what appears to be flames—but the hair never burns. This phenomenon is known as St. Elmo's Fire.

The third table of "Degrees or Comparisons of Heat" lists instances where the same item may differ in temperature. For example, plants are not warm to human touch but may become so if they are enclosed in a box or allowed to decay. The heat of animals is increased by exercise, wine, feasting, Venus, fever, and pain.

Induction, Hypothesis, Deduction

Now comes the truly extraordinary part of Bacon's analysis. It seems impossible that anyone could consider all these varied, often irrelevant, and dubious bits of data listed in his three tables and, by induction, arrive at an understanding of heat. First he eliminates some possibilities. For example, light cannot be the basis of heat, since light from the moon and stars is not hot even though light from flames and the sun may be. Color cannot be the cause, since a red hot iron and the relatively cool flame of burning alcohol differ so much. Heat cannot be a substance, since iron and other materials may be made hot and not lose substance.

After ruling out many possibilities, Bacon reaches this astonishing conclusion:

> From instances taken collectively, as well as singly, the nature whose limit is heat appears to be *motion*. This is chiefly exhibited in flame, which is in constant motion, and in warm or boiling liquids, which are likewise in constant motion. It is also shown in the excitement or increase of heat by motion and by bellows and draughts . . . It is also shown by the extinction of fire and heat upon any strong pressure, which restrains and puts a stop to motion . . . (thus is with tinder, or the burning snuff of a candle or lamp, or even hot charcoal cinders, for when they are squeezed by snuffers, or the foot, and the like, the effect of the fire instantly ceases) . . . It is further shown by this circumstance, namely, that every substance is destroyed, or at least materially changed, by strong and powerful heat: whence it is clear that tumult and confusion are occasioned by heat, together with a violent motion in the internal parts of bodies, and this gradually tends to their dissolution . . . It must not be thought that heat generates motion, or motion heat, (though in some respects this be true,) but that the very essence of heat . . . is motion and nothing else. (*Novum Organum*, Book 2, Aphorism 20)

The data available to Bacon were wholly inadequate for him to reach this view of the nature of heat that we now accept as correct. Furthermore, induction—the philosophical process that Bacon so valued—

could not alone sort among all of the observations Bacon thought to be relevant and conclude that heat is a form of motion. In this case a fine mind had made a lucky guess.

This is not an isolated example. In the years following Bacon many important scientific discoveries turned out to be based on erroneous data. In these instances the acceptable evidential basis was provided later.

Perhaps the greatest weakness in Bacon's system was the lack of a clear indication on how to make the intellectual step from the isolated facts to the general statement—the hypothesis. That remains the central difficulty of inductive reasoning to this day. It is here that genius, intuition, inspiration, serendipity, and luck—one or several—must assume control of the analysis.

Two and a half centuries later the great English scientist John Tyndall (1863) had this to say: "From the direct contemplation of some of the phenomena of heat, a profound mind is led almost instinctively to conclude that heat is a kind of motion, Bacon held a view of this kind." But the *sine qua non* is that profound mind.

Induction means no more than that one begins a study with observation and experimentation relating to some natural phenomena and uses the data to reach an understanding of the fundamental causes or the association of seemingly unrelated events. Selected data are used to frame provisional hypotheses, and from these hypotheses deductions are made and tested. Deduction remains a powerful adjunct of analysis, but the deduction of modern scientists is not the same as the deductive reasoning that Bacon found so repugnant. In science today deductions from a hypothesis are necessary conclusions from that hypothesis. Their value is to suggest what observations or experiments can be done in order to confirm or deny the hypothesis, nothing more. In this manner the original hypothesis is improved or replaced. The new hypothesis then becomes the basis for new deductions and they, in turn, refine or replace the working hypothesis of the moment. This constant interplay of induction, hypothesis formation, deduction, and testing leads to improved understanding of the phenomenon being explored.

The deductions of the early philosophers and theologians were often regarded as eternal conclusions drawn from eternal truths, but in reality they were based on shared faith or bold imagination. Thus philosophers and theologians began with a statement accepted as true, in contrast with scientists who begin with a statement, a hypothesis, which they seek to confirm or disprove through tests.

To this day scientists strive to start only with the most reliable and confirmable data, and thereafter employ a constant interplay of inductive and deductive procedures to reach a more fundamental level of understanding of the natural world. That understanding can be no more than "this is the most accurate statement that can be made with the evidence at hand." It must be emphasized that this does not mean that the science of the day is "wrong." It means that the science of today is provisional and will be replaced by better science tomorrow. Our analysis of the development of genetic concepts will provide an excellent example. The genetics of Mendel in 1865 was not wrong. It was expanded to be the better genetics of Sutton in 1903 and then to the still better genetics of Morgan in 1912, and finally, to the vastly better genetics of today.

Some philosophers of science have maintained that Bacon was seriously inadequate in not appreciating the value of deductive reasoning. To be sure, he was not as explicit as the philosophers of today; but considering the pioneer nature of his effort, one can argue that he comes off fairly well. For example, "The signs for the interpretation of nature comprehend two divisions: the first regards the eliciting or creating of axioms from experiment, the second the deducing or deriving of new experiments from axioms" (*Novum Organum*, Book 2, Aphorism 10). Today we would use "hypotheses" as a synonym for "axioms," so we find here not only a description of what philosophers of today recognize as important components of scientific methodology but about as accurate a statement as can be made of what working scientists actually do. And far from expecting scientific understanding to come solely from those ants collecting facts, Bacon suggests that with his approach, "we have good reason, therefore, to derive hope from a closer and purer alliance of these faculties, (the experimental and rational) than has yet been attempted" (Book 1, Aphorism 95).

The fundamental difference between Bacon's approach and the approaches that he attacked was that scientific statements must be based on the data derived from observations and experiments of natural phenomena and not on preconceived principles, or beliefs of classical authors, or imagination, or superstition. As Bacon advises, we "should not arrogantly search for the sciences in the narrow cells of human wit, but humbly in the greater world."

Thus it is incorrect to say that Bacon believed that induction is the only effective procedure for arriving at acceptable scientific statements. His emphasis on induction was to counter the seemingly total reliance of philosophers and theologians on deductive reasoning from broadly inclusive *a priori* beliefs. Induction is not an automatic procedure for

advancing science. It depends absolutely on the brilliance, perseverance, knowledge, and luck of the scientist. And deduction is an effective
and powerful procedure when one uses it to make testable deductions
from provisional hypotheses.

The legacy of Sir Francis Bacon, Lord Chancellor of England, is that
we must study nature, not books alone, and cease the worship of those
four idols. Scientists of the seventeenth century—Andreas Vesalius,
Galileo Galilei, Johannes Kepler, William Harvey, and Sir Isaac
Newton—were attempting to do these things. Bacon was most influential in being a publicist and codifier of the Great Instauration—a new
way of obtaining reliable information about the natural world.

The Very Small—Animalcules

The followers of Vesalius in anatomy and Harvey in physiology were
to continue with their predecessors' approaches and would contribute
much new information. In the late sixteenth century, however, studies
of entirely different sorts, which seemed to have little likelihood of
making major contributions to the way we look at the animal kingdom,
were under way. One, which we will look at in this chapter, was the
study of the very small; the other, taken up in the next chapter, was
the study of the very old. Without the former, the study of genetics
would have been extremely limited; without the latter, there would
have been no modern theory of evolution.

Over most of history the human view of nature was restricted by the
capabilities of the sense organs—mainly the eyes. This view included
creatures as large as great trees, whales, and elephants at one end of
the scale and as small as tiny flies and cheese mites at the other. The
limits of resolution of the human eye—about 1 minute of arc—meant
that the smallest creatures that could be seen would be about 0.1 millimeter in diameter.

Was there life apart from these visible members of the animal and
plant kingdoms? Larger life was unlikely, but smaller life was distinctly
possible. Such life was about to be discovered—the world of microorganisms. We now know that they are the most abundant forms of life,
probably accounting for more than 99 percent of all individuals. A gram
of fertile soil may contain a million microorganisms whose activities are
essential for the life of all the macroorganisms. Our knowledge of them
awaited the invention of magnifying lenses.

A single double convex magnifying lens, or simple microscope, seems to have been known by the middle of the fifteenth century. It magnified a few diameters and provided the viewer with some of the fascinating details of the structure of insects and other living and nonliving objects.

The simple microscope reached a considerable degree of perfection in the hands of Anton van Leeuwenhoek (1632–1723) of Delft, Holland. His instrument consisted of a tiny double convex lens held in a frame and a movable needle to which a specimen could be attached and viewed. The greatest magnifications he was able to obtain appear to have been about 200 times. This was sufficient to permit him to see bacteria of about 0.002–0.003 millimeter in size.

In 1673 the Secretary of the Royal Society of London, Henry Oldenberg, began correspondence with Leeuwenhoek. For 50 years Leeuwenhoek wrote the members of the Society about his discoveries. They were impressed and supportive. He was made a Fellow of the Royal Society in 1680, a very great honor indeed. Many of Leeuwenhoek's letters appeared in the *Philosophical Transactions of the Royal Society.*

The following letter of September 7, 1674, records the discovery of a previously hidden world (these and other quotations are modified from Dobell 1960):

> About two hours distant from Delft there is a lake with marshy or boggy places on the bottom. Its water is very clear in the winter but in the summer it becomes whitish and there are then little green clouds floating in it; which, according to the country folk, are caused by the dew. Passing the lake recently, and seeing the water as just described, I took a little vial of it. When I examined it next day, I found earthy particles floating in it and some green streaks, spirally wound and orderly arranged. The whole circumference of each of these streaks was about the thickness of the hair of one's head [Dobell identifies these as the green alga *Spirogyra*]. Among these were also very many little animalcules, some of which were roundish [protozoans]. Others, a bit bigger, were oval and I saw two little legs near the head and two little fins at the hind end of the body [rotifers]. Others were somewhat longer than an oval and these were moved very slowly and were few in number [probably ciliates]. The motion of most of these animalcules in the water was so swift and so varied—upward, downward, and round about—that 'twas wonderful to see. I judge that some of these little creatures had a volume less than one thousandth that of the smallest ones I have ever seen upon the rind of cheese, in wheaten flour, mould, and the like.

Over the course of years, Leeuwenhoek sent descriptions of a rich variety of objects he saw with his simple microscope, organisms now recognized as protozoans, single-celled algae, rotifers, and bacteria. He found these in rainwater after it had stood for several days, in sea water, in pond water from many sources, and in infusions of pepper, ginger, cloves, nutmeg. He observed "eels" (nematode worms) in vinegar. He estimated that 27,000,000 of his animalcules were equivalent to a grain of sand and that a cubic inch could hold 13,824,000,000,000 of them.

Leeuwenhoek examined bile and the contents of the gut of various animals and human beings and found bacteria and protozoans. He found bacteria in the human mouth, red blood corpuscles in blood, and animalcules in semen of humans and other animals. The last he called spermatozoa, "sperm animals."

He tried to test Redi's hypothesis that spontaneous generation of organisms could not occur in closed containers but found that bacteria appeared. He remarked that maybe Redi was talking about "only worms or maggots, which you commonly see in rotten meat, and which ordinarily proceed from the eggs of flies, and which are so big that we have no need of a good microscope to descry them" (p. 198). However, he doubted that spontaneous generation could occur and suspected contamination as the source of the bacteria.

Although Leeuwenhoek had revealed a new world, one beyond the capabilities of the unaided human eyes, he felt that "all we have yet discovered is but a trifle, in comparison of what still lies in the great treasury of nature" (p. 192). And why did he do it?

> The learned Secretary of the Royal Society writes with so many expressions of satisfaction at my discoveries, which I sent to your Fellows, that I stood all abashed when the letter was read to me. Nay, my eyes filled with tears at all the great expressions and respect that they have for my work. My work has been accomplished alone and by my own impulse and inclination for no money could even have driven me to make discoveries. I am only working out as 'twere an impulse that was born in me. I do not believe I have ever met with others who would spend so much time and work in searching into the things of Nature. (pp. 88–89)

A few others were driven with the same impulse to search for that hidden treasure, and slowly, over the decades, they too observed and described the world of microorganisms. Such a person was Robert Hooke, a contemporary of Leeuwenhoek. He was a most active member of the Royal Society and thus knew of Leeuwenhoek's work, repeated some of it, and made many new microscopic observations of his own.

Robert Hooke and the Discovery of Cells

Robert Hooke (1635–1703) was more interested in the microscopic structure of the parts of visible animals and plants than in microorganisms. For these studies he designed a compound microscope, quite different from the single lens instruments used by Leeuwenhoek.

The compound microscope consists of two lenses, one at each end of a tube of leather, wood, cardboard, or, later, brass. They were probably first made between 1590 and 1609 in Holland. These early instruments with uncorrected lenses gave very poor images. Nevertheless they were widely used in England and Italy as well as Holland—any image is better than no image. In 1830 Joseph Lister (not the one of antisepsis fame) showed how lenses corrected for both chromatic and spherical aberration could be made. These achromatic lenses were the best available until apochromatic lenses were produced late in the nineteenth century.

Hooke began to demonstrate objects with his microscope to the Fellows at the very first meetings of the Royal Society in the 1660s, and he was requested to continue to do so at all meetings—he was appointed Curator of Experiments. In 1665 his discoveries were published in *Micrographia or Some Physiological Descriptions of Minute Bodies, Made by Magnifying Glasses; with Observations and Inquiries Thereupon.*

He began with "Observ. I. Of the Point of a sharp small Needle" and observed that, "If view'd with a very good *Microscope*, we may find that the *top* of a Needle (though as to the sense very *sharp*) appears a *broad, blunt,* and very *irregular* end; not resembling a Cone, as is imagin'd, but only a piece of a tapering body, with a great part of the top remov'd, or deficient" (pp. 1–2). Hooke examined a great many familiar nonliving objects: the edge of a razor, linen cloth, silk, pieces of glass, sand, ice crystals. Objects of biological origin were gravel in urine, charcoal, petrified wood, cork, mould, moss, sponges, seaweed, leaves, spines on nettles, seeds of many plants, hair from many animals, fish scales, bee stings, feathers, as well as the external structure of many species of arthropods. A wealth of detail was revealed that was unknown without magnification.

Hooke's *Micrographia,* with its splendid woodcuts and descriptions of the microscopic world, made an enormous impression on his contemporaries. Samuel Pepys, the English Admiralty official, diarist, and subsequently President of the Royal Society, entered in his diary for January 20, 1665: "To my bookseller's, and there took home Hooke's book of Microscopy, a most excellent piece, and of which I am very proud." And the next day he wrote: "Before I went to bed I sat up till two o'clock in my chamber reading Mr. Hooke's Microscopical Observations, the most ingenious book that I ever read in my life."

Hooke is usually remembered for describing those boxlike structures, which he called "cells," in a slice of cork. That observation was the beginning of a line of research that established the concept that cells are the fundamental units of structure and function of living organisms. The key episodes in that drama were Hooke's description of the cellular nature of cork and a few other plants; the work of many microscopists who concluded that many plants were cellular; Schleiden and Schwann's hypothesis that all organisms are cellular; and Virchov's hypothesis that cells originate only from other cells. Thus cells became

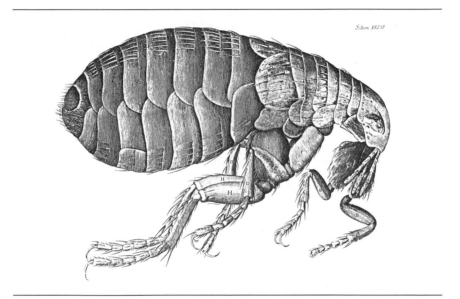

15 *The famous flea from Robert Hooke's* Micrographia *(1665). "The strength and beauty of this small creature, had it no other relation at all to man, would deserve a description . . . As for the beauty of it, the* Microscope *manifests it to be all over adorn'd with a curiously polish'd suit of* sable *Armour, neatly jointed, and beset with multitudes of sharp pinns . . . the head is on either side beautifi'd with a quick and round black eye." Hooke's careful description, together with the huge illustration of so humble and bothersome a creature, was found amusing or ridiculous by many of his contemporaries.*

the units not only of structure and function in embryos and adults but in inheritance as well. The sum total of information about them came to be called the cell theory.

The microscope, which eventually became the distinctive tool of biologists, began to decline in importance in the late seventeenth century. The instruments gave poor images and the methods for preparing biological materials for examination were inadequate. Writing in 1692, Hooke remarked that few grinders of lenses were still at their benches, and Nicolson (1972, p. 22) comments that

> Only Leeuwenhoek remained, and he had become a lonely figure, without many scientific followers, the butt of satire, a fantastic who propounded what to the layman seemed like absurdities . . . For a time the microscope ceased to be an important scientific instrument and became a plaything of the aristocracy—most of all, of the "ladies" . . . As it ceased temporarily to influence the history of science, it became an important influence in the history of literature.

And of satire there was aplenty. Thomas Shadwell's drama, *The Virtuoso*, opened in 1676 with its hero, Sir Nicholas Gimcrack, a takeoff on Hooke. Somehow a deep concern for the detailed structure of louse and flea was not understood by the public at large—nor for that matter would it be today.

Vesalius, Harvey, Leeuwenhoek, and Hooke made differing contributions to the biological side of the Scientific Revolution, but not one of these remarkable men was responsible for a concept of major significance or a true paradigm shift. There emerged no theory of protozoans, theory of human anatomy and physiology, or theory of insect structure—nor could any theories emerge since no fundamental questions were being asked, no major hypotheses being tested. The work of these scientists was mainly descriptive, not theoretical, biology. In fact, the naturalists of those days were suspicious of speculation, being influenced by that guiding light of the Royal Society, Sir Francis Bacon, who was vigorous in his attacks on the speculating scholastic philosophy, which to him was the antithesis of science. Hooke came the closest to a fundamental biological theory with his discovery of cells, yet even this was of trivial importance at the time and remained so for more than two centuries. The concept that cells are the fundamental units of life became firmly established only in the nineteenth century.

This is not to depreciate the work and significance of those sixteenth- and seventeenth-century biologists but to emphasize again the enormous amount of labor that has been required to provide the conceptual framework of biology that today organizes our view of the world of life. Biology had to await the nineteenth century for a true revolution.

chapter 6

Figur'd Stones and Plastick Virtue

The earth's crust is a cluttered, complex place. In areas populated by human beings, artifacts such as houses, roads, beer bottles, and wrecked automobiles are clearly associated with human beings, and there is no problem in accounting for their origin. Leaves and seeds, feathers, empty shells, and bleached bones are recognized as having an organic origin. Then there are objects with no obvious relation to organisms or any evidence of their origin—the rocks and minerals that come in various shapes, colors, and consistencies. But a fourth class of objects—confusing to naturalists before the seventeenth century—were rocks that had some resemblances to organisms or parts of them. Even ancient observers knew of what appeared to be shells of marine organisms and the bones of huge animals. They were called "figured stones."

Were they related in some way to organisms or were they merely stones with strange designs? The eventual understanding of their nature was to revolutionize the way we look upon the kingdoms of life. Just as Leeuwenhoek, Hooke, and others had discovered the world of the very small, those who sought to understand figured stones discovered the world of the very old. The discovery of that world was to involve science and the Judeo-Christian worldview in yet another confrontation—one that lingers to this day.

In the seventeenth century this much could be said about figured stones:

(1) They varied from objects having an obvious resemblance to organisms—shells of mollusks and leaves of plants—through a complete gradation to others resembling no known organism.
(2) Their consistency was that of the stone in which they were situated and not that of living or recently dead plants and animals.
(3) The figured stones that resembled sea creatures occurred not only close to the shore but also high in mountains far from the sea.
(4) The organismlike structures might occur inside hard stones.

The naturalists of the seventeenth century had various explanations for figured stones. Some of them proposed the hypothesis that figured stones represented the range of variation normally encountered in purely inorganic materials and that they had no relation to organisms. The hypothetical cause of the figured stones was "latent plastick virtue," which was present in rocks. That hypothesis was not easy to test.

Other naturalists formulated the hypothesis that figured stones were petrified organisms—fossils, as we call them today. Although this hypothesis eventually proved correct, consider some of the problems with it. If the object was once a living creature composed of soft structures, how could it change to stone? If the fossil resembled a marine organism, a clam let us say, how could one explain its occurrence in rocks high up in mountains and far from the sea? Could one imagine a clam working its way overland for hundreds of kilometers and, once it got to those inland mountains, boring its way into the middle of a huge stone and leaving no trace of its entrance passage?

Robert Hooke was one of those naturalists who accepted the hypothesis that fossils had an organic origin. The following long excerpt from *Micrographia* is a fascinating example of how he attempted to show that petrified wood had indeed begun as wood:

Observ. XVII. of *Petrify'd wood*, and other *Petrify'd bodies*.

Of this sort of substance, I observ'd several pieces of very differing kinds, both for their outward shape, colour, grain, *texture*, hardness, etc. . . . . That which I more particular examin'd, was a piece about the bigness of a mans hand, which seem'd to have been a part of some large tree, that by rottenness had been broken off from it before it began to be *petrify'd*.

Hooke examined a piece of rotten oak wood and charcoal with his microscope and compared them with the petrified wood. The three objects resembled one another closely.

> This *Petrify'd* substance resembled Wood, in that, First all the parts of it seem'd not at all dislocated, or alter'd from their natural Position, whil'st they were Wood, but the whole piece retain'd the exact shape of Wood, having many of the conspicuous pores of wood still remaining pores, and shewing a manifest difference visible enough between the grain of the Wood and that of the bark . . . Next (it resembled Wood) in that all the smaller and . . . *Microscopical* pores of it appear (both when the substance is cut and polish'd *transversly* and *parallel* to the pores of it) perfectly like the *Microscopical* pores of several kinds of Wood.

So much for the resemblances, now for the differences.

> The [petrified wood] differed from Wood: First, in *weight* being to common water as 3 1/4 to 1. whereas there are few of our *English* Woods, that when very dry are found to be as heavie as water . . . Secondly, in *hardness*, being very neer as hard as Flint . . . Thirdly, in the *closeness* of it, [for the *microscopical* pores of the petrified substance] appear'd darker then the rest of the body, onely because they were fill'd up with a more duskie substance, and not because they were hollow . . . Fourthly, in its *incombustibleness*, in that it would not burn in the fire . . . Fifthly, in its *dissolubleness*; for putting some drops of distilled *Vinegar* upon the Stone, I found it presently to yield many Bubbles . . . Sixthly, in its *rigidness* and *friability*, being not at all flexible but brittle like a Flint . . . Seventhly, it seem'd also very different from Wood to the *touch, feeling* more cold then Wood usually does, and much like other close stones and Minerals.

In this fascinating analysis of the problem of the nature of the petrified object, Hooke gave compelling data for its resemblance to rotting wood. The microscopical evidence was especially convincing, and note that he made a comparison of the petrified object as viewed in two planes. That was an especially demanding test. Hooke also did what all careful scientists should do—he listed the negative evidence. Six of the seven differences listed are precisely what would be expected from a petrified object, that is, one consisting of stone. For the other difference, the third listed, Hooke provided a probable explanation. Thus the hypothesis that the petrified object was altered wood seemed highly probable. His task then was to account for the extraordinary events that saw a piece of rotten wood changed into solid stone.

The Reasons of all which Phenomena seem to be, That this *petrify'd* Wood having lain in some place where it was well soak'd with *petrifying* water (that is, such water as is well *impregnated* with stony and earthy particles) did by degrees separate [an] abundance of stony particles from the permeating water, which stony particles, being by means of the fluid *vehicle* convey'd, not onely into the *Microscopical* pores, and so perfectly stoping them up, but also into the pores or *interstitia* . . . of that part of the Wood, which through the *Microscope*, appears most solid . . . By this *intrusion* of the *petrifying* particles, this substance also becomes hard and *friable*.

It may seem like a blatant *deus ex machina* to invoke "petrifying water." How could water change something to stone? This proposal might seem as dubious as "latent plastic virtue." Not at all—it was well known that caves in limestone regions have long tapering formations, stalactites on the roof and stalagmites on the floor. These were known to be formed by the evaporation of salt-laden water, which had percolated through the soil and dripped from the roof of the cave.

Hooke extended his hypothesis about the organic origin of fossil wood to other kinds of fossils. After a very thorough examination, by eye and microscope, he concluded:

I cannot but think, that all these, and most other kinds of stony bodies which are found thus strangely figured, do owe their formation and figuration, not to any kind of *Plastick virtue* inherent in the earth, but to the Shells of certain Shel-fishes, which, either by some Deluge, Inundation, Earthquake, or some such other means, came to be thrown to that place, and there to be fill'd with some kind of Mud or Clay, or *petrifying* Water, or some other substance, which in tract of time had been settled together and hardened in those shelly moulds into those shaped substances we now find them . . . And he that shall thoroughly examine several kinds of such curiously form'd stones, will (I am very apt to think) find reason to suppose their generation or formation to be ascribable to some such accidents as I have mention'd, and not to any *Plastick virtue*.

Today we read Hooke's hypothesis for the origin of fossils with pleasure and approval because subsequent events proved that he was essentially correct. Not all of his fellow naturalists were convinced, and many questions remained to be answered.

Marine Life on Mountain Tops?

More than a century before Hooke published his observations on petrified wood, Leonardo da Vinci had entered in his notebooks the following solution to the problem of fossils of marine organisms in high places:

> When the floods of the rivers which were turbid with fine mud were deposited upon creatures which have their bones on the outside, such as cockles, snails, oysters, scallops, and the like, these creatures became embedded in this mud. Being entirely covered under a great weight of mud they perished for lack of a supply of the creatures on which they were accustomed to feed. In the course of time the level of the sea became lower and, as the salt water flowed away, this mud became changed into stone; and those shells that had lost their inhabitants became filled with mud, which was changed to stone. (modified from MacCurdy 1938, p. 330)

As was the case with his studies of human anatomy, Leonardo did not publish this hypothesis and hence it did nothing to advance science. This quotation does show how a sixteenth-century genius could propose what was to become the accepted hypothesis of modern geologists: organisms covered with mud, petrification of mud and organisms, and uplifting of the layer of rock with the fossil organisms. Leonardo's hypothesis also accounted for the problem of an organism boring its way into a stone. It didn't. Instead, the mollusk was covered by the soft mud and both were turned to stone.

Robert Hooke, in his remarkable *Discourse of Earthquakes*, which was published posthumously (1705), greatly extended his observations on fossils beyond those in *Micrographia*. His hypothesis for the occurrence of marine fossils far from the sea was similar to that of Leonardo:

> A great part of the Surface of the Earth hath been since the Creation transformed and made of another Nature; namely, many Parts which have been Sea are now Land, and divers other Parts are now Sea which were once a firm land; Mountains have been turned into Plains, and Plains into Mountains, and the like.
>
> Most of those Inland Places, where these kinds of Stones [the fossils] are, or have been, have been heretofore under the Water; and that either by the departing of the Water to another part or side of the Earth, by the alteration of the Center of Gravity of the whole Bulk, which is not impossible; or rather by the Eruption of some kind of subterraneous Fires, or Earthquakes, whereby great quantities of Earth have been rais'd above the former Level of those Parts, the Waters have been forc'd away from the Parts formerly cover'd, and many of those surfaces are now raised above the Level of the Water's Surface many scores of Fathoms.
>
> It seems not improbable, that the tops of the highest and most considerable Mountains in the World have been under Water, and that they

themselves most probably seem to have been the Effects of some very great Earthquake, such as the *Alps* and Appennine Mountains, *Caucasus,* the Pike of *Tenariff,* the Pike in the *Terceras,* and the like. (pp. 290–291)

Figured Stones of Unknown Creatures

Hooke's *Discourse on Earthquakes* described many strange fossils. Some were huge coiled objects, weighing as much as 400 pounds and being upwards of 2.5 feet in diameter. He had been told of others much larger. The common name for these coiled fossils was "snake stones," or *Sceleta Serpentum,* because of their resemblance to a coiled snake. It was difficult to claim that these fossils had an organic origin because there were no known living creatures that they resembled. So the question arose, "How could something that did not exist become a fossil?"

Hooke described the snake stones carefully and decided they were some sort of mollusk. He compared them with the living *Nautilus,* which later work has shown to be their closest living relative. They are now known as ammonites—a group of cephalopods that lived in late Paleozoic and Mesozoic seas and then became extinct.

Then there were the puzzling "tongue stones" on the island of Malta. Some of them were a few inches long. Again, there was no known living creature that resembled a tongue stone. Their correct interpretation, due to Nicolaus Steno (1667), was one of the early triumphs of paleontology. He noticed that the teeth of a modern shark were identical, except for size, with the tongue stones. He concluded that tongue stones were not fossils of whole organisms but were the fossil teeth of sharks. He had no problem with their occurrence on land:

> If we believe the accounts, new islands have emerged from the sea; and who knows where Malta's cradle was situated? Perhaps formerly when this place was submerged in the sea it was the haunt of sharks, whose teeth in times past were buried in the muddy sea-bed; then, afterwards it had changed its level by a sudden ignition of subterranean emanations, these sharks' teeth are found in the middle of the island . . .
>
> Since then the bodies resembling parts of animals that are dug out of the ground may be assumed to be parts of animals; since the shape of the tongue stones is like the shark's teeth as one egg to another; since neither their number nor their position in the ground speaks against it; it appears to me that they cannot be far from the truth who assert that tongue stones are sharks' teeth. (Garboe 1958, pp. 43, 45)

16 *A nautilus shell, helmet stones, and button stones as illustrated by Hooke in*
A Discourse on Earthquakes *(1715). The nautilus is recent and the others are*
fossils related to present-day echinoderms.

Thus during the seventeenth century some naturalists were becoming convinced that life has had a long history and that the rocks of the earth's crust contain the evidence for that former life. It also appeared that some fossils were of creatures that existed no longer. This is what Hooke had to say: "There have been many other Species of Creatures in former Ages, of which we can find none at present; and that 'tis not unlikely also that there may be divers new kinds now, which have not been from the beginning." That was only possible, not probable, since perhaps in some faraway place there lived creatures, yet to be discovered, of the same sorts as those enigmatic fossils.

17   A "tongue stone" magnified two times. Pliny believed that these puzzling stones rained down from the heavens during the period when the moon was waning. Steno suggested they were the teeth of huge extinct sharks, which turned out to be correct. This is a tooth from the shark Isurus hastilus, *from the Miocene of California.*

The hypothesis that some species had become extinct was a serious problem not only for naturalists but also for theologians. Extinct species would seem to imply that the Creator had made a mistake—designed a species that could not survive.

Baron Cuvier

Léopold Chrétien Frédéric Dagobert, Baron Cuvier, known also as Georges Cuvier, was born in 1769 and died in 1832. He came from a humble background and rose to a commanding position in French science and society. He lived through a most stressful period in French history that spanned monarchy, revolution, republic, and Napoleon— and survived. In the years following the Revolution there was great intellectual ferment in France; the importance of science was recognized and scientists were supported. Cuvier held the post of professor of anatomy at the Muséum d'Histoire Naturelle in Paris for most of his adult life.

Cuvier was an exceptional person. Throughout life he engaged in a series of activities that would have taxed the combined talents of several ordinary people. He was arrogant, vain, highly susceptible to flattery, quick to stifle criticism, authoritarian to the many he considered his inferiors, and somewhat fawning to the very few he considered his superiors. He was suspicious of ungrounded speculation and a great admirer of Aristotle. There is no doubt whatsoever that he was a truly great scientist.

Cuvier had an extraordinary memory and was said to know in considerable detail the contents of the nearly 20,000 volumes in his library. This characteristic was ideally suited for his research interests: comparative anatomy and paleontology. For each of these fields one has to command an incredible amount of detail about the structure of organisms. The amount is awesome for one to be a gifted comparative anatomist. To be a gifted paleontologist even more is required. Fossil vertebrates, his favorite material, almost always come as isolated bones or fragments of bones, making the task of identification difficult. Thus a successful paleontologist must have a recallable mental image of the individual bones of innumerable animals, living and extinct. When a new fossil bone or fragment is encountered there will be the possibility of identification.

Cuvier was tireless in building up the collections of the Muséum d'Histoire Naturelle in Paris, and this reference collection, so basic to his research interests in paleontology, included materials from all over the world. This made it possible for him to compare any fossil bones he might obtain with the bones of present-day species in the collections.

One of his first important contributions was to establish that some species had become extinct. That might seem a strange thing to have to prove since long before it had appeared highly probable to Hooke, Steno, and others in the seventeenth century that those huge shark's teeth and ammonites belonged to species no longer living. Later research has shown that their hypothesis is true beyond all reasonable doubt. But there *was* reasonable doubt until the twentieth century.

The problem was that if a fossil species has no known living repre-

18 *A fossil ammonite from the Lower Cretaceous at natural size. Compare with the nautilus of figure 16.*

sentatives, it may really be extinct, or individuals might still be living but not yet discovered. Critics of Hooke and Steno could maintain, quite rightly, that just because no huge sharks had been collected in the Mediterranean, or no ammonites found in the seas around the British Isles, they do not exist. They might be so rare that they were yet to be collected; or, alternatively, they might exist in some faraway ocean or in the depths where they would be safe from collectors. It is impossible to prove a "negative" of this sort in science; the most that can be said is, "As of now, no living ammonites or huge sharks have been encountered."

Subsequent history was to prove the importance of being cautious. A common fossil in some deposits had a buttonlike columnar structure. These fossils resembled no known organism. Only later when entire specimens were discovered was it realized that the puzzling objects were no more than the fragmented stems of crinoids. Another example was a strange fish caught off the east coast of Africa in 1939. It was named *Latimeria* and found to be a coelacanth fish, a group thought to have become extinct about 75 million years ago. A more recent discovery of an "extinct" species is *Neopilina*, found off the coast of Costa Rica. As in the case of *Latimeria*, the first specimens were collected at great depths. *Neopilina* belongs to a group of very primitive mollusks, otherwise known only from the Cambrian—a half billion years ago.

So there was no satisfactory answer to the question of whether or not any species had become extinct when Cuvier began to study the problem around the beginning of the nineteenth century. Not only was this question of great interest to biologists and paleontologists but there were religious overtones as well. One prominent school of thought in the Judeo-Christian tradition interpreted Genesis as saying that all species had been created during Creation Week and that all survive to this day. Extinction would imply that some species were imperfect, which might imply that some of God's work had been inferior. Furthermore, since the date of creation had been established by theologians to be about 4004 BC, there would not have been much time for extinction.

What sorts of fossil species would be the best suited to test the hypothesis of extinction? Cuvier excluded fossils of marine organisms because, as noted before, it would be difficult to prove that a putative extinct species might not be still living and lurking in the briny deep. So he selected a type of fossil species most likely to give an unambiguous answer: large terrestrial quadrupeds, or mammals. Cuvier was able to demonstrate that many fossil quadrupeds were similar to, but not identical with, living species. The bones of fossil elephants differed from those of elephants living in Africa and Asia. Since these fossils were of large terrestrial quadrupeds that had no living counterparts, Cuvier assumed that they had become extinct. Although he realized that the interiors of Africa and South American had not been fully explored, it seemed unlikely that the dozens of fossil species he described as new remained sequestered in those unknown lands. Surely the largest quadrupeds would be the first to be known to the European explorers.

But Cuvier was interested in far more than identifying ancient bones or in the anatomical details of living vertebrates. What were the rules, the causes, and the history of the phenomena he studied? His broad, firsthand experience with the anatomy, and especially the bones of terrestrial vertebrates, led him to realize that all was not chaos—there was a rule that applied to structure, and by extension to function. The general principle—a unity of structural plan and a correlation of parts —had been known to Aristotle, but it remained for Cuvier to show its predictive power.

> Comparative anatomy possesses a principle that when used properly can solve these problems. It is that the parts of organized creatures are so adjusted to one another that each species can be recognized on the basis of each part. Every organized creature forms a unique and perfectly integrated whole. None of its parts can change without the whole changing so all parts can be known on the basis of one.
>
> Thus, if the intestines of an animal are so organized that it can digest only flesh, it follows that its jaws must be constructed to devour prey, its claws to seize and tear it, its teeth to cut and divide it, the organs of locomotion be such as to enable the animals to pursue prey and catch it, its sense organs to detect distant prey, and it must possess a brain that has the necessary instincts to enable the animal to conceal itself and lay snares for its victims.
>
> Such will be the general conditions of all carnivores. Every carnivore will always possess this complex of characteristics for it could not exist without them. But the different species of carnivores may differ in many ways but these are but modifications of the general carnivore condition. Thus among the carnivores not only the class, but the order, genus, and even the species have the characteristic formation of each part. (Cuvier 1831, pp. 58–60)

This unity of plan and correlation of parts can be seen in the details of form and function as well:

> If the jaw of a carnivore is to be able to seize prey, it must have a certain type of jaw articulation with certain relation between the position of the resisting power and that of the force on the fulcrum. The temporal muscle must be of a certain volume and that requires a specific construction of the zygomatic arch. (pp. 59–50)

There is a similar unity of plan and correlation of parts among another great group of terrestrial quadrupeds—those with hoofs.

> We see very plainly that hoofed animals must all be herbivorous because they have no means of seizing prey. Having no use for the front feet other than support of the body, they have no need for a powerfully built shoulder. This accounts for the absence of the clavicle and the acromion [of the scapula] and the straightness of the scapula. Since there is no need to rotate the front leg, the radius will be solidly united to the cubitus, or at least articulated by a hinge joint, instead of a ball and socket, with the shoulder. The herbaceous diet will require teeth with a broad surface to crush the seeds and herbs. The surface of the teeth must be irregular with the enameled and bony parts alternating. This sort of surface requires a horizontal motion in chewing and this grinds the food. The articulation of the jaw cannot form a hinge so close as in carnivorous animals. Instead it must be flattened and adjusted to the temporal bones. The temporal cavity, which will contain only a very small muscle, will be small and shallow. All these things are necessary deductions, one from another. (pp. 61–62)

An extraordinary amount of study of a wide variety of species was required before Cuvier could reach these general conclusions. His approach is a model of scientific investigation. On the basis of observation, a large amount of data is accumulated. These data suggest broad principles, which serve as hypotheses. These hypotheses allow deductions to be made and tested by further observations.

> The least prominence of a bone, the smallest apophysis, have a specific character relative to the class, order, genus, and even species. Even if we have only the extremity of a well-preserved bone we can, by examining it carefully, applying analogical skills, and comparing it with other materials, determine as much as if we had the whole animal. I have often experimented with portions of known animals before applying this method to fossils. This has always led to such infallible success that I no longer doubt the accuracy of the results that will be obtained. (pp. 64–65)

Cuvier's claim of being able to construct a whole animal from a bit of bone has embarrassed paleontologists ever since. It cannot be done. Yet Cuvier seemed able to do so, and there is a special reason for his success. Most of the large fossil mammals he studied were from very recent times—the Pleistocene. Thus the bones of the fossil elephants he studied were so similar to the bones of present-day elephants that once he identified a piece of fossil bone as elephantlike, he could extrapolate to an entire elephant.

The predictive value of Cuvier's approach was given a fascinating public test. He had received a fossil embedded in a rock. Only a small part was exposed but he hypothesized it was a marsupial. That identification was possible since skeletons of living American and Australian marsupials were in the museum's collections. Therefore when the skeleton was exposed, it should have the characteristics of marsupials. Now marsupials share with the monotremes a unique structure, the epipubic or marsupial bones. These support the marsupial pouch. Since Cuvier predicted that the specimen was a marsupial, it should have these strange bones that existed in no placental mammal. He dug down in the rock and there they were—epipubic bones—confirming his method and hypothesis.

The extraordinary integration of the parts of quadrupeds calls for an explanation; there must be an underlying reason. Cuvier was unable to supply one and suggested that by continued observation answers would emerge. Although he lived at the same time as the famous evolutionist Lamarck, Cuvier was not an evolutionist. To Cuvier, species were constant. One interesting way he tried to prove that they were was to make a careful comparison of the skeleton of a mummified sacred ibis, collected in Egypt, with those living today. They were identical. Thus, if they had not evolved in more than 4,000 years, maybe they never had or would.

Quarries of the Paris Basin

In the early nineteenth century geologists were seeking the significance of those rocks that were found in roughly horizontal layers, parallel to the earth's surface. These layers appeared to have been derived from silt that had been deposited in water and eventually turned to stone. They were known as stratified rocks, or strata, in contrast with igneous rocks, which were not stratified and appeared to be formed by volcanic action.

Examples of stratified rocks occurred near Paris, which is situated in a flat area surrounded by low hills—the whole forming a basin. Throughout this Paris Basin there were numerous commercial quarries that gave geologists a glimpse of a slice through the upper portions of the earth's crust.

Cuvier and a fellow scientist, Alexandre Brongniart (1770–1847), used these quarries to study the relation between the layers of rocks and the fossils they contained. They recognized five major layers on the basis of the type of rocks.

(1) The first and uppermost layer, immediately below the soil, consisted of marls. It contained many fossils of animals and plants characteristic of lakes and freshwater marshes.

(2) The second layer, at places more than 80 feet in thickness, consisted of marls and sandstone. The fossils consisted mainly of shells of marine organisms.

(3) The third layer consisted of gypsum, marls, and limestone, and the fossils appeared to be of freshwater species.

(4) The fourth layer consisted of a coarse-grained limestone with an abundance of fossils belonging to about 400 different species, mainly marine but with a few freshwater forms.

(5) The fifth and lowest layer, resting on a chalk formation, consisted mostly of clay and sand. The fossils were freshwater shells and driftwood.

Several observations, fundamental to paleontological analysis, were made by Cuvier and Brongniart in those quarries. First, these same five layers, arranged in an identical sequence, were found in numerous quarries throughout the Paris Basin. Second, each formation contained a specific assemblage of species. Thus, the five layers could be identified by their composition, relative position, and the specific fossil species they contained.

These layers allowed Cuvier and Brongniart to infer something of the geological history of the Paris Basin. A layer containing fossils of marine species was evidence that sediments had been deposited at the bottom of a bay or inlet of the sea. If the layer contained freshwater species, it was concluded that the Paris Basin had, at a different time, been a huge freshwater lake. In the case of the fourth layer, which had many marine and a few freshwater species, it was concluded that the region had been part of the sea close to where a river entered.

It seemed reasonable to conclude that each layer represented a specific time-period in which the sediments that formed it were deposited. If this is so, the most important conclusion of all is reached: the relative position of a layer is an indication of its relative age. This conclusion is made necessary by a very simple observation: if material is being deposited at the bottom of a freshwater lake or marine bay, the material deposited first will be covered by material deposited later. No layer could be deposited below one already there.

The hypothesis is that layer five, resting on the chalk, is the oldest, and layer one the youngest. This hypothesis was made more probable by the following analysis. Layers two and four both contained marine species. These fossil species were compared with those still living on the coast of France. Many in layer two were found to be identical with them, but others were not known from the French coast. The fossil species of layer four, on the other hand, contained almost no living

19 *Geological strata. Erosion of an ancient seabed formed these cliffs in the Petrified Forest National Park of Arizona. The differently colored layers represent a sequence of deposits, the oldest being at the bottom and the youngest on top.*

species. The argument, then, was that a layer containing mainly extinct species must be older than one consisting mainly of species that still exist.

It was beginning to look as though the stratified, sedimentary rocks of the earth's crust contained a running diary of earth's history. Seemingly great changes had occurred in the Paris Basin over the years: at times it was dry land, at others a freshwater lake, and at still others an arm of the sea.

But what could account for the different species in the different strata? Cuvier developed a hypothesis based on three major conclusions: species are "fixed," that is, they do not evolve, contrary to what his countryman Lamarck maintained; species may become extinct; and different assemblages of fossils occur at different times in the past, as shown by the layers of sedimentary rocks.

Cuvier's solution was to propose another hypothesis—periodical geological catastrophes that saw the ocean sweeping over the land and exterminating all life. When the waters would finally recede, terrestrial organisms from unaffected regions would repopulate the devastated area. His observations in the quarries of the Paris Basin demonstrated such ingressions from the sea.

Cuvier believed the earth, in its present form, to be of recent origin, and he offered evidence from history, astronomy, and geology that this was so. He concluded that if there is anything that has been learned from geology, it is that the surface of our globe has been subjected to a vast and sudden revolution, the "general deluge," as the sea swept over the land. This occurred not more than from five to six thousand years ago. When the waters receded, the newly dry land was repopulated by any creatures that were spared or by migration from regions not covered by the sea. Prior to this last revolution there may have been two or three other eruptions of the sea.

Such revolutions, each being followed by a specific assemblage of organisms, suggested another important principle to Cuvier: there seemed to be a trend toward greater structural complexity of the fossils found in the geological formations of successively younger age. The fossils of the oldest formations known to him were coelenterates, other invertebrates, and possibly fishes. In the next oldest, the coal layers, there were ferns and palms but no terrestrial vertebrates. The next major formation had unique terrestrial reptiles along with many fish species no longer living. Still later deposits, such as the limestone of Jura (mountains of eastern France), had many different kinds of strange reptiles, some of gigantic proportions, and monocotyledonous plants. The mammals appeared only in much more recent layers. The oldest mammalian fossils were entirely different from the mammals of today. No evidence of fossil remains of human beings were known to Cuvier.

> Did the last and the most perfect work of the Creator exist no where? And where were the animals now present on the earth that are unknown as fossils? Have the lands inhabited by human beings together with these other animals been swallowed up by the sea? Was the land now inhabited by human beings and modern animals previously swept by a great inundation that destroyed the creatures previously inhabiting it? On this subject the study of fossils gives us no information and in this Discourse we must not seek an answer to our question from other sources.
>
> It is certain that we are now at least in the fourth succession of terrestrial animals. The age of reptiles was followed by that of the palaeotheres [primitive mammals], then the age of mammoths, mastodons, and megatheria. Finally we arrive at the age of the human species together with domestic animals. It is only in the deposits subsequent to the beginning of this age, in turf-bogs and alluvial deposits, that we find bones all of which belong to animals now existing . . . None of these remains belong either to the vast deposits of the great catastrophe or to those of the ages preceding that wonderful event. (pp. 220–221)

So that last catastrophe, the deluge that Cuvier believed to have occurred five or six thousand years ago, had cleared the earth for the advent of human beings.

Catastrophism and Uniformitarianism

Cuvier's hypothesis that successive deluges—his geological catastrophes—were the cause of the existing state of the earth's crust and its inhabitants was part of a vigorous, ongoing debate. One school of geologists—the catastrophists—felt that the present state of the earth was the consequence of violent catastrophes of short duration, such as Cuvier's deluges. Another school—the uniformitarians—thought that the present state of the earth could be accounted for by the slow, uniform action of the same geological forces that exist today but having acted over vast eons of time.

There was ample evidence for the deluges described by Cuvier. His and Brongniart's studies of the strata of the Paris Basin indicated a succession of inundations. There was other evidence of violent geologically recent floods. Numerous regions in Western Europe were covered by thick alluvial deposits that appeared to have been deposited by the movement of tremendous amounts of water.

These observations and the deduction were perfectly correct. We now know that the alluvial deposits were the result of the great volume of water released by the melting glaciers of the Pleistocene Ice Age, about 10,000 years ago. The idea that glaciers could modify the terrain had been clearly expressed by John Playfair in 1802, but it was not widely accepted:

> For the moving of large masses of rock, the most powerful engines without doubt which nature employs are the glaciers, those lakes or rivers of ice which are formed in the highest valleys of the Alps, and other mountains of the first order. These great masses are in perpetual motion, undermined by the influx of heat from the earth, and impelled down the declivities on which they rest by their own enormous weight, together with that of the innumerable fragments of rock with which they are loaded. These fragments they gradually transport to their utmost boundaries, where a formidable wall ascertains the magnitude, and attests the force, of the great engine by which it was erected. The immense quantity and size of the rocks thus transported, have been remarked with astonishment by every observer. (pp. 388–389)

It would require another few decades before Louis Agassiz was to study the glaciers of the Alps, observe the evidence of ice actions in other regions, and propose that a vast ice sheet had covered much of Western Europe in fairly recent geological times.

So Cuvier and other catastrophists did have evidence of great changes. There was also evidence that the last deluge was of recent date. The Rhone River flows into and out of Lake Lausanne, and the Swiss naturalist Jean-André De Luc (1727–1817) studied the delta formed at the river's ingress. He estimated that the delta was increasing at such a rate that it should have filled Lake Lausanne long ago unless it was of recent origin. This observation and interpretation also proved to be true: lake and river were formed after the last Pleistocene ice sheet had melted.

There were equally vigorous proponents of the view that the same sorts of geological processes acting today have been those acting in the past. One of those uniformatarians was James Hutton (1726–1797), a geologist of Scotland. He believed that the present state of the earth's crust could be explained by three main geological processes.

(1) Sediments that are deposited, mainly in the ocean, become the layers of stratified sedimentary rocks, or strata.
(2) Volcanic action uplifts the strata to form mountains.
(3) Once the strata have been elevated, they are subject to erosion by rain, rivers, and wind.

Whereas the catastrophists thought of major geological events occurring in short periods of time, the uniformitarians demanded vast periods of time for the postulated geological forces to act. In contrast with Judeo-Christian tradition, Hutton believed that the past was of immense duration. His *Theory of the Earth* (1788) closes: "The result, therefore, of our present enquiry is, that we find no vestige of a beginning,—no prospect of an end." The poetic beauty of that sentence did not appeal to all. Hutton was severely criticized for holding a belief contrary to scripture. Times had changed, however, and he did not die catastrophically at the stake but, being a true uniformitarian, of natural causes.

William Smith and the Geological Column

During the years when Cuvier and Brongniart were studying the Tertiary strata in the Paris Basin, an Englishman, William Smith (1769–1839), was studying the older Secondary strata in England. His large-scale geological map covering most of England and Wales was published in 1815. He, too, was aware that the strata might cover a large horizontal distance and that each contained a characteristic assemblage of fossils.

Smith's work was but a spectacular example of similar studies being done in many parts of Europe, and the notion was developing that, once all the strata had been studied, it would be possible to arrange them in a vertical sequence, the geological column, with the oldest at the bottom and the most recent at the top. The relative position of the strata in the column would correspond with the relative age of their formation.

The sequence of strata is never complete for any given region, nor should we expect it to be. Sedimentary rocks are formed in places where sedimentation of materials is occurring, and such situations are almost always restricted to shallow seas, river deltas, swamps, and freshwater lakes. Such places are widely scattered, and sedimentation in any one of them might be for a relatively short period of geological time.

The geological column, therefore, is constructed from bits of information coming from many localities. When all these bits are superimposed, the sum of the different strata totals more than a hundred kilometers. It becomes the geological timescale with worldwide applicability and is the reference point for all studies in paleontology. In Smith's day, one could only guess about their absolute age. Reliable measures of absolute time, using radioactivity, came later.

Early in the nineteenth century, Sir Charles Lyell, a friend of Charles Darwin, provided a monumental synthesis of what was known of geology. In fact, his *Principles of Geology* established geology as a science. Lyell, who was born in Scotland in 1797 and died in 1875, was a uniformitarian. The *Principles* was first published in a three-volume set in 1830–1833. It was to go through 12 editions during Lyell's life. The elegance of his writing style and his almost constant revisions to take into account new data and interpretations meant that *Principles* did much to advance the geological sciences, though it added little to the general conclusions of Hooke, Steno, Hutton, Smith, and Cuvier, except for emphasis on the immensity of geological time.

Understanding Nature in 1850

In Chapter 2 there is a list of the sorts of questions about living organisms that were suggested by the observations and speculations of Aristotle. In the more than 2,000-year interval between Aristotle and 1850 there was a vast increase in biological information, so it is of interest to ask to what degree had those ancient questions been answered. The answer has to be that, in spite of a great increase in data, there was little increase in understanding. There had been no fundamental conceptual breakthroughs, no shift in basic paradigm.

Nevertheless, by mid-century biologists knew a great deal more about the questions that Aristotle had sought to answer. The two unique improvements in knowledge had been that microscopes revealed the world of the very small and geology revealed the world of the very old. Both were basic to the conceptual advances that were to occur after 1850.

In reviewing Aristotle's questions, we find that there had been no real improvement in our understanding of what it means to be alive. At the beginning of the nineteenth century one hypothesis was that chemical composition of organisms is unique. They and they alone appeared to be composed of complex chemical substances, difficult to study, and rich in carbon. These substances were known only from living organisms or their products. Thus, life was more than an assemblage of molecules—it was an assemblage of molecules with a "life force." The hypothesis of the uniqueness of these organic substances was destroyed in 1828 when Fredrich Wöhler synthesized the organic compound urea from inorganic molecules.

Nothing new could be added to the observation that being alive seemed to be much the same in all animals. One had to accept "that this is just the way it is."

Nothing new had been added to the list of requirements for life. They remained food, water, air, and a suitable ambient environment.

The most important new information about requirements for life had to do with the nature of *pneuma*. That essential component of air was discovered by Joseph Priestly (1733–1804) and named oxygen by Antoine Laurent Lavoisier (1743–1794). Lavoisier established the importance of oxygen in combustion and its role in the respiration of animals and plants. So the answer to why animals must breathe is to obtain oxygen. Lavoisier was not so fortunate as his countryman Georges Cuvier, and was sent to the guillotine during the French Revolution.

The great increase in biological knowledge since Aristotle was mainly related to the diversity of life. The European Age of Discovery, from the fifteenth century to the nineteenth, brought naturalists in contact with a rich, diverse, and previously unknown flora and fauna from all parts of the world. Those who stayed home worked on the local species. There was great interest in describing new species, adding specimens to private or public collections, and maintaining animals as pets or in zoos.

This expanding knowledge of diversity, however, was not accompanied by scientific explanations of its cause. Most people did not even recognize diversity as a problem. A perfectly acceptable explanation had always been part of the Judco-Christian tradition: the species of animals and plants had been created by God on the third, fifth, and sixth days of Creation Week.

The resemblances of both structure and function among similar organisms that Aristotle had noticed were confirmed by later workers and reached a climax with Cuvier. These relationships could be used to arrange animals in groups that shared common characteristics. But what could possibly be the explanation for these basic plans? There was no scientific answer.

And finally there was the obvious rule that like produces like. But how? The answer in 1850 was no better than Aristotle's.

Despite an overall lack of conceptual advance, we must not ignore the substantial increases in information about diversity, structure, and to a lesser degree function. These data were to fuel the biological revolution that was to begin in the last half of the nineteenth century. They can be summarized as follows:

Microstructure

The most important new insight since Aristotle was the concept that cells are the units of structure and function in all organisms. By 1850 this was suspected by an increasing number of biologists, but the data were far from conclusive, especially the evidence that cells are units of function. Although in 1858 Virchow was to propose the hypothesis that cells can only originate from other cells, *omnis cellula e cellula*, an understanding of how this came about was not to be available for another generation.

This hypothesis was antithetical to that of spontaneous generation, the bugaboo that had accompanied biological thought from earliest times. Acceptance of the possibility of spontaneous generation was decreasing, but the notion would last even after it was made improbable by the experiments of Louis Pasteur.

Macrostructure

The most notable advance in understanding structure was at the level of organs and systems. In fact, morphology was to be the dominant field of inquiry throughout the nineteenth century—the Queen of the Biological Sciences, so to speak. Animals belonging to all groups were investigated, and patterns of structure were discovered. Seemingly all animals had the same general problems associated with life: obtaining food, water, and oxygen; digesting food; transporting these necessities to all parts of the body; eliminating waste products; reproducing; developing; avoiding predators; moving; and so on. Structure was studied in relation to these functions. It was found that the structure of the organ systems of different animals could vary enormously while subserving the same functional needs.

Comparative anatomists were finding also that the species of a given group, originally defined on external characteristics, shared a common plan of internal structure. Even the species of major groups, such as the vertebrates, seemed to be variations on a basic structural plan.

The concepts of homology and analogy were explored with vigor. In pre-Darwinian days, homology meant the "same structure" as modified in different species. Thus the wings of birds and the forelegs of mammals were seen as variations on the same basic structure. Analogy meant "different structures" that have a similar function, such as wings of insects and wings of birds. The concepts of homology and analogy proved to be powerful tools in helping to understand the relationships of parts and how seemingly "new" parts could be formed as variations of what was already present.

Another important principle of comparative anatomy was Cuvier's view that the parts of the body are so closely adjusted to one another that the whole essentially determines the form of the parts.

Physiology

Few questioned that form and function must be considered parallel phenomena: form is what the structure is, function is what it does. Progress was rapid in the first but painfully slow in the second, and would remain so until better methods, instruments, and knowledge of chemistry became available. Some functions are obvious to the senses: eating, walking, speaking, beating of the heart, movements of the chest, elimination of wastes, and so on. But the body remained a "black box" for functions without an external manifestation. Most studies were of the human body, and those of a few other mammals, because the information gained was thought to be of use in medicine.

The difficulties in studying internal physiology are nowhere made more obvious than in our evaluation of Harvey's discovery of the circulation of the blood. It had taken millennia to establish that single fact—which in the overall scheme of things may seem fairly trivial. Yet that was the breakthrough that started physiology on a course of sustained discovery and understanding.

Physiologists obtain their information by experimentation, but in 1850 experimentation was not widely employed in the other biological sciences.

Ecology

In 1850 this was not yet an organized field of inquiry, but some basic observations had been made. It was recognized that most species are restricted to a specific habitat and way of life. The complementary lifestyles of green plants and animals were understood in broad outlines —one using the sun's energy to combine carbon dioxide, water, and simple salts to produce complex food molecules and the other to break down these food molecules to release energy, with water and carbon dioxide being end products. The meaning of "soil fertility" was becoming apparent. There was very great interest in the geographic distribution of animals—strange lands and seas were found to have strange faunas. As these new species were catalogued, a discovery of exceptional interest was made: most of them fit into the existing scheme of classification. A loose relation was also found between geographic distribution and classification. A given species was usually found on a single landmass and often on only a part of it. There were many examples of all of the species of a genus occurring on the same continent. What could this mean?

Developmental biology

There were two competing hypotheses to explain development. Epigenesis held that new structures appear in the course of development. Preformation held that all structures are present in the fertilized egg and that development is no more than the growth of these structures. Aristotle had accepted epigenesis as the pattern of development, but later embryologists squabbled over the relative merits of epigenesis and preformation until the end of the eighteenth century.

Embryologists had also been busy describing the structural aspects of development in all available organisms. Aristotle's finding that the patterns of development are similar in similar species was confirmed and extended.

Classification

The observation that different species could be grouped on the basis of characteristics held in common was used as a means of classifying them. Aristotle and naturalists of classical times had proposed schemes of classification, and that of Aristotle was as good as any until the eighteenth century. It was then that Carolus Linnaeus (1707–1778) attempted to arrange all the known plants and animals into a binomial hierarchial system, which became the basis of the system we use today.

The vertebrates had long been recognized as a natural group, but the invertebrates presented far more problems. Not only were there many more kinds but they could be small, seemingly amorphous, and totally different in their internal structure from the better known vertebrates. Lamarck had contributed greatly to the morphology of invertebrates, and after him much progress was made. By 1850 the major groups, as we know them today, were being slowly delimited. Gone were the days when invertebrates were classified as either "insects" or "worms."

The notion of a linear scale of nature—the idea that all species could be arranged in a sequence of increasing complexity from the simplest microorganisms to *Homo sapiens*—was becoming progressively less important as more was learned about the structure of the invertebrates. Cuvier found no evidence for a single sequence of increasing complexity, but suspected that one might be found within each of the four major groupings of animals that he recognized.

For many naturalists, especially those in England, many of whom were clergymen, there was no deep philosophical problem with the seemingly endless diversity of organisms, or with their possession of common characteristics that allowed the recognition of categories in the system of classification, or with the resemblance of human beings to other animals, or common patterns of development, or the rules of geographical distribution. For those adopting the Judeo-Christian paradigm, the living world is the way it is because that is the way the Creator made it. For some, however, it was becoming increasingly doubtful that Genesis provided a complete and reliable description of the natural world.

Even more disturbing for the Judeo-Christian paradigm were the answers coming from the geological strata. More and more it seemed that the earth's history might have covered an enormous span of time. And there was the question of the fossil evidence of life in the past. Cuvier was convinced that some species had become extinct. By 1850 it was becoming clearer that the oldest strata tended to have fossils of creatures less structurally complex than those of more recent strata.

A dim vision of a coherent scheme to account for the data of biology was beginning to emerge. Biological facts were being related to general ideas, but almost never was there a clue as to why this could be so. One could see that animal diversity was not all chaos, but the underlying causes for the apparent order were elusive.

part two

THE GROWTH OF EVOLUTIONARY THOUGHT

chapter 7

The Paradigm of Evolution

C onceptual biology shifted gears in 1859, the year that Charles Darwin published *On the Origin of Species*. A vast amount of data about living organisms began to make sense to a degree never achieved before. As is the case today with the attempt in physics to seek a way of uniting all the basic forces in one grand scheme, biology after 1859 seemed to be on the road to developing its own Theory of Everything (TOE).

The new paradigm was that life has had a complex, ever-changing history. John Ray's belief that "the Works created by God at the first, and conserv'd to this Day in the Same State and Condition in which they were first made" was replaced by the hypothesis of descent with change; that is, life had evolved.

Darwin's hypothesis of descent with change as a consequence of the natural selection of natural variations was a very simple idea:

A struggle for existence inevitably follows from the high rate at which all organic beings tend to increase. Every being, which during its natural lifetime produces several eggs or seeds, must suffer destruction during some period of its life, and during some season or occasional year, otherwise, on the principle of geometrical increase, its numbers would quickly become so inordinately great that no country could support the product. Hence, as more individuals are produced than can possibly survive, there must in every case be a struggle for existence, either one individual with another of the same species, or with individuals of distinct species, or with the physical conditions of life. It is the doctrine of Malthus applied with manifold force to the whole animal and vegetable kingdoms . . . There is no exception to the rule that every organic being naturally increases at so high a rate, that if not destroyed, the earth would soon be covered by the progeny of a single pair. (*Origin*, pp. 61–62)

Owing to this struggle for life, any variation, however slight and from whatever cause proceeding, if it be in any degree profitable to an individual of any species, in its infinitely complex relations to other organic beings and to external nature, will tend to the preservation of that individual, and will generally be inherited by its offspring. The offspring, also, will thus have a better chance of surviving, for, of the many individuals of any species which are periodically born, but a small number can survive. I have called this principle, by which each slight variation, if useful, is preserved, by the term of Natural Selection, in order to mark its relation to man's power of selection. We have seen that man by selection can produce great results, and can adapt organic beings to his own uses, through the accumulation of slight but useful variations, given to him by the hand of Nature. But Natural Selection . . . is a power incessantly ready for action, and is as immeasurably superior to man's feeble efforts, as the works of Nature are to those of Art. (p. 61)

One of the most astonishing things, in retrospect, is that the argument of the *Origin* was not proposed in a systematic manner until 1859. It seems so simple—and obvious. All that is required is to observe that the individuals of a species vary, suspect that some variants may be better adapted than others, observe that the power of reproduction can exceed the carrying capacity of the environment, and that the better-adapted individuals have a statistically better chance of surviving and leaving offspring.

One might seek to blame the Judeo-Christian dogma of special creation for inhibiting thoughts about descent with change, and to some extent this blame is valid. One cannot, however, hold this tradition accountable for inhibiting those Ionian Greeks. There is no reason why they could not have put together the elements of Darwinian evolution two and a half millennia before Darwin did.

Even more surprising is that any one of the early nineteenth-century naturalists, by considering the data already provided by John Ray, William Paley, and many others, could have realized that evolution would explain the data in a more satisfying manner than the concept of special creation—more satisfying to a scientist because its basis was natural rather than supernatural processes. There were even theories of evolution being discussed—those of Jean Baptiste Lamarck and Robert Chambers. They were not taken very seriously, however. Even

Darwin did not begin to think of evolution as a useful hypothesis until he had made personal observations about the fossils of Patagonia and the fauna of the Galapagos Islands while on the voyage of HMS *Beagle* in 1831–1836. But we forget a fundamental aspect of the history of science: an idea that seems obvious to many once it is proposed was not so previously.

Evolution through natural selection as outlined by Darwin was not immediately accepted as a new paradigm. It has had a rocky history over the last century and a half. In fact, evolutionary biology is the only discipline of biology, indeed of any science, that in our day has encountered considerable, even violent, opposition from some segments of society. The opposition comes from those who see the theory of evolution as incompatible with a literal interpretation of the account of the origin and history of life as given in Genesis. And, of course, it is. This is a current episode in the long-enduring religion-versus-science conflict and presents a problem for teaching science in the schools. Biological evolution is the only scientific theory to have reached the Supreme Court.

This donnybrook can be a source of incredulity or amusement for some sophisticates, but it is a very serious matter. The central issue is whether understanding of the phenomena of science is to be based solely on observation and experiment or whether supernatural events and processes are to be included. In these years when the use of science dominates our civilization, it is important that we understand what it is and what its strengths and limitations are. Evolutionary biology provides a fine case history of how science is done.

In 1958 the distinguished geneticist H. J. Muller gave his famous speech "One Hundred Years without Darwinism Are Enough." Muller was concerned that evolution was taught either ineffectually or not at all in the schools, and that this lack results in a citizenry ignorant of one of the essential concepts of modern biology—and, even worse, a citizenry that accepts supernatural explanations for natural phenomena:

> It ill befits our great people, four generations after Darwin and Wallace published their epochal discovery of evolution by natural selection, to turn our backs on it, to pretend that it is unimportant or uncertain, to adopt euphemistic expressions to hide and soften its impact, to teach it only as one alternative theory, to leave it for advanced courses where the multitudes cannot encounter it, or, if it is dealt with at all in a school or high school biology course, to present it as unobtrusively and near the end of

the course as possible, so that the student will fail to appreciate how every other feature and principle found in living things is in reality an outgrowth of its universal operation. (Muller 1959)

Muller was talking mainly about the biology of the precollege years where Christian fundamentalists were a powerful force in preventing the teaching of evolution in the schools. But, as Muller wrote, "We have no more right to starve the masses of our people intellectually and emotionally because of the objections of the uninformed than we have a right to allow people to keep their children from being vaccinated, and thus endanger the whole community physically." Muller saw the intellectual starvation of our masses as terribly dangerous.

In an equally well known essay, "Nothing in Biology Makes Sense Except in the Light of Evolution," Theodosius Dobzhansky (1973) summarized his position after surveying the major fields of the biological sciences: "Seen in the light of evolution, biology is, perhaps, intellectually the most satisfying and inspiring science. Without that light it becomes a pile of sundry facts—some of them interesting or curious but making no meaningful picture as a whole."

But why were two such eminent biologists called upon to make such pronouncements? The answer is that, beginning in the late 1950s after a long period of quiescence, the Christian fundamentalists were again vigorously trying to block the teaching of evolutionary biology in the schools or, if that proved impossible, to demand equal time for the teaching of "creation science," which was an attempt to bend scientific data to conform to one of the biblical accounts of creation and throw doubt on the discoveries of evolutionists. Hard-line creationists insist that the species of today are no more than trivially different from when they were first created and that divine creation was a very recent affair—not more than about 10,000 years ago.

Evolution is the only scientific theory that has been found repugnant to sections of society in recent years. School boards are not petitioned to suppress teaching the cell theory, the theory of gravitation, or Mendelian heredity. A few centuries ago there would have been serious protests against teaching the Copernican theory and the theory of a spherical earth, but over the years most of us have come to accept what the astronomers tell us even though our firsthand experience is that the sun does circle a flat earth.

Biological evolution will be discussed frequently in the chapters that follow, and this theme will continue to provide understanding for a variety of biological phenomena. But here it will serve as a vehicle to emphasize how scientists come to hold the views they do and why they reject all explanations of natural events that rely on the supernatural. I do not know of a single professional biologist who does not regard evolution as the basic concept of the biological sciences. It *is* the way we think. But that is not the way some others may think.

When discussing evolution, we must take special care to indicate how terms such as *theory, hypothesis, proven, fact,* and *truth* are used. This is important since there is considerable variation in usage. For example, *theory* for a scientist may represent the grandest synthesis of a large and important body of information about some related group of natural phenomena. For nonscientists the term may be pejorative: "Evolution is just a theory," meaning that it is a dubious notion. Both of these very different meanings, together with others, are fully countenanced by lexicographers. *Theory* is being used here for a body of knowledge and explanatory concepts that seek to increase our understanding of (explain) a major phenomenon of nature. Thus, the cell theory would consist of the many sorts of morphological, biochemical, and physiological observations relating to the basic units of structure and function of most living creatures.

Used this way, a theory cannot be disproven. There is no way that the cell theory can be falsified. Some of the things we think we know about cells may be shown to be false, but the consequence of this will be that some of the data or concepts included in the cell theory will be eliminated or replaced by other data and concepts that are more probable. General theories are never disproved, only improved.

Theories are so important and so nearly coextensive with the fields of knowledge for which they provide the conceptual framework that, quite frequently, the term *theory* is abandoned. Thus, the books appearing today are not entitled *The Cell Theory*. More likely the title will be something like *Cells* or *The Biology of Cells*.

One can acknowledge the usefulness of a theory even though the ultimate causes of the phenomena to which it applies are unknown. Thus, one can accept the theory of gravitation as a useful synthesis of available knowledge relating to the mutual attraction of bodies, yet have no adequate understanding of what "pulls" them together.

Hypothesis will be used as a tentative explanation of some phenomenon. It is an "educated guess" to be tested. Many use "theory" and "hypothesis" as synonyms. I will not. In the formative years of a science, hypotheses may grow into theories. Some of the early speculators about the underlying causes of organic diversity and of adaptation adopted a hypothesis of evolutionary change to account for these phenomena. As data and certainty increased, one could begin to recognize a large body of information and verified hypotheses as the "theory of evolution."

A scientific statement is *true* if all attempts to falsify it have failed. Truth in science is not some final statement that is correct for all time. It means "true beyond all reasonable doubt." It is important not to overdo this tentativeness of scientific statements. Good form, I know, but it can be carried to foolish extremes. Cannot we accept as "true" that water is composed of hydrogen and oxygen? Then again, some statements are true because they are definitions. An experienced ornithologist, after having a good look at a medium-sized bird with a grayish back and a reddish breast, does not have to be so careful as to say that the bird is "a robin beyond all reasonable doubt." We have agreed that a bird with certain well-defined features is to be known as "robin." If the experienced ornithologist has seen a bird with those features it is, by definition, a robin.

We can also accept that historical events that have left adequate records need not be verified by experiment or by observation of a rerun of the event. One can accept the reality ("truth") that dinosaurs once existed without demanding that they evolve through the many steps from primitive amphibians through primitive reptiles to their magnificent selves. There are many things in biology that cannot be repeated experimentally or verified by direct observation. This does not mean that we exclude them from the domain of science.

For a biological phenomenon to be *proven* will mean that it has been tested extensively and elegantly and has not been falsified. Once again this means "proven beyond all reasonable doubt."

First Questions

The road to the theory of evolution has been long. The hypothesis of evolutionary change smoldered in the human intellect from the days of the early Greeks until it came into full flame during the nineteenth century, when interest in natural history was intensified. Travelers, scientists, and professional collectors brought back to Western Europe a bewildering variety of new kinds of animals and plants from all parts of the world and even from the depths of the ocean.

As familiarity with living organisms increased, so did questions about them. The sorts of questions that eventually led to the concept of evolutionary change were varied, but the main ones seem to have been these three.

(1) How could one account for the extraordinary amount of organic diversity? What was the explanation for all these species, hundreds of thousands and possibly millions of them, and with their amazing differences in structure, physiology, behavior, and way of life?

(2) How could one explain the remarkable adaptations of living creatures? The more one learned about the structure and behavior of organisms, the more remarkable were their exquisite adjustments to the life they led. The intricate structure of a bird's feather, as revealed by the microscope, showed it to be a contrivance perfected for lightness and strength. Bones are generally thick and heavy, but some of the bird's bones are hollow and the walls thin. The lungs of birds extend into air sacs, which both lightens the body and improves respiration. The entire bird seems to be a perfect solution to the problem of flight. No human engineer could duplicate the bird's remarkable achievement, and until recently all human attempts to fly had ended in death or ridicule.

(3) What was the basis of the *scala naturae*, or scale of nature, that saw all species of animals or plants as part of a continuum, which, with small gaps, appeared to extend from the simplest to the most complex species? One could pass in small steps from typical lizards, through lizards with ever-shorter legs, to snakes with no legs at all. There seemed to be patterns in organic diversity.

But over most of human history, for the last two millennia at least, these three phenomena did not stimulate questions in the Western world. That is because the answer was provided by the Judeo-Christian religion: the world and all its living inhabitants are the way they are because that is the way God made them. It was divine creation, as described in Genesis. God had created everything, animals, plants, the earth, and the cosmos at some remote time in the past—some said as recently as 4,004 BC. The many sorts of animals and plants of today were each a consequence of an act of divine creation. The remarkable adaptations were an example of the care exercised by the Creator. That the species could be arranged in an almost continuous scale of variation meant only that they were created that way. Whatever one observed in nature was the consequence of the specific events of creation. There were no real problems. Every question relating to cause had the same answer—that is the way God did it.

One might have supposed that the belief that what organisms were and did depended directly on divine creation would have stopped all serious study. Far from it. It was reasoned that one might learn about the Creator from studying what He had created. This approach was known as natural theology.

The Paradigm of Natural Theology

Natural theology was not traditional theology but an attempt to see if something could be learned of the attributes of the Deity by studying his handiwork—nature. In contrast, traditional theology was based on received dogma that traced itself back to revelation, that is, God's message and attributes as revealed to someone who had lived in the distant past. Natural theology, on the other hand, was a do-it-yourself theology. Anyone could study the creation in the hope of glimpsing something of the attributes of the Creator. These two theologies were not in conflict since both assumed that living and nonliving nature were the handiwork of God.

An early exponent of natural theology in England was the clergyman-naturalist John Ray (1627–1705), who in 1691 published *The Wisdom of God Manifested in the Works of the Creation . . . viz the Heavenly Bodies, Elements, Meteors, Fossils, Vegetables, Animals (Beasts, Birds, Fishes, and Insects), more particularly in the Body of the Earth, its Figure, Motion, and Consistency; and the Admirable Structure of the Bodies of Man and other Animals; as also in their Generation, etc. With Answers to some Objections.* He further defined the contents in the Preface:

> That by the Works of the Creation in the Title, I mean the Works created by God at the first, and by him conserv'd to this Day in the same State and Condition in which they were at first made; for Conservation (according to the *Judgement both of Philosophers and Divines*) is a *continu'd Creation*.

There follows a long account of animate and inanimate nature that reveals John Ray as the finest English naturalist of his age. Much of what he says is just plain good science. He described the works of creation as accurately as he could. The advance over Gesner and Topsell is astonishing.

Ray's point of view became widely known in nineteenth-century England in a book published in 1802 by the clergyman William Paley (1743–1808). It was entitled *Natural Theology: or, Evidences of the Existence and Attributes of the Deity Collected from the Appearances of Nature.* Paley relied heavily on information supplied by Ray, though it was not adequately acknowledged.

Paley's book passed through many editions and was enormously popular. It was well known to naturalists, and Charles Darwin stated in a letter written in 1859, nine days before the publication of *On the Origin of Species,* "I do not think I hardly ever admired a book more than Paley's 'Natural Theology.' I could almost formerly have said it by heart." In his *Autobiography* he wrote, "I was charmed and convinced by the long line of argumentation." Thomas Henry Huxley remembered it as the most interesting book he was allowed to read on Sunday when he was a boy. Until the early twentieth century it was part of the readings required for those who wished to enter Cambridge University. There was no similar requirement for Oxford, it probably being assumed that anyone wishing to enter Oxford would already be well steeped in Paley.

Paley begins his proof for the existence and attributes of the Deity with the argument from design, his famous metaphor of the watch:

> In crossing a heath, suppose I pitched my foot against a *stone*, and were asked how the stone came to be there, I might possibly answer, that for any thing I knew to the contrary, it had lain there forever; nor would it perhaps be very easy to show the absurdity of this answer. But suppose I had found a *watch* upon the ground, and it should be enquired how the watch happened to be in that place, I should hardly think of the answer which I had before given, that, for any thing I knew, the watch might have always been there. Yet, why should not this answer serve for the watch, as well as for the stone? Why is it not as admissible in the second case, as in the first? For this reason, and for no other, viz. that, when we come to inspect the watch, we perceive (what we could not discover in the stone) that its several parts are framed and put together for a purpose, e.g. that they are so formed and adjusted as to produce motion, and that motion so regulated as to point out the hour of the day; that, if the several parts had been differently shaped from what they are, of a different size from what they are, or placed after any other manner, or in any other order, than that in which they are placed, either no motion at all would have been carried on in the machine, or none which would have answered

the use that is now served by it . . . This mechanism being observed (it requires indeed an examination of the instrument, and perhaps some previous knowledge of the subject, to perceive and understand it; but being once, as we have said, observed and understood,) the inference, we think, is inevitable; that the watch must have had a maker; that there must have existed, at some time and at some place or another, an artificer or artificers who formed it for the purpose which we find it actually to answer; who comprehended its construction, and designed its use. (pp. 1–3)

We should note that the geologists such as Charles Lyell were busily trying to find out about that stone Paley pitched his foot against. They suspected that the stone might *not* always have been there. Paley then catalogs some of the remarkable structures found in living organisms and shows how admirably they are adapted to serve the purpose for which they are used. The more one studies the natural world the more one realizes that organisms show clear evidence of design. He argued that no chance association of atoms could result in even the simplest of organic structures. Some structures, such as the human eye, are far more complex than a watch, so what conclusions must we reach?

Were there no example in the world of contrivance except that of the *eye*, it would be alone sufficient to support the conclusion which we draw from it, as to the necessity of an intelligent Creator. It could never be got rid of: because it could not be accounted for by any other supposition, which did not contradict all the principles we possess of knowledge; the principles according to which things do, as often as they can be brought to the test of experience, turn out to be true or false. Its coats and humours, constructed, as the lenses of a telescope are constructed, for the refraction of rays of light to a point, which forms the proper action of the organ; the provision in its muscular tendons for turning its pupil to the object, similar to that which is given to the telescope by screws, and upon which power of direction in the eye, the exercise of its office, as an optical instrument, depends; the further provision for its defence, for its constant lubricity and moisture, which we see in its socket and its lids, in its glands for secretion of the matter of tears, its outlet or communication with the nose for carrying off the liquid after the eye is washed with it; these provisions composed altogether an apparatus, a system of parts, a preparation of means, so manifest in their design, so exquisite in their contrivance, so successful in their issue, so precious and so infinitely beneficial in their use, as, in my opinion, to bear down all doubt that can be raised upon the subject. And what I wish . . . to observe, is, that, if other parts of nature were inaccessible to our inquiries, or even if other parts of nature presented nothing to our examination but disorder and confusion, the validity of this example would remain the same. If there were but one watch in the world, it would not be less certain that it had a maker . . . Of this point each machine is a proof, independently of all the rest. So it is with the evidences of a divine agency. The proof is not a conclusion, which lies at the end of a chain of reasoning, of which chain each instance of a contrivance is only a link, and of which, if one link fail, the whole falls; but it is an argument separately supplied by every separate example. An error in stating an example affects only that example. The argument is cumulative in the fullest sense of that term. The eye proves it without the ear; the ear without the eye. The proof in each example is complete; for when the design of the part, and the conduciveness of its structure to that design, is shown, the mind may set itself at rest: no future consideration can detract any thing from the force of the example. (pp. 75–77)

These selections emphasize Paley's argument for the existence of a Deity and not the large amount of natural history that forms the bulk of the material discussed. *Natural Theology* provides a fine overview of what was known of nature, especially animals, in the late eighteenth and early nineteenth centuries. It is well worth reading on Sunday. In contrast with the above quotations, most of it is clearly written. Paley becomes turgid only when he is defending his thesis, and then his prose is nearly as complex as the contrivances of nature he describes.

Paley's famous argument from design was convincing to many individuals, including Darwin at an early period in his life. It is, however, yet another case of deductive reasoning. Paley accepted on faith that there is a Deity and then arranged his observations about living organisms in such a fashion to suggest that life was so complex that it could not "have just happened." Anything as complex as life must have required a Creator. Having been raised in the Judeo-Christian tradition, and being a clergyman, his conclusion was preordained. If one first accepts that a Creator exists, no matter what his handiwork might be it would be evidence of his design.

But any notion of a deity belongs to the realm of the supernatural. One cannot directly study, observe, or experiment with this realm— one accepts it on faith. If one assumes the existence and operation of supernatural forces, nothing is impossible. Remarking about the eye, Paley states "it could not be accounted for by any other supposition." Such a position closes all avenues of deeper inquiry and scientific analysis. Furthermore, one cannot revise or correct an assumption which is accepted as absolutely true and "no future consideration can detract any thing from the force of the example."

Deductive reasoning, based on resolutely held assumptions of supernatural forces, has been singularly unproductive in advancing our understanding of the natural world. When one maintains that "no future consideration can detract any thing from the force of the example," science comes to a standstill. Paley's countryman, Sir Francis Bacon, had warned about this type of reasoning long before, but Paley did not abide by his advice. For most people, in most situations, deductive reasoning is the more comfortable mode—it is better to know what the answer is before you start to look for it.

Paley and Darwin dealt with the same biological phenomena—mainly structure, function, diversity, and adaptation. The deductive approach of one was sterile and remains so to this day. The inductive approach of the other would give us modern biology with its vast explanatory and predictive powers.

Natural theology reached a remarkable peak with the publication of the *Bridgewater Treatises* between 1833 and 1836. The Right Honourable and Reverend Francis Henry, Earl of Bridgewater, left a sum of 8,000 pounds sterling to be invested and the interest used to have a series of books written

> On the Power, Wisdom, and Goodness of God, as manifested in the Creation; illustrating such work by all reasonable arguments, as for instance the variety and formation of God's creatures in the animal, vegetable, and mineral kingdoms; the effect of digestion, and thereby of conversion; the construction of the hand of man, and an infinite variety of other arguments; as also by discoveries ancient and modern, in arts, sciences, and the whole extent of literature.

In all there were eight treatises covering subjects in biology, geology, mineralogy, meteorology, chemistry, physics, and astronomy. They provide a synthesis of science for the early years of the nineteenth century, and those in biology were especially successful in covering the field. They included many observations of the sort that Charles Darwin was to use in support of his hypothesis of evolution by natural selection. In fact, one cannot read the *Bridgewater Treatises* devoted to biology without being struck by how slight a change would be required to switch from the hypothesis of divine creation of species to a hypothesis of creation of species by evolution. So far as obtaining the data is concerned, Darwin need not have made that voyage of the *Beagle*—he could have stayed home and obtained enough facts for his theory from Paley and the *Bridgewater Treatises*. Or then again, maybe not, because some seemingly trivial observations made on the voyage of the *Beagle* were the clues to his hypothesis.

Within three decades after publication of the *Bridgewater Treatises*, the divine creation hypothesis began to be replaced in the minds of most scientists by the hypothesis of evolution. An approach that explains everything, as did natural theology, in the end was seen to explain nothing. The inquisitive mind is not satisfied with the same answer for all natural phenomena—it's God's way. Thomas Kuhn (1970) employs the term "paradigm" to mean a generally accepted way of explaining a scientific phenomenon of first magnitude. Divine creation was one paradigm for explaining the phenomenon of organic variation, adaptation, the *scala naturae*, and much more. The paradigm of evolution was a different way of explaining the same phenomena. Kuhn observes that one paradigm may replace another in what he calls a "scientific revolution." A scientific revolution will occur when the old paradigm ceases to be a useful and acceptable way of explaining the phenomena to which it applies. His scientific revolution represents one way of knowing being replaced by another way of knowing.

First Answers

But if one rejects supernatural explanations, how does one seek naturalistic ones? It is not obvious. Even with our knowledge today of what the answers proved to be, can we imagine any direct research program that might be expected to provide answers to the three questions listed above relating to diversity, adaptation, and a *scala naturae*? What observations would one make, and what experiments would one perform? Not one of our three questions has been formulated in a manner precise enough to enable us to use the procedures of science to seek an answer. One cannot begin to make observations or conduct experiments with a question. One must start with a provisional answer—a testable hypothesis—if there is to be any hope of gaining some understanding of the phenomenon. In science, we might almost say that we seek answers to "answers." We cannot just ask why there is incredible variety in the world of life. We have to make a guess—hypothesis—for what the reason might be and then set about seeing if it is so.

It was during the voyage of the *Beagle* that Charles Robert Darwin (1809–1882), became interested in answering the three sorts of questions that have been our concern. He was rather successful in doing so, and this is because he hit upon a way of asking questions that could be answered. Our three questions are so broad and vague that they defy analysis. How could one possibly study a question so broad and vague as "Why are there so many species?" or "Why are individuals so well adapted to their environment?" Such questions are about as useful as "Why are some things heavy?"

One has to start small, and Darwin did so. He did not start by saying, "I want to study evolution" but, on the basis of some very specific observations, came to the conclusion that a hypothesis of evolution would account for what he had observed. In fact, Darwin's analysis is a case study of the procedures that scientists are supposed to use— mainly the hypothetico-deductive method. He tells us what made him consider the hypothesis of evolution at a time when it was held in low esteem. Then he tells us what he did: "The line of argument often pursued throughout my theory is to establish a point as a probability by deduction and to apply it as hypotheses to other points to see whether it will solve them."

This putting together of clues and observations to form a hypothesis—induction—is, as Sir Francis Bacon observed, a logical step from specific statements to more general statements, in contrast with deduction, where one begins with a general statement (the hypothesis) and moves to the more specific deductions. There were three key observations, which are casually referred to in the first sentence of the Introduction to the *Origin* and in more detail in Darwin's *Autobiography*.

(1) While the *Beagle* was engaged in charting the coasts of Argentina, Darwin conducted considerable fieldwork on land, stimulated in part by his susceptibility to *mal de mer* (seasickness). He observed many species that were new to him. Some were very strange, such as the armadillo. Darwin was also collecting fossils, and among these he found the remains of some extinct armadillos—the glyptodonts. The clue is this: two very strange animals of the same general sort, one living and one extinct, were observed in the same part of the world.

(2) Darwin visited many localities on the east coast of South America, from Brazil to southern Argentina. He noted that some of the species encountered in one locality might be present at other places, yet the individuals in the various localities might not be exactly the same. Thus, the clue is that what appeared to be the same species was made up of populations that varied with the locality. Individuals from populations close to one another might differ almost imperceptibly, while populations more distantly separated might be almost as different as two

species. This phenomenon is known as *geographic variation*. It was not the sort of thing Darwin had observed in England. There all of the individuals of a species were very much alike. But England is small and the climate does not vary greatly from place to place.

(3) A similar phenomenon was encountered in the Galapagos, a group of volcanic islands off the coast of Ecuador. One of the most striking kinds of animals there were the giant tortoises. It was pointed out to Darwin by one of the residents that each island had its own variety of tortoise. The differences were such that an experienced person could tell the island from which any individual originated. Here was geographic variation occurring on islands within sight of one another. He noted similar examples in some species of plants and birds. The Galapagos finches remain to this day a notable example.

The standard explanation in the 1830s would have been that the living and extinct armadillolike creature and the different populations of living species had been separately created. If so, Darwin asked, isn't it surprising that both the fossil form and the living form are found in precisely the same locality? The living animals were scampering over the land that held the entombed fossils. Wouldn't it be simpler to assume that the extinct armadillos evolved into the living ones? And for geographic variation, Darwin wondered just how precise divine creation had to be. John Ray had said that living species are exactly the same as the day they were created. That would require that each local population of a species, no matter how slightly it differed from the neighboring populations, represents a distinct act of creation. Thus a slightly different sort of tortoise had been created for each little island in the Galapagos.

Alternatively, the phenomenon of geographic variation could be explained as the result of evolution. The hypothesis would go something like this. A widespread species on the South American mainland would find its local populations in a variety of environmental situations. Possibly the local populations would change because of the local conditions and, in time, evolve into distinctive populations. The Galapagos finches could have had a similar history. Long ago a few birds might have been accidentally carried by winds or storms from the west coast of South America, where their closest relatives live today, to the Galapagos. There they could have spread slowly from island to island. This may have been a rare event since at the present time there seems to be little or no interisland migration. Thus, each island would have an almost completely isolated population, and possibly it would slowly evolve so that each island would come to have its own variant—as with the tortoises. But what could be the mechanism of these changes?

Darwin was offering speculation and, for that period, rather wild speculation. He was saying that if evolution could occur, it could account for the armadillos, tortoises, and finches. A skeptical biologist would have observed that neither the hypothesis of creation nor of evolution could be validated. Both might seem reasonable—and be wrong. There is, however, a fundamental difference between the two hypotheses: creationism is based on supernatural events that could never be studied by the methods of science; evolution is assumed to be based on natural phenomena that could, in theory at least, be studied by scientists. For those individuals deeply interested in answers to our original three questions, the research program seemed clear: study what can be studied with the procedures of science; ignore creationism, not because it is wrong but because it cannot be studied.

Darwin remained doubtful. A hypothesis of evolution was certainly plausible, but the statements of science must be based on probability, not plausibility. He was fully aware that the writings of Jean Baptiste Lamarck and Robert Chambers on evolution were thoroughly discounted. Charles Lyell was the foremost authority in geology, and he had much to say on species as well. He had rejected evolution. The influences of fellow scientists are powerful, and essentially all rejected the hypothesis of evolution; moreover, Darwin could not imagine what could be the mechanism of evolutionary change. How could one species possibly change into another? Nearly all experience suggested the "fixity of species," that is, that species do not change.

If one suspects that species are not fixed, proof must be given that one species can change into another. No one had ever observed such a thing, and Darwin believed that the reason was that evolution is an exceedingly slow process. Yet if one was to consider seriously the hypothesis of evolution, a way to test it must be found. Otherwise, the hypothesis of evolution was as useless as creationism.

But it proved possible to suggest a hypothesis for evolutionary change—*natural selection*. Interestingly enough, the basic idea came not from fellow biologists or geologists but from an economist. The Reverend Thomas Robert Malthus (1766–1834) published his important study, *An Essay on Population*, in 1798. Malthus was greatly distressed by the prevalence of human misery and poverty that he saw in England during those years of the industrial revolution. Why was there such misery?

He suggested that the answer was to be found in the relation between the rate of human population growth and the rate of increase of the human food supply. He suspected that people were outbreeding the crops, and the inevitable consequence would be not enough food for all. Thus, misery and hunger are inexorable consequences of being human. There has always been starvation and probably always would be unless human beings changed their reproductive behavior. Malthus suggested that all life has the same problem:

> Through the animal and vegetable kingdoms, nature has scattered the seeds of life abroad with the most profuse and liberal hand. She has been comparatively sparing in the room, and the nourishment necessary to rear them. The germs of existence contained in this spot of earth, with ample food, and ample room to expand in, would fill millions of worlds in the course of a few thousand years. Necessity, that imperious all-pervading law of nature, restrains them within the prescribed bounds. The race of plants, and the race of animals shrink under this great restrictive law. And the race of man cannot, by any efforts of reason, escape from it. Among plants and animals its effects are waste of seed, sickness, and premature death. Among mankind, misery and vice.

Darwin tells us that this suggestion of Malthus was the key that he sought—a mechanism for evolutionary change. He was well aware of variation among the individuals of a species—in minor ways to be sure. Some might be large, others small; some with one color pattern, others with different patterns; some with long hair, others with short hair. So far as he could tell, there is no reason to suspect that every character (a word meaning "attribute" or "characteristic") might not vary. If some of the variant characters better adapt the individual to survive and leave offspring, wouldn't it be reasonable to suppose that, over the course of time, individuals of the better adapted type will make up most and eventually all of the population? They would simply outbreed the less adapted types of individuals.

Thus the normal variation that all naturalists agreed characterized species in nature would allow nature to select the "better" variants, and slowly the species would change. Darwin referred to this phenomenon as *natural selection* and hypothesized that it could be the long sought mechanism of evolutionary change.

chapter 8

Testing Darwin's Hypotheses

The hypothesis that organic diversity could be understood as the consequence of natural selection acting upon genetic variability dealt only with natural phenomena. And so it could be tested by the usual procedures of scientific methodology: formulate deductions and test them. This chapter will present ten deductions and then assemble Darwin's data to test them.

Have Life Forms Changed over Time?

The critical information required for proving the hypothesis of evolution is data on the conversion of one species into another. Evolution means descent with change, so it would be necessary to show that ancestors and descendants are in fact different. On the face of it, that sounds like a nearly impossible task. Remote ancestors are remote and hence not available for direct study. Nevertheless, the hypothesis of evolution, no matter how logical it might seem, would remain unproven unless some means of reconstructing the past history of life exists. So one of our most critical deductions will be:

Deduction 1: If the hypothesis of evolution is true, the species that lived in the remote past must be different from the species alive today.

This deduction may be logically correct, but in practice how does one obtain information on organisms that died thousands or millions of years ago? As Cuvier and others discovered a century before Darwin, the rocks of the earth's crust hold the secrets of many events that occurred in the past. Long-cooled lava flows tell of ancient volcanic eruptions. Limestone tells of ancient sea bottoms where the sediments were slowly changed to rocks. Smooth horizontal rocks, with parallel grooves, tell of ancient glaciers that had ground across them. The depths of canyons give some idea of the length of time that it must have taken rivers to cut them.

Evidences of past life occur in one of the two basic kinds of rocks. The granites and lavas are igneous rocks, formed by the outpouring of molten rock from the interior of the earth. They contain no fossils. But sedimentary rocks—the horizontal layers of rock that form such a conspicuous feature of cliff faces in many parts of the world—may contain evidences of former life. As Cuvier and Brongniart discovered in the quarries of the Paris Basin, these layers, or strata, start as material settling at the bottom of a lake, inland sea, or ocean or are deposited on land by wind or eroding waters. Gradually these deposits are covered and slowly change to stone. This method of formation means that the topmost deposits of sediments are the most recent and the lowest layer the oldest. If these deposits happen to include the remains of animals or plants, these remains may be fossilized—changed to stone. Thus the sedimentary rocks are time capsules of earth history. Each stratum contains information about events occurring at the time of its formation.

The possibility of the remains of a dying organism being fossilized is exceedingly remote. Most dead organisms are food for some creatures and are quickly consumed. If the organism has hard parts—shells, teeth, bones, or an exoskeleton—there is a better chance of these structures surviving destruction, and hence the possibility of fossilization increases.

Fossils make it possible to learn something about life in the past that could throw light on whether or not evolution has occurred. Thus, we can test the basic idea in evolution that natural populations slowly change over long stretches of time and, therefore, ancestors and descendants will be different from one another. The data to evaluate the deduction were available to Darwin and other naturalists: fossils had been known for a very long time. A few fossil species proved to be very similar, or even identical, to living species, but the vast majority were very different. Especially striking and interesting were the remains of some of the vertebrates, whose bony skeletons increased their chances of being fossilized. There were huge elephantlike creatures that had lived on land and monstrous reptiles that had lived in the swamps and seas. There were spectacular plant remains such as giant tree ferns and many other species entirely different from any alive today.

Thus it was true beyond a reasonable doubt that few of the species alive today occur also as fossils in the remote past. The vast majority of the fossil species are no longer with us. The deduction is shown to be true, so the hypothesis of evolution is made more probable.

We can continue with a similar but more sophisticated deduction.

Deduction 2: If the hypothesis of evolution is true, the older the sedimentary strata, the less the chance of finding fossils of contemporary species.

Tests of this deduction involve not only finding fossils but knowing something about the time they lived. In the middle of the nineteenth century there were no accurate methods for determining the ages of the strata of sedimentary rocks. Nevertheless, it was possible to obtain data that allowed a test of this deduction. If we are looking at the face of a cliff of sedimentary rocks, the layers will generally be more or less horizontal and the bottommost layer will be the oldest and the topmost the youngest.

As we have seen, during the first half of the nineteenth century, geologists had developed a *relative time scale*—the geological column— that forms a framework for discussing the past. The basic data consisted of measuring the thickness of the sedimentary strata and arranging them in a sequence from the oldest to more recent. Of course, this huge pile of rocks, many miles thick, could not be observed in any one locality. Sedimentary rocks are not formed as a uniform layer across the earth's crust; at any specific period of time, deposition will occur in one restricted region, and later elsewhere. By examining all the available cliffs and railroad cuts, it was eventually possible to put the pieces of the puzzle in place and to arrange all of the strata in an imaginary column of strata.

When first devised, this geological column measured only relative time. That is, the lower the stratum, the older it was. Absolute geological time became known only after the radioactive methods for dating became available in the early years of the twentieth century.

The hypothesis that the species in progressively older strata would be ever less like living species was tested by Lyell and by a French geologist, Deshayes. They collected fossil shells from different strata that had been formed in a geological era known as the Tertiary (figure 20). The oldest Tertiary strata had been given the name Eocene and the most recent, the Pliocene. Nearly all of the Recent Pliocene fossils belong to species still alive, whereas almost none of the Eocene fossils do (table 1). Again, the deduction has been tested and found to be true, so we can say that the hypothesis of evolution becomes even more probable.

The finding that progressively older strata have progressively fewer species that are alive today is not what one would expect from the hypothesis of divine creation. If all species of animals and plants had been created within four days, as a literal interpretation of Genesis demands, all strata might be expected to have the same array of fossils. That is most definitely not what the geological record shows. Nevertheless, Lyell's data do not disprove divine creation—if supernatural forces are invoked, anything can be explained. What one could say, however, is that if we accept as valid evidence only what we can observe, and employ only the naturalistic methods of science, the hypothesis of evolution is a more probable hypothesis than the hypothesis of divine creation.

Deduction 3: If the hypothesis of evolution is true, then we would expect to find only the simplest organisms in the very oldest fossiliferous strata and the more complex ones only in more recent strata.

Some cautionary remarks must be made about this deduction. Evolution means change—not necessarily becoming more complex. Some species alive today, many parasites for example, may be structurally simpler than their ancestors. Nevertheless, Darwin and other evolutionists assumed, as we do today, that the very earliest forms of life must have

Table 1. Percentages of Tertiary species still living. Data from Lyell (1854, pp. 389–395).

	Fossil species	Alive today	Percent of fossil species still alive
Recent Pliocene	226	216	96
Older Pliocene	569	238	42
Miocene	1,021	176	17
Eocene	1,238	42	3

been small and simple and that slowly, very slowly, more complicated species evolved.

Even in the early years of the nineteenth century geologists realized that the oldest strata contained simpler forms—invertebrates alone were found. Only later did higher vertebrates—reptiles, birds, and mammals—appear. The same was true for the plant kingdom. Algae, mosses, ferns, and similar plants were very old; the angiosperms were more recent. In 1824 Lyell had this to say:

> An opinion was entertained soon after the commencement of the study of organic remains, that in ascending from the lowest to the more recent strata, a gradual and progressive scale could be traced from the simplest forms of organization to those more complicated, ending at length in the class of animals most related to man [the mammals]. (1826, p. 513)

In the 1820s one had to be cautious—after all, very little paleontological work had been done. Much more was known in the 1830s when Lyell's *Principles of Geology* was first published. His *Manual*, which contained the data on paleontology that had been removed from later editions of the *Principles*, appeared in its fourth edition in 1852—a few years before the *Origin*. By then the data were much more convincing.

Darwin and others at mid-century spoke of the inadequacy of the geological record. In many ways it was inadequate since the strata in only a small part of the world had been studied. Nevertheless, the record was sufficiently adequate to test Deduction 3. It was true beyond a reasonable doubt that there had been a progression of forms of life, with the less complex species appearing before the more complex. The various kinds of invertebrates were well represented at the then-known beginning of the fossil record in the Silurian; the mammals were first encountered in the much younger Jurassic strata. Thus the hypothesis of evolution becomes more probable.

Do Species Evolve into Different Species over Time?

Essentially all of the prominent geologists in the pre-*Origin* years were creationists, including Lyell, who made the first grand synthesis of geological data and saw the progression of life. He fully realized that each major group of strata had its own distinctive life forms—many not known in either younger or older strata. How could a creationist explain such facts? Cuvier had suggested that there had been a series of creations and extinctions. Darwin's hypothesis was radically different. For

MILLION YEARS AGO	EON	ERA	PERIOD	EPOCH	DESCRIPTION
0.01	PHANEROZOIC	CENOZOIC	Quarternary	Recent	Earth warms. Neolithic revolution. Civilization.
1.6				Pleistocene	Glaciation in northern hemisphere. Human beings worldwide. Extinction of large mammals.
5 24 34 53 65			Tertiary	Pliocene Miocene Oligocene Eocene Paleocene	Continued diversification of modern birds, angiosperms, and placental mammals. By the end of the Tertiary the genus *Homo* had appeared. Grasses and grazing mammals become abundant.
135		MESOZOIC	Cretaceous		Beginnings of radiations of flowering plants. Dinosaurs abundant then extinct. Many other extinctions possibly associated with impact with meteor at end of period.
205			Jurassic		First known birds and mammals. Dinosaurs abundant.
250			Triassic		First dinosaurs. Cycads and conifers abundant. Continents moving apart.
290		PALEOZOIC	Permian		Land masses form single continent, Pangea. Frigid conditions and massive extinctions at end of period. Mammal-like reptiles.
355			Carboniferous		Warm humid conditions result in huge forests of primitive plants, which formed extensive coal deposits. Reptiles appear at end of period. Insects present. Trilobites, brachiopods, and crinoids abundant. Labyrinthodonts and jawed fishes abundant.
405			Devonian		Age of fishes. First amphibians. Land plants and land arthropods abundant. Atmospheric oxygen at present levels or higher. Continents moving toward one another. Brachiopods, echinoderms, and cephalopods abundant. First insects.
435			Silurian		Brachiopods, trilobites, and eurypterids common. First jawed fishes. Plants and animals invade land. Atmospheric oxygen about 20 percent.
510			Ordovician		Earliest known vertebrates. Brachiopods, trilobites, cephalopods, graptolites common.
570			Cambrian		Probably all metazoan phyla present. Trilobites and brachiopods abundant. Atmospheric oxygen reaches about 2 percent.
670			Ediacarian		Oldest known metazoans. Coelenterates, annelids, and arthropods may have been present.
2,500	PROTEROZOIC				Abundant prokaryotic life. Eukaryotes may have appeared by 2,000 million years ago. Atmospheric oxygen about 0.2 percent.
3,800	ARCHEAN				Oldest known rocks and prokaryotes.
4,600	HADEAN				Earth forms. No geological record.

20 *A modern version of the geological time scale.*

Darwin the species at any one period of earth history were the progenitors of the species of subsequent periods. There was no break in the lineage of life. Presumably all forms alive today are the remote descendants of the first sorts of life that appeared on earth. Thus the most critical deduction of all will be:

Deduction 4: If the hypothesis of evolution is true, it must be possible to demonstrate the slow change of one species into another.

Here Darwin failed. There were no critical data in the 1850s that showed the evolution of one species into another. Darwin thought that this was but a reflection of the inadequacy of the fossil record. In a sense he was correct, but today, when we know so much more about speciation, we realize why it is exceedingly difficult to obtain convincing fossil evidence of the evolution of one species into another. Fossilization is a highly improbable event, and critical evidence would consist of a series of fossils of a single lineage from adjacent strata over a long period of time. Even today there are not many such examples.

Darwin and others at the time were more interested in a related question: Were there fossils that were intermediate between major groups of animals? Were there *missing links*, which if discovered, could document the evolution of life? This leads to another deduction.

Deduction 5: If the hypothesis of evolution, which assumes that all of today's species are the descendants of a few original forms, is true, there should have been connecting forms between the major groups (phyla, classes, orders).

> Why then is not every geological formation and every stratum full of such intermediate links? Geology assuredly does not reveal any such finely graduated organic chain; and this, perhaps is the most obvious and gravest objection that can be urged against my theory. The explanation lies, as I believe, in the extreme imperfection of the geological record. (*Origin*, p. 280)

If that explanation was correct, one must assume that greater efforts by collectors of fossils would eventually supply the data. It was also possible to predict what sorts of fossils might be the most promising. The answer is the vertebrates, for two main reasons. First, the bones and teeth of vertebrates are excellent candidates for fossilization. Sec-

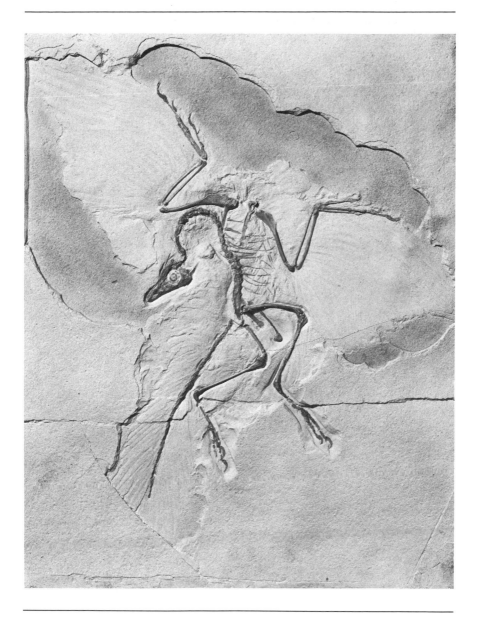

21 *A fossil skeleton of the early bird* Archaeopteryx. *The importance of this specimen is that it combines characteristics of birds and reptiles. Its skeleton has many reptilian features but* Archaeopteryx *is birdlike in being covered with feathers and being capable of flight. Thus it was seen as a "missing link" between birds and reptiles, confirming the hypothesis that birds evolved from reptiles.*

ond, the phylum Chordata, which includes the vertebrates, is the most recently evolved of the major phyla. Many of the invertebrate phyla were already present in the then oldest known strata from which fossils had been collected—the Silurian rocks. This means that the common ancestors of any two of the phyla already present in the Silurian would have lived at an earlier time when no fossil record was then available. On the other hand, one could work with the bony chordates—the vertebrates. Of the major groups—fishes, amphibians, reptiles, birds, and mammals—only the most primitive, the fishes, were present in earliest fossil-bearing strata. If, as Darwin imagined, all the other vertebrates were evolved from fishes, there would be a chance of documenting his belief.

The first dramatic proof for the existence of fossils intermediate between major groups came in 1861—two years after the publication of the *Origin*. In a quarry near Solenhofen in Bavaria, a feather was discovered in strata of the Jurassic age. Shortly thereafter a well-preserved specimen of a strange fossil with feathers was discovered and named *Archaeopteryx*. It was a confusing mixture of structural features of birds and extinct reptiles known as archosaurians. The skull was somewhat birdlike but the jaws contained teeth, characteristic of reptiles but absent in all modern birds. There was a long reptilianlike bony tail, in contrast with the modern birds, which have the tail vertebrae fused in a short projection—the "tail" being feathers only. In fact, the skeleton was more similar to that of the archosaurians than to modern birds. Nevertheless, *Archaeopteryx* had wings and the body was covered with feathers. It could be debated whether *Archaeopteryx* should be classed as a reptile or bird, but, using the presence or absence of feathers as the main diagnostic feature, paleontologists chose to classify it as a bird.

Of course this is exactly what would be hoped for in a "missing link"—a species so perfectly intermediate between two major groups that one could debate to which group it should be assigned. Darwin had assumed that birds must have evolved from primitive reptiles, and here was the evidence that should have convinced many scientists that they had. *Archaeopteryx* remains today as one of the best examples of a link no longer missing (figure 21).

Here then was the "smoking gun" that Darwin and the other evolutionists required—a fossilized link between two major groups of vertebrates. This gave strong credibility to the hypothesis of evolution. One would assume, therefore, that intense enthusiasm would follow such a discovery. Surprisingly, this was not the case.

The specimen discovered in 1861 was sent to the British Museum (Natural History) in 1862 and was studied immediately by the famous English anatomist Richard Owen (1863). It was also studied by Thomas Henry Huxley (1868). Neither saw in *Archaeopteryx* the importance we ascribe to it today. In the fourth edition and later editions of the *Origin*, Darwin gave a single entry in the index for *Archaeopteryx*, and we read only that a "strange bird, the Archaeopteryx, with a long lizard-like tail, bearing a pair of feathers on each joint, and with its wings furnished with two free claws, has been discovered in the [Jurassic] slates of Solenhofen" (*Origin*, 1892, p. 266). There is another mention, however, that first appeared on page 284 of the fifth edition. "Even the wide interval between birds and reptiles has been shown by [Huxley] to be partially bridged over in the most unexpected manner, on one hand by the ostrich and extinct Archaeopteryx, and on the other hand by [one of the dinosaurs]."

Slowness in recognizing the importance of a new discovery in science is far from unusual. As in other fields of knowledge, it is exceedingly difficult to put two and two together unless one already knows the answer is four.

Thus the critical deduction that there must be forms intermediate between major groups of organisms could be satisfied in this one case. Is one enough? Only one was available to Darwin in his lifetime. Subsequently, and especially in the twentieth century, other dramatic and more complete examples have been discovered. Some of these will be mentioned later.

Has There Been Time Enough for Evolution?

Deduction 6: If the hypothesis of evolution is true, the age of the earth must be very great, possibly many millions of years old.

Darwin's *Origin* appeared during a period when most educated people in the West assumed that the earth was not very old. Bishop Ussher's date of creation, 4004 BC, was so widely accepted that it was printed in the margins of the first chapter of Genesis in the King James Version of the Bible. Nevertheless, geologists were coming more and more to the opinion (following Hutton) that the earth is very old. Darwin said that "a man must for years examine for himself great piles of superimposed strata, and watch the sea at work grinding down old rocks and making fresh sediments, before he can hope to comprehend anything of the lapse of time" (*Origin*, p. 282). But even with this firsthand knowledge of geology, that statement is hypothesis, not fact.

Darwin knew that he needed vast ages if his hypothesis of evolution through natural selection was to be supported. The first crude estimates of the earth's age were based on the thickness of the strata. Geologists had become convinced that sedimentary rocks came from material that had been deposited, generally under water. If one knew the rate of deposition and the thickness of a group of sedimentary rocks, one could compute how long it had taken them to form. Then, if one knew the total thickness of all the sedimentary rocks, it would be possible to give a figure for how long it took for all sedimentary rocks to form. That would give a minimum estimate of the age of the earth.

None of these estimates could be made with any exactness. The rate of deposition of materials would depend on the rainfall, slope of the land, and the nature of the material being eroded—all of which might vary over long periods of time. A given stratum of rocks might be very thick in one place and thin in another. One could never be sure that all strata were known.

Nevertheless, geologists came to believe that the earth must be very old. The then-known strata in Great Britain alone totaled 72,584 feet, or 14 miles in thickness. Darwin knew of an estimate for deposition by the Mississippi River: 600 feet in 100,000 years. Thus each foot would represent 166.67 years. That would mean that the known sedimentary rocks in Great Britain would have taken 12 million years to form. Yet an even longer time might have elapsed, as one might imagine that the great Mississippi River, draining much of the continent, would have a very high rate of deposition.

Various other ingenious methods were used to determine the age of the earth. The salinity of rivers flowing into the ocean was measured. Knowing that and the total amount of salt in the ocean, one could estimate how long it had taken the ocean to receive its salt. Another method consisted of making an assumption about the temperature of the earth at the time it was formed and then estimating how long it would take for it to cool to its present temperature. Whatever methods were employed, it seemed that the earth must be very old. No method, however, was capable of giving a reliable answer.

It was not until the twentieth century, and especially since 1940, that reliable methods for determining the age of rocks have been perfected. All of these depend on the rate of radioactive decay of materials in the rocks. One can now estimate the age of rocks with an error of only several percent.

Although an acceptably accurate method for dating rocks was not available to Darwin, it seemed true beyond all reasonable doubt that an almost inconceivable length of time had elapsed between the present and when those organisms that lay buried in the oldest fossiliferous rocks had lived.

Is Natural Selection the Mechanism of Change?

Evolution was an ancient idea; natural selection was a new idea. Darwin realized that he had to make a case for natural selection, and if he could, the hypothesis of evolution would become more likely.

Deduction 7: There must be variation among organisms if the hypothesis of evolution by natural selection is true.

This deduction can be tested by finding out whether or not there is variation among the individuals of a species. The *Origin* begins not with the phenomenon of variation of organisms in nature but under domestication. Darwin could be confident that most of his readers would be familiar with the dramatic variation shown by most cultivated plants and domesticated animals. In the nineteenth century there was a great deal of interest in England in selecting new varieties of domesticated plants and animals. The results could be striking. Darwin emphasized that the selected varieties "generally differ much more from each other, than do the individuals of any one species or variety in a state of nature." Darwin was planting the notion that selection could be both quick and powerful in producing new varieties. His readers might not be aware that there was variation, albeit less, in wild populations. Thus he was wise to begin his argument with the familiar.

Darwin was also planting the idea that all differences need not be the consequence of divine creation. He could be sure that no reader would believe that creation was the reason for the new breed of sheep or for a better beet. It would be accepted that the cultivated varieties were the product of *artificial selection* by the farmer. Darwin proposed that artificial selection is a model for natural selection, the difference being the agent responsible. In one case nature selected what was "better" for survival and the production of offspring. In the other, the farmer selected what was "better" for his purposes: cows that produced more milk, hens that had more breast meat or laid more eggs, roses that were more beautiful or fragrant.

There was no doubt that variation under domestication is a fact. But what about variation in nature? After all, Darwin was interested mainly in the possibility of evolution in nature, not in the barnyard, but he hypothesized that the two were basically the same. There is a tremendous amount of variation in nature. Darwin had been impressed by this while on the *Beagle* expedition. Later, combining the publications of fellow scientists for additional facts, he found abundant examples of variation not only in superficial external features but in internal structures as well.

The most important point about variation of populations in nature was that there seemed to be a continuous array of cases: from two populations being essentially identical, to being as different as varieties, or subspecies, or even "good" species.

> Certainly no clear line of demarcation has as yet been drawn between species and sub-species—that is, the forms which in the opinion of some naturalists come very near to, but do not quite arrive at the rank of species; or again, between sub-species and well-marked varieties, or between lesser varieties and individual differences. These differences blend into each other in an insensible series; and a series impresses the mind with the idea of an actual passage. (*Origin*, p. 51)

Thus, Darwin found ample and convincing evidence of variation in populations of organisms, both in nature and in domestication. Therefore the deduction is correct and the hypothesis of evolution by natural selection becomes somewhat probable—at least it has not been falsified.

Deduction 8: Natural selection can be operative only if more offspring are born than survive.

The idea that only a few of the many offspring of animals, or of the many seeds of plants, survive was again something that Darwin's readers would know or suspect. They might also have observed that the population size of species seems to remain about the same year after year. A single oyster produces millions of eggs each year yet the sea does not fill with oysters. A single oak tree can produce thousands of seeds each year yet, in natural areas, the number of oak trees remains about the same.

There is no exception to the rule that every organic being naturally increases at so high a rate, that if not destroyed, the earth would soon be covered by the progeny of a single pair. Even slow-breeding man has doubled in twenty-five years, and at this rate, in a few thousand years, there would literally not be standing room for his progeny . . . The elephant is reckoned to be the slowest breeder of all known animals, and I have taken some pains to estimate its probable minimum rate of increase: it will be under the mark to assume that it breeds when thirty years old, and goes on breeding till ninety years old, bringing forth three pair of young

22 *The female balloon fly accepts a freshly killed insect from the male that is courting her. In many species of the more complex animals, especially arthropods and vertebrates, females choose mates from a number of suitors. Their criteria of choice may range from the practical, as in this case, to the highly impractical, as in the case of the female bird of paradises's preference for males with enlarged and brilliant tailfeathers. In other species the male's greater size, song, or courtship dance might attract the female. Darwin recognized the importance of female choice in evolution, a phenomenon he called* sexual *selection.*

in this interval; if this be so, at the end of the fifth century there would be alive fifteen million elephants, descended from the first pair." (*Origin*, p. 64)

Yet, in all likelihood, an average original pair would have only two surviving offspring—14,999,998 would have perished. The struggle for existence is a tremendously impressive fact of nature. Thus, the deduction that more offspring are produced than can survive can be accepted as true. The hypothesis of evolution by natural selection is not falsified by this test and, hence, becomes more probable.

Deduction 9: If the hypothesis of evolution by natural selection is true, there must be differences between the offspring that survive and reproduce and those that do not.

Selection implies that some individuals are chosen and others are discarded. An animal breeder cannot develop a new variety of sheep, with heavier fleece for example, if he lets all the individuals in his flock reproduce. Success could only be assured if he kept the lambs of parents with the better fleece as his breeding stock and sent all the other little lambs to market.

Similarly in nature, if it is a matter of chance or luck which perish and which survive, there would be no selection. What evidence did Darwin have to offer on this important deduction, which is the very heart of the concept of natural selection? He could make an effective case for the reality of artificial selection and suggested that the same general principle could hold in nature as well.

> It may be said that natural selection is daily and hourly scrutinizing, throughout the world, every variation, even the slightest; rejecting that which is bad, preserving and adding up all that is good; silently and insensibly working, whenever and wherever opportunity offers, at the improvement of each organic being in relation to its organic and inorganic conditions of life. We see nothing of these slow changes in progress, until the hand of time has marked the long lapse of ages, and then so imperfect is our view into the long past geological ages, that we see only that the forms of life are now different from what they formerly were. (*Origin*, p. 84)

Darwin rested his case on this logical argument, not data. He had no critical evidence that demonstrated a difference between the offspring that survived and those that perished. Compelling as the logical argument might be, the hypothesis of evolution by natural selection would be iffy until evidence for the reality of natural selection could be offered. Nevertheless, the inherent logic of the argument was sufficient for many nineteenth-century naturalists. With the field so lacking in rigor, the possible became probable and the probable became acceptable.

Even to this day, the demonstration of natural selection remains impossible except in certain special situations. Darwin himself realized that demonstration would be most difficult for he believed that "we see nothing of these slow changes in progress, until the hand of time has marked the long lapse of the ages." And he felt sure that those ages were considerably greater than the life span of any human observer. So it would never be possible for a scientist to observe one species evolving into another.

Today this is how we view the situation: If natural selection has been operating on natural populations continually, any genetic variant that could be classed as "good" would surely have appeared at some time in the past and have been screened by selection. Thus, the populations that we see today are the best that selection has been able to accomplish with the genetic variants that have appeared to date. The chance of any observer being lucky enough to detect a truly new genetic variant in a natural population under natural conditions that would be "better" is extraordinarily slight.

The deduction was eventually tested by other means and shown to be correct. The most dramatic evidence comes from situations in which a population is presented with an environmental challenge never before encountered and hence never before selected for. When the population encounters this new environment, for which most members are clearly not adapted, most individuals are killed, but any which have genes that confer some adaptation survive and reproduce.

The peppered moth (*Biston betularia*) of England and many other species of moths living in the more heavily industrialized regions of the world evolved melanic forms following the industrial revolution. This can be documented in many cases because naturalists have collected moths for many years and deposited the specimens in museums. Therefore the recent history of the population is known. In addition, the genetics of melanism can be determined by crossing the original and the melanic forms.

Biston rests on tree trunks. In nonindustrialized regions of England the tree trunks have heavy growths of pale-colored lichens. The moths in such places are also pale. In industrialized regions, polluting gases kill the lichens and expose the dark tree trunks. In the last century melanic forms of the moths have become common in these areas.

Mutations to the melanic form occur throughout the range of the species. If they happen to appear in the nonpolluted areas where there are lichens on the trees, they are conspicuous on the pale trunks and thus more easily seen and captured by birds. If the melanic forms appear in an industrial area where the lichens have been killed and the tree trunks are dark, they will be protectively colored and have a better chance of surviving. Laborious field observations have shown that this actually occurs. Natural selection, in this case predation by birds, causes notable changes in the frequencies of the alleles that control the moth's pigmentation.

It should be emphasized that industrial melanism is not an example of Lamarckian evolution, even though common sense might suggest that the polluted environment is somehow directly causing the change in pigmentation. The data seem conclusive that melanic mutations occur at random throughout the range of *Biston*. The polluted environment is not the cause of the mutation to melanism, but if such a mutation occurs the melanic form will be favored by natural selection.

Other modern-day examples of natural selection are the very large number of insects and other arthropods that have become resistant to pesticides, and the many kinds of microorganisms that have become resistant to antibiotics and other drugs. These examples of evolution to resistance are of enormous importance in agriculture, in the control of vectors that transmit diseases, and in the practice of medicine. In those cases that have been carefully analyzed, the explanation is the same as for industrial melanism. Mutations appear by chance. If they happen to increase the survival of the individuals in their new environment— one with pesticides or with drugs—they increase in frequency. But this kind of evidence for natural selection came a century after Darwin and hence was not available for him to answer his critics.

Deduction 10: If the hypothesis of evolution by natural selection is true, only those variations that are inherited will be important.

In Darwin's time the relation between the characteristics of individuals and their inheritance was poorly understood. It was obvious, of course, that much was inherited: the offspring of chickens were chickens, not pigeons. On the other hand, some of the minor differences among a flock of chickens might be passed to the offspring and some not. Some of these minor characteristics might then reappear after several generations. If that was the case, was the variant new or had it been transmitted in a dormant state—whatever that might mean?

In 1859 there was simply not enough information about inheritance to provide the necessary genetic foundation for the hypothesis of natural selection. Neither was Darwin, or anyone else, able to account for the origin, or apparent origin, of new variations. Darwin assumed that new variations were related in some way to the environment but not in a Lamarckian sense—that is, he did not believe that the environment induced specific adaptations (long hair in cold climates, for example). The best that Darwin could do was to conclude:

> Whatever the cause may be of each slight difference in the offspring from their parents—and a cause for each must exist—it is the steady accumulation, through natural selection, of such differences, when beneficial to the individual, that gives to all the more important modifications of structure, by which the innumerable beings on the face of this earth are enabled to struggle with each other, and the best adapted survive. (*Origin*, p. 170)

Although Darwin accepted that there could be new variations appearing in populations, there was great uncertainty on the part of many scientists that these variations could be of sufficient magnitude to enable one species to evolve into another. All sorts of bizarre traits had been selected to produce a bewildering variety of breeds of pigeons—*but they were all pigeons*. Even the most extreme breeds could be crossed and produce fertile offspring, and the ability to cross and produce fertile offspring was the generally accepted test for good species at that time. Darwin had emphasized earlier that some of the breeds of dogs, for example, are as different in structure and behavior as different wild canine species, but they still could be crossed. Nevertheless, no artificial selection had produced a "new species."

One of Darwin's critics, Fleeming Jenkin (1867), made a most compelling argument that new variants could never produce any lasting effect. If a new variant arose in a population, according to Jenkin, it would of necessity have to cross with the other members of the population. When different varieties are crossed it was accepted in those days that *blended inheritance* was the rule. That is, the offspring would be intermediate. So, the offspring of the variant and a regular individual would be intermediate. Their offspring would, most likely, breed with the regular forms again so the next generation would be even more like the regular individuals. Thus, any new variations would be expected to be diluted and disappear. That would be the consequence if blended inheritance was the rule—and in Darwin's time that was accepted as one of the few sure things that could be said about inheritance. Thus, Jenkin's argument was a telling blow to the Darwinian hypothesis. It was telling, however, because it was thought to be valid. Now we know better (see below). Inheritance is not a matter of blending the differences of the parents—but Mendel's work was unknown at the time.

Thus Darwin was not able to confirm Deduction 10—that the variations which are important in evolution are inherited. This deduction did seem highly probable, and eventually convincing data were accumulated, but this, coupled with the fact that neither could Deduction 9—natural selection in nature—be confirmed, led many scientists to have serious doubts about the hypothesis that natural selection is the mechanism for evolutionary change. Yet the number of scientists and others who came to accept evolution increased. To them Mr. Darwin had made his point about the possibility of evolution; he just did not know what made it happen.

The Genetic Basis of Natural Selection

If inheritance had proved to be as vague and capricious a phenomenon as Jenkin and many others supposed, natural selection could not be the agent of evolutionary change—there would be nothing to select. For natural selection to work, the mechanisms of inheritance would have to be based on a close relation between whatever is inherited and the characteristics of the individual. Furthermore, those mechanisms must be almost constant from generation to generation (to account for offspring resembling their parents), yet be subject to rare alterations (to provide the genetic variability on which natural selection could operate).

A hypothesis of inheritance that would fulfill these basic requirements began to take shape in 1900, long after Darwin's death, with the discovery of some experiments by Gregor Mendel performed in the 1860s. This was the effective beginning of the science of genetics, which will be treated in detail in Part Three. However, a thumbnail sketch of enough genetics to see how it can support Darwinism will be given here.

(1) The characteristics of an individual are controlled by thousands of different genes, which are composed of DNA (deoxyribonucleic acid) and are parts of chromosomes.

(2) Every cell has two pairs of each kind of chromosome, and each kind of chromosome has a unique group of genes. Thus there will be two of each kind of gene in each cell. Each gene occupies a specific place, its locus, in the chromosome.

(3) The same kind of gene usually has two or more variations, called alleles. The two alleles in a cell may be the same, the homozygous condition, or they may be different, the heterozygous condition.

(4) There is no blending of genes in inheritance. In heterozygous cells, different versions of a gene, as in blue or brown eyes, are not altered by their association, as shown by their reappearance in subsequent generations in unmodified form. The old notion of blended inheritance was based on perfectly valid observations: if two markedly different individuals are crossed, their offspring are usually more or less intermediate. It was only when Mendel and his followers concentrated on individual characters, not the sum of all characters, that it became obvious that blending does not occur.

(5) The change of a gene from one kind of allele to another is a mutation. Mutation is a rare event—a given allele may change to a new form about once in a million cell generations but usually far less frequently. Nevertheless, since there are thousands of kinds of genes, at least a few loci may experience a mutation every generation.

(6) To the best of our knowledge, mutation is random with respect both to the locus involved and to the change that occurs. The neo-Lamarckian view that the type of mutation is directly related to the environmental stimulus remains unproven. The neo-Darwinian view that mutation is random and that natural selection is responsible for selecting some mutants and eliminating others accords with available data.

(7) Although the rate of mutation is very low for any one locus on the chromosome, given the very large number of loci, new variations are appearing constantly. Although the individuals of most species resemble one another closely, there is nevertheless a huge amount of hidden genetic variability. It has been estimated that as many as a third of all loci may be polymorphic, that is, they have two or more different alleles. Furthermore, in sexual reproduction there is a considerable shuffling of the alleles—genetic recombination.

(8) Thus mutation plus genetic recombination result in highly diverse offspring. It has been suggested, for example, that every human being, apart from identical twins, has a unique genetic makeup that neither existed before nor will recur in human history. There is ample variation on which natural selection can act.

(9) The entire individual is the target of selection—not the individual genes. Genes survive or not because they are parts of individuals that survive or do not. Those individuals that survive will have, on the whole, genes that provide a better chance of surviving and leaving offspring. But the surviving individual might even have some deleterious genes. For example, a given gene might be so much "better" that the individual with it would survive even though it has other genes that are slightly deleterious. The not-so-good genes would be hitchhiking.

Accounting for the Diversity of Life

Given the reality of inherited variability in populations and the efficacy of natural selection, one can by logical induction predict the slow change of a population over time. Thus one can begin over 3 billion years ago with an ancient prokaryote population and have a model for how genetic variation and natural selection can lead, species by species through time, to human beings.

But the interaction of variation and natural selection alone cannot explain the presence of the *millions* of species on the earth today or the vastly greater numbers that occurred throughout the past. As Fleeming Jenkin's criticisms made clear, that was Darwin's main problem—what is the origin of species? What accounts for the vast diversity of life at a given moment in time?

Darwin and the naturalists who followed him were much impressed with the frequent difficulty in deciding whether two populations should be considered a single species or two. The two populations might run the gamut from seeming identity, to differing in slight ways, to differing enough to be regarded as subspecies, or differing so much that they could be accepted as different species. There was a clear reason for the difficulty, in Darwin's mind—the gamut reflected different degrees of evolutionary divergence. Again this is an example of a perplexing phenomenon making sense in terms of a theory.

Darwin and later naturalists noted frequently that the extent of the differentiation between two populations was associated with barriers to

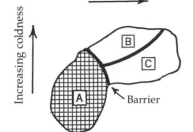

Increasing dryness →

Increasing coldness ↑

Barrier

Time 1. Single species.
The population is restricted to Zone A.

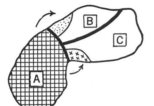

Time 2. Single species.
A few individuals have migrated to Zone B and others to Zone C, where they are geographically isolated from the original population. Selection will promote the development of two new populations, one adapted to a cold, dry environment in B, the other to a warm, dry environment in C.

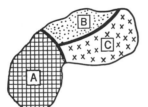

Time 3. Three subspecies.
Evolution through the selection of spontaneous mutations in the three isolated populations has reached the point where each zone has its own adapted population. These populations, though slightly different from one another, could still interchange genes should the geographical barriers between populations disappear. Consequently, they are subspecies, not yet true species.

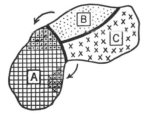

Time 4. Three species.
The three isolated populations have diverged to the point where physiological or behavioral isolating mechanisms prevent the interchange of genes. If individuals from either Zone B or Zone C migrate back to Zone A, they can occupy this zone together with the Zone A species without interbreeding and losing their specific identity.

23   *A model for geographic speciation, which is thought to be the dominant pattern for the evolution of diversity.*

dispersal: two subpopulations might be separated by a mountain range or other condition inhospitable to the organisms and difficult to cross. This seemed to be a clue to speciation. It was not possible to understand how a single population in which all individuals had the opportunity to interbreed freely could split into two species, that is, *speciate*. But if an interbreeding population became physically divided by some barrier to free dispersal, one could imagine how the isolated subpopulations might slowly evolve differences from one another. The evidence now seems overwhelming that this pattern, *geographic speciation*, is the principal mechanism for the origin of new species (figure 23).

If only a few individuals become geographically isolated from the rest of the population, these few individuals will have only a sampling of the alleles of the entire population, so any new population that develops from these "founders," as Ernst Mayr calls them, will differ from the original population and will diverge more quickly. Genetic drift, as this phenomenon is called, has interested evolutionists for decades, and its importance is that it could produce a new genetic situation that is then subject to natural selection. In any event it is but part of the more general phenomenon of genetic change through geographic speciation. Once a portion of a population becomes isolated from the bulk of the population, or a population is greatly reduced in size, it must inevitably have a different gene pool. That new gene pool will continue to be molded by natural selection. Natural selection is continuous; genetic drift is rare and episodic.

chapter 9

In the Light of Evolution

The great appeal of the theory of evolution for scientists in the last part of the nineteenth century was that it made so much sense of an otherwise bewildering mass of data in biology and geology. The theory of evolution also proved of great value in directing the course of research in comparative anatomy, cytology, inheritance, paleontology, and, to a lesser degree, in systematics and animal behavior. All important theories in science have this dual role—to explain the data already at hand and to suggest ways of acquiring additional information. Science is not only a way of knowing; it is also a way of discovering.

It is time for a change in terminology. By the last quarter of the nineteenth century, it was accepted by the majority of biologists and geologists that evolution was true beyond all reasonable doubt even though its mechanisms were not well understood. It is proper, therefore, to change its designation from "hypothesis" to "theory."

Dobzhansky's 1973 statement that "Nothing in biology makes sense except in the light of evolution" is likely to be accepted by scientists familiar with biology. By implication, no other concept except evolution provides a satisfying explanation of biological phenomena. But surely the biology of Ray, Paley, and the authors of the *Bridgewater Treatises* made perfect sense to them. For them "Nothing in biology makes sense except in the light of Divine Creation." Natural theology can account for all the phenomena of life, but to do so it must invoke supernatural phenomena.

Natural theology came to be viewed as a sterile enterprise. The answer was always the same: "What is, is what was created." Supernatural powers and processes, far beyond the ken of scientists to detect or employ, explained all. After 1859 the more intellectually satisfying answer became: "What is, is what evolved." Nothing in evolutionary theory was beyond the power of scientists to try to study directly or indirectly. Some problems might be very difficult to study—for example, the history of life. For such information one had to be content mainly with fossils, fortuitously formed and fortuitously found. An additional glimpse of the past was possible when it was recognized that the species alive today have built into their gross and microscopic structure evidences of their past history.

Comparative Anatomy

The modern mind, even of a fervent creationist, has the notion of biological relatedness so firmly embedded that we must remind ourselves of a fundamental difference between species that have been created and species that have evolved. If each species had been created and remained in essentially the same form until the present, there could be no biological, or as we would now say genetic, relatedness among different species. Even the untutored mind recognizes a relatedness among dogs, wolves, foxes, and coyotes. Yet if each had been separately created, they could be no more related to one another than any one of them is related to a hippopotamus, dogfish, dodo, or oak tree. The biological, and usual, meaning of "related" is "connected by common ancestry." Separately created canine species could not share a common ancestry. Each species could have only creatures of its own kind for ancestors.

The new view that evolution brought was that canine species must have shared a common ancestor at some time in the remote past and that, over the ages, the ancestral populations had split into separate lineages that led to the species we know today. On a larger scale, an evolutionist could suggest that all members of the phylum Chordata share a common ancestor—some long extinct protochordate living in a time long gone. If we accept this notion as worthy of analysis, we can view it as a hypothesis to be tested and formulate the following deduction:

Deduction 11: If the members of a taxonomic unit, such as the phylum Chordata, share a common ancestry, that fact should be reflected in their structure.

This notion—that the structure of all descendants from a common ancestor will show evidences of this descent—is based on the belief that modifications of the original structure would evolve very slowly. After all, the structural differences between parent and offspring are usually trivial unless they are pathological. Hence, some of the original characters, or obvious modifications of them, should be present in the descendants.

Comparative anatomy was the discipline that first recorded the common characteristics of organisms in a systematic manner. This is not surprising since, at an earlier time, knowledge of physiology was rudimentary and cellular structure was unknown. It was a major activity for biologists from the sixteenth to well into the nineteenth century. Skeletons received special attention. Not only were they more permanent than the soft tissues but, when the study of fossils became an active enterprise in the eighteenth century, the only possible comparisons of living and fossil vertebrates were of their bones.

Twelve years after Vesalius put human anatomy on a firm footing with his *De Humani Corporis Fabrica* of 1543, a French naturalist, Pierre Belon (1517–1564), published a book on birds in which he showed, side by side, the skeletons of a bird and of a human being. The dramatic thing he did was to give the same names to the bones of the bird that had human counterparts. There was no difficulty, for example, in showing that two such apparently different structures—the bird's wing and the human arm—had corresponding bones.

The recognition of similar parts in different organisms is the concept of *homology.* For Belon in the sixteenth century and for later workers, the proximal bone of the human arm and of the bird wing are modifications of the "same thing." A skeptic could ask, "How could they be the same thing when arms and wings differ so greatly in appearance and function?" The answer could only have been vague and unsatisfactory. But the concept of evolution made it obvious how the parts of birds and mammals could resemble one another: it would be held that the arm and wing are derived, through evolutionary modifications, from the anterior paired appendage of a common ancestor. Homology, then, is similarity based on common descent.

This explanation could be extended to suggest that the anterior paired appendages of all vertebrates—the pectoral fins of fishes, whales, and dolphins; the wings of birds, bats, and pterodactyls; the digging foreleg of a mole; the hoofed foreleg of horses and cattle; the prehensile arm of a sloth; and the tool-using human hand—are examples of the same

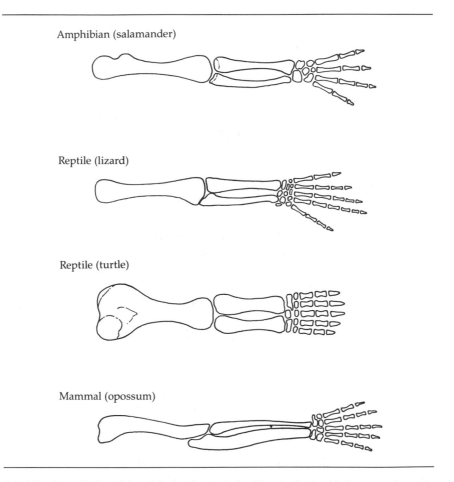

Amphibian (salamander)

Reptile (lizard)

Reptile (turtle)

Mammal (opossum)

24 *The front limbs of four kinds of tetrapods. The similarity of the appendages in these living vertebrates is an example of homology, the limb of each species having evolved from the same appendage in a common ancestor—in this case an early amphibian. The salamander (top) is an aquatic amphibian; the lizard is a structurally primitive reptile; the turtle is a more specialized reptile; and the opossum is a terrestrial mammal.*

ancestral structure modified in different ways in different animals (figure 24). The driving force, according to theory, is natural selection acting on variation.

The phylum Chordata has numerous splendid examples of how the concept of evolution made sense of the details of comparative anatomy. The data of comparative anatomy alone suggest that the living chordates can be arranged in ten subphyla and classes that indicate, in a very general way, increasing complexity: Hemichorda, Urochorda, Cephalochorda, Agnatha, Chondrichthyes, Osteichthyes, Amphibia, Reptilia, Aves, and Mammalia. The first three are marine protochordates; the last seven are vertebrates. Among the vertebrates the Agnatha, Chondrichthyes, and Osteichthyes are fishes and the Amphibia, Reptilia, Aves, and Mammalia are tetrapods—four-legged creatures.

The sequence could also be viewed as the consequence of evolution from very simple protochordates to the more complex birds and mammals (figure 25). The hypothesis was that different groups had split off the main line (the "main line" being that leading to us, naturally) and to varying degrees had retained some of the characteristics of their ancestors at the time of the split. Thus the famous cephalochordate *Amphioxus* was looked upon as a relic of a prevertebrate stage of chordate evolution. The lamprey and hagfish of today, which differ so much in superficial appearances from those early vertebrates, the ostracoderms, are so like them in fundamental ways that all are placed in the same class—the Agnatha. One can study living Agnatha, therefore, to learn something of a remote agnathan period of vertebrate evolution.

Slowly the paleontological data were obtained to show the broad outlines of vertebrate evolution. To a gratifying degree, the sequence that had been suggested by the data of comparative anatomy turned out to be the sequence of evolution.

As we will see in more detail in Part Four, many bewildering phenomena of development made sense in terms of evolution. The ear bones of mammals could be understood as modifications of the bones associated with the jaws of lower vertebrates (figure 26). The variations in the circulatory systems of all vertebrates could be understood as evolutionary modifications of a basic plan not unlike that of a shark embryo (figure 27). And finally the different types of vertebrate kidneys made sense in terms of descent with change (figure 28).

If each species had been created, why would there be these unduly complex ways of making ears, circulatory systems, or excretory systems? Wouldn't it be simpler to begin with the adult plan? On the other hand, one could account for the data in a satisfactory manner if the concept of evolution was invoked as an explanatory hypothesis. In evolution, what is already in existence can be modified.

Before there was fossil evidence from the mammallike reptiles, one could say only that the transformation of the reptilian articular and quadrate bones into the mammalian malleus and incus could be explained on the basis of evolution. That putative relationship was a

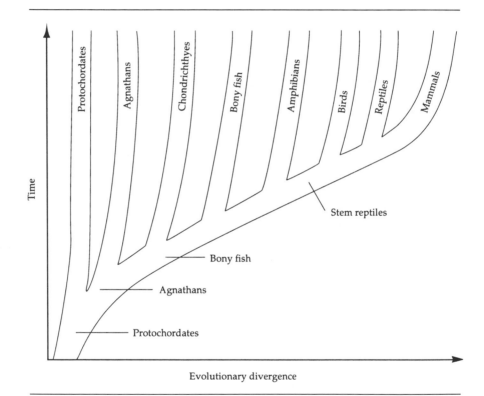

25 *The evolutionary relationships of the major chordate groups. The earliest chordates were protochordates, whose key features are shown in Figure 29. Some of their descendants, such as amphioxus and tunicates, are still alive. Others evolved into the earliest vertebrates, the agnathans. These were primitive fishes without jaws; their descendants, such as the lamprey and hagfish, are still with us. Agnathans were ancestors to the cartilaginous chondrichthyes (sharks) and to bony fish. Amphibians evolved from the primitive bony fish, and the stem reptiles from amphibians. Birds, modern reptiles, and mammals all evolved from the stem reptiles.*

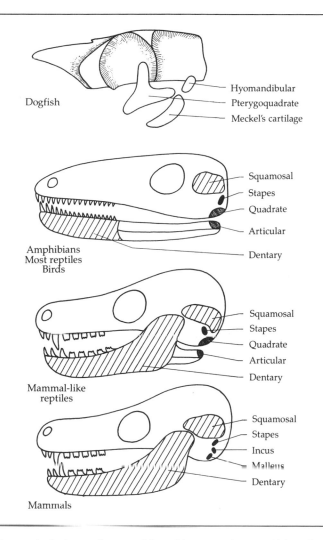

Dogfish

Hyomandibular
Pterygoquadrate
Meckel's cartilage

Squamosal
Stapes
Quadrate
Articular

Amphibians
Most reptiles
Birds

Dentary

Squamosal
Stapes
Quadrate
Articular
Dentary

Mammal-like
reptiles

Squamosal
Stapes
Incus
Malleus
Dentary

Mammals

26 *The jaw articulation and ear ossicles of four vertebrates. Although novelty does arise in evolution, an exceedingly common pattern is the alteration of an existing structure. Not one of the three ear bones of mammals—stapes, incus, and malleus —is really new. The stapes, which began as part of the hyomandibular in the primitive fish, became the single ear bone in amphibians, reptiles, and birds. The incus was a prominent upper jawbone (pterygoquadrate) in primitive fish and became reduced to the quadrate in amphibians, reptiles, and birds. It formed a jaw joint with the articular in the lower jaw, the latter a modification of Meckel's cartilage. In the mammal-like reptiles the dentary of the lower jaw begins to enlarge; and, finally, in the mammals it forms a new jaw joint with the squamosal of the upper part of the skull. Of the two bones of the old joint, the quadrate becomes the incus and the articular becomes the malleus.*

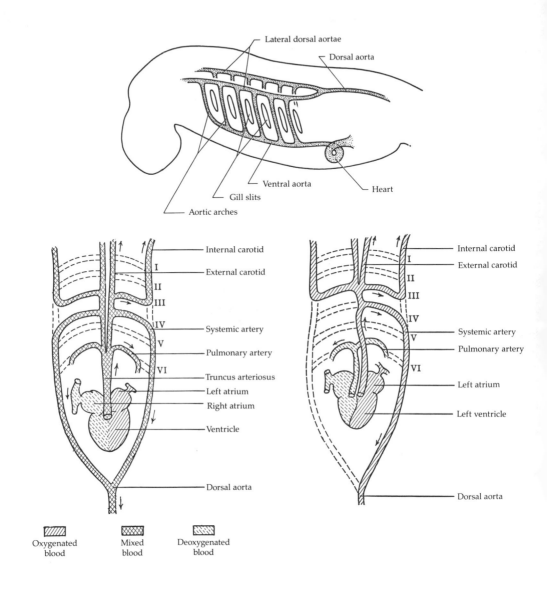

27 (Top) *Lateral view of the main arteries of the anterior trunk region of a vertebrate embryo. Note the six pairs of aortic arches. This basic pattern of the aortic arches is repeated in the embryos of all vertebrates and is variously modified in the adults as shown in two examples:* (left) *a frog and* (right) *a mammal.*

hypothesis. Only later would the fossil record show that this deduction was true beyond all reasonable doubt. The fossils, therefore, provided direct evidence for the correctness of the hypothesis of evolution. The case with the circulatory and excretory systems, although they "make

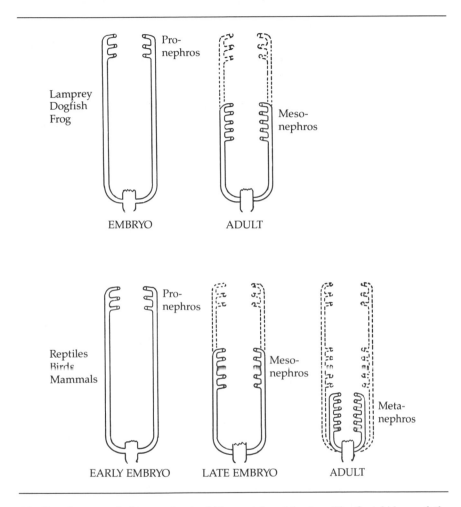

28 *Developmental changes in the kidneys of vertebrates. The first kidney of the vertebrate embryo is the pronephros, one on either side of the body cavity. In the lamprey (agnathan), dogfish (chrondrichthyes), and frog (amphibian) the definitive kidney, the mesonephros, develops later. The embryos of reptiles, birds, and mammals recapitulate the pronephros and mesonephros and then form the metanephros, which is the adult kidney.*

sense" as modifications brought about by evolution, must remain hypotheses. Circulatory and excretory systems are not usually preserved in the fossil record.

Thus the concept of evolution allows us to "make sense" out of much otherwise inexplicable data—and that is the grand purpose of a good theory. This is not an easy point of view to understand, for as Medawar points out:

> The reasons that have led professionals without exception to accept the hypothesis of evolution are in the main too subtle to be grasped by laymen. The reason is that only the evolutionary hypothesis makes sense of the natural order as it is revealed by taxonomy and the animal relationships revealed by the study of comparative anatomy . . . In biosystematics and comparative zoology the alternative to thinking in evolutionary terms is not to think at all. (1981)

Embryonic Development

The discussion of these examples from comparative anatomy emphasizes the close relationship between adult structure and embryology. This is not surprising—the structure of adults is derived directly from embryos. The next deduction, therefore, will come as a natural outgrowth of the last.

Deduction 12: If the members of a major taxonomic unit share a common ancestry, that fact should be reflected in their embryonic development.

Early in the nineteenth century embryologists noticed that embryos of different classes of chordates resemble one another much more closely than do the adults (figure 29). Darwin mentioned that Louis Agassiz had inadvertently failed to label an embryo when he first obtained it and then could not tell "whether it be that of a mammal, bird, or reptile" (*Origin*, p. 439). What could be the basis of this extraordinary similarity?

After 1859 one of the main preoccupations of embryologists was to study embryos for whatever light they could throw on evolution. (This will be discussed at length in Part Four on developmental biology.) In those days when the fossil record was even more inadequate than it is today, the changes in the course of development (ontogeny) were thought to reflect the changes that had occurred in the course of evolution (phylogeny). There could almost never be any direct evidence for the evolutionary changes in the soft tissues that fail to fossilize.

The reciprocal relationship of the structure of embryos and the theory of evolution is, again, both subtle and powerful. The fact is that the diversity of embryonic patterns of development not only "makes sense" as a reflection of evolution but, in turn, makes the hypothesis of evolution more probable. One of the most dramatic discoveries of comparative embryology was that the structure of all chordates can be understood as modifications of a basic form. All pass through an early stage with a notochord, dorsal nerve tube, and pharyngeal gill pouches. Only later do the embryos diverge to become adult tunicates, sharks, bony fishes, amphibians, reptiles, birds, and mammals. Derivation from a common embryological pattern suggested derivation from a common ancestor.

Classification

Classification is a device of utmost importance in enabling us to deal with everyday life. It consists of combining objects and ideas because of characteristics held in common. Classification is the most powerful

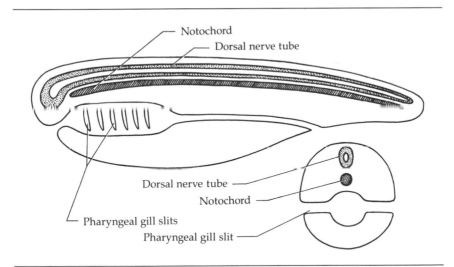

29 *An idealized chordate showing the three diagnostic characteristics of all members of the phylum: the presence at some stage in the life history of a notochord, dorsal nerve tube, and pharyngeal gill slits. Lateral view above, cross section in the gill region below.*

method we possess for packaging information. It is so much a part of our lives that we forget how basic it is.

Suppose that someone begins to speak about a VW, one that you have never seen. Nevertheless, you automatically recognize "VW" as "automobile." With that recognition you will immediately have a vast amount of available information—about motors, wheels, seats, batteries, gasoline, role in accidents, relation to pollution, oil and the nation's balance of payments, and a host of other details.

Classification is the same information-packed and predictive device in biology. Let us suppose that you have carefully dissected and studied one mammal—a fetal pig, rabbit, white rat, whatever. You are then presented with a new mammalian species never before studied by a scientist. Without touching the creature, you would be able to make a host of predictions about its anatomy, physiology, reproduction, and development. When you checked the predictions, they would be found to be correct or almost correct. To paraphrase a well-known dictum: "When you have studied one mammal, you know them all."

Different systems of classification for biological species have been used for different purposes. The ancient Hebrews made an important distinction between animals that were clean or unclean. The clean could be eaten, the unclean not. Aristotle, the father of so many fields in biology, is also acclaimed as the father of the classification of animals. The principal features that he employed were the presence or absence of blood and the mode of reproduction.

The modern period of biological classification begins with Carl von Linné (1707–1778), who regularized the system of giving Latin names to all plants and animals (including himself—"Carolus Linnaeus"). His *Systema Naturae* of 1758 is the starting point for our taxonomy. Linnaeus, following Aristotle, employed a very few characteristics for classifying organisms: the nature of the heart, type of reproduction, warm or cold blooded, and so on. This procedure of using only a few characteristics is called an *artificial system.*

As more and more was learned about the anatomy and embryology of animals, the artificial system proved to be inadequate: it appeared to unite species that differed in fundamental ways. It was suggested that a *natural system* of classification would better reflect the state of nature. The natural system differed from the artificial system in employing many more characteristics, including those of internal anatomy and development.

The way the natural system worked in practice can be understood by seeing how two of the "errors" made by a mid-nineteenth century taxonomist were corrected. In the Radiata, the hydroids were placed in one class and the medusae in another. That is a reasonable step if one uses only external morphology. The hydroids are plantlike species that are usually sessile. The medusae, or jellyfish, are free-swimming cup- or plate-shaped animals—with no apparent similarities to the hydroids. Later it was discovered that most species of hydroids reproduce by asexual budding and the buds become medusae. Furthermore, many of the medusae reproduce sexually, and the offspring are hydroids. Before this was known, there were cases where two very different forms were placed in separate classes when they were none other than parent and offspring of the same species.

A similar error occurred in the class Acephala, which combines the tunicates and the bivalve mollusks. A bivalve mollusk without its shell is a rather amorphous creature—recall the last oyster on a half shell that you ate. The tunicates are, likewise, usually rather featureless lumps growing on wharf pilings. So bivalve mollusks and tunicates were put in the Acephala because they lacked well-marked heads. The story changed when the embryos of tunicates were discovered. They proved to be tiny tadpolelike larvae with a notochord, dorsal nerve tube, and pharyngeal gill slits. The adult retains only the gill slits. Thus it seemed more natural to include the tunicates with the other kinds of animals that have those three characteristics—the chordates.

But what can possibly be the basis of there being groups of related species? We noted before that if species are separately created they cannot be genetically related. Yet there are groups of organisms: groups of individuals make up a species; groups of related species are united in a genus; related genera in a family; related families in orders, orders in classes, and so on.

Before the theory of evolution became available, it was difficult to provide a naturalistic explanation for how there could be natural groups. Some saw no need to. Agassiz recognized the problem and offered a personal answer.

> The divisions of animals according to branch, class, order, family, genus, and species . . . constitute the primary questions respecting any system of Zoology, [and] seem to me to deserve the consideration of all thoughtful minds . . . To me it appears indisputable, that this order and arrangement . . . [are] in truth but translations into human language of the thoughts of the Creator. (1859, pp. 8–9)

The purpose of the natural system is to ascertain relationships. It proved much easier in the nineteenth century to delimit natural groups in the lesser categories, such as genera and families, than in the major categories, such as classes and phyla. Thus one could easily imagine a common ancestor for all canine species but not for all Radiata. When the hypothesis of evolution was suggested, there was no change in the system of classification. What had changed was that, for the first time, there was a naturalistic explanation for natural groups: the data could be explained by the hypothesis that all the members of any well-defined natural taxonomic group shared a common ancestor. Thus,

Deduction 13: If evolutionary divergence is the basis of organic diversity, that fact should be reflected in the system of classification.

Naturalists found that it was.

Microstructure

Deduction 14: If there is a unity of life based on descent from a common ancestor, this should be reflected in the structure of cells.

And it is—to an astonishing degree. Common ancestry demands a unity of life, and nowhere is this more obvious than in the fact that the bodies of nearly all organisms are built of the same units of structure and function—cells.

During the last half of the nineteenth century, considerable knowledge of cells at the level of resolution of the light microscope was obtained. Cells were found to exist in great variety, but all shared a basic structure consisting of an outer limiting membrane, a nucleus, and associated cytoplasm. And all cells came from other cells (see Part Three). There were questions about the microorganisms—were bacteria really cells and did they have a nucleus? These debates lasted until the availability of the electron microscope and improved techniques settled many issues of fine structure. We now know that there are two fundamentally different types of cells—prokaryotic and eukaryotic. Bacteria and blue-green algae consist of single prokaryotic cells, which lack nuclei. Their DNA is free in the cytoplasm. The bodies of all higher organisms—true plants, fungi, and animals—have eukaryotic cells, which have nuclei.

Evolutionists in the nineteenth century more or less accepted the hypothesis that life must have started with organisms at least as simple as a bacterium, yet few of them would have dared to be so optimistic as to believe that a useful fossil record existed for such tiny and fragile creatures. In recent years, however, new methods of collecting and preparing specimens are providing us with information about organisms that lived 3.5 billion years ago (figure 30). The oldest so far discovered are from Western Australia and South Africa. The Australian

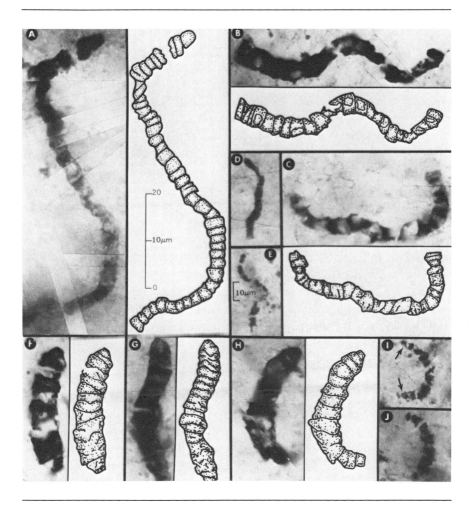

30 *The oldest known fossils. These microscopic organisms lived about 3.5 billion years ago in Pre-Cambrian rocks in Australia. They are related to present-day blue-green algae.*

material comes from a site with the improbable name North Pole. There, in the Warrawoona Group rocks, specimens identified as blue-green algae with a probable age of 3.5 billion years have been discovered. The Fig Tree Chert of South Africa has yielded what seem to be bacteria and blue-green algae. These first life forms are prokaryotes, which are much like the prokaryotes living today.

Living prokaryotes are all small—usually less than 10 micrometers and some only 0.2 micrometers. In common with all cells, there is an outer limiting membrane or plasmalemma. The contents of the cell are not compartmentalized. There are ribosomes but no mitochondria, Golgi, chloroplasts, or lysosomes. There is no nucleus, that is, a membrane-bound vesicle containing chromosomes. There is, of course, DNA (or RNA). This is in the form of a double helix, which is not associated with proteins, and is free in the cell. The DNA may form a tangled mass visible with the electron microscope as a nucleoid.

The characteristic features of the prokaryotic cell, therefore, are a plasmalemma, RNA-containing ribosomes, and naked DNA. We now realize that this was the limit of evolutionary complexity for most of life's span on earth. The generally accepted date for the formation of the earth is 4.5 billion years ago. The oldest known prokaryotic organisms date to 3.5 billion years ago.

Life took a momentous step 1 billion, or possibly even 2 billion, years ago, when eukaryotic cells came into existence. The important difference between a prokaryotic and eukaryotic cell is that the latter has some of its contents compartmentalized. That is, related sets of molecules concerned with the same function are surrounded by membranes that provide some isolation and regulate the passage of materials in and out of the compartments. The DNA, now associated with protein, is surrounded by a nuclear membrane with pores. The energy-yielding reactions are compartmentalized in mitochondria. The photosynthetic reactions of green plants are isolated in chloroplasts. Some of the secretory processes are restricted to the Golgi. Other membrane systems, such as the endoplasmic reticulum and the associated ribosomes, provide the protein-synthesizing machinery of the cell.

In recent years an interesting hypothesis for the origin of certain organelles of eukaryotic cells has gained increasing support. The suggestion is that intracellular structures such as mitochondria and chloroplasts began as free-living prokaryotic cells. These entered other cells, where they became symbionts. The result was the eukaryotic cell. Examples of symbiosis have long been known: algae and fungi cooperate to form lichens; algae live in the cells of several kinds of animals, including the giant clam, to which they supply food from their photosynthetic reactions.

Once the functions of cells became compartmentalized at the eukaryotic level of organization, there could be a true division of labor. With time this permitted the evolution of multicellular organisms with some cells different from others. Groups of specialized cells then took over the separate physiological functions. Yet even in the largest and most complex organisms the fundamental unit of structure and function remains the cell. The cells of eukaryotic organisms form an integrated whole. The functioning of the individual cells of the complex eukaryotic organisms can be seen as necessary for the life of the organisms, just as the functioning of the organism seems to be concerned mainly with the life of its individual cells.

Most of the details of structure and function in the cells make sense because what cells are and what they do are reflections of their evolutionary history.

Molecular Processes

Deduction 15: If there is a unity of life based on evolution, that fact should be reflected in the molecular processes of organisms.

There is a tremendous diversity in structure, physiology, and behavior among the species of the five kingdoms of life: The Monera (bacteria, blue-green algae), Protista (protozoans and green algae), Fungi (yeast, molds, mushrooms), Animalia (multicellular animals), and Plantae (red and brown algae, mosses, vascular plants). There is far less diversity in the cells of which their bodies are composed. There is still less diversity in the basic molecular processes that occur in all cells. This has become clear only in recent years with the wealth of new techniques available for studying molecular events in cells.

The fact that the basic metabolic reactions in the cells of all organisms are so much alike can be interpreted to mean that they developed in the first stages of life and have been retained, with relatively little modification, in the descendants. It is difficult to find another naturalistic explanation. We cannot study the molecular reactions of those ancient blue-green algae and bacteria from the North Pole deposits of Australia or the Fig Tree Cherts of South Africa. Nevertheless, these ancient progenitors have closely similar living counterparts and, by studying them, we can make reasonable inferences about what happened in the past.

Some of the more striking of these molecular uniformities are:

(1) All cells consist of the same classes of organic compounds: nucleotides, proteins, lipids, and carbohydrates.

(2) Reactions involving these organic compounds are controlled in all cells by the same class of proteins—the enzymes.

(3) DNA contains the coded information that regulates the life of the cell and transmits the information to the next generation. In a few types of viruses the closely similar RNA performs the same functions.

(4) In all organisms, DNA is composed almost entirely of combinations of six kinds of molecules: 2 pyrimidines (thymine and cytosine), 2 purines (adenine and guanine), a sugar (deoxyribose), and phosphate. The purine and pyrimidine bases are sometimes chemically modified.

(5) In all organisms, RNA is composed of combinations of six kinds of molecules: 2 pyrimidines (uracil and cytosine), 2 purines (adenine and guanine), a sugar (ribose), and phosphate. Again, the bases are sometimes chemically modified.

(6) The DNA code, which transmits hereditary information from generation to generation and provides the information for intracellular synthesis and control, consists of the sequence in which the purine and pyrimidine bases are arranged. Each specific triplet of bases is related to a specific amino acid. The code is universal for all forms of life—a simply astonishing fact.

(7) In all organisms the specific information encoded in DNA serves as a template for the synthesis of specific messenger RNA. Enzymes organized by specific messenger RNA molecules, in cooperation with transfer RNA, join amino acids to form specific proteins.

(8) In all cells, proteins are synthesized from about 20 kinds of amino acids. Except for the 5 carbon sugars of DNA, RNA, ATP, and other specialized molecules including some co-enzymes, the major carbohydrates of cells consist of 6 carbon sugars, such as glucose, and polymers of 6 carbon sugars, such as starch and cellulose. Important lipids in all cells consist of glycerol plus 3 molecules of fatty acids or 2 molecules of fatty acid and 1 molecule of a phosphorus-containing compound.

Nucleic acids, proteins, carbohydrates, and lipids account for nearly all of the organic compounds of life, of which there are enormous numbers of specific kinds within these major groups. Nevertheless, this great diversity is based on a relatively few building blocks—much as the rich English language combines only a few of the 26 letters to fill an unabridged dictionary.

(9) All cells obtain energy by oxidizing reduced compounds, as in glycolysis—the anaerobic fermentation of glucose to pyruvic acid. In glycolysis, chemical energy is transferred to convert ADP to ATP. ATP is the immediate energy source for cells. Some prokaryotes and most eukaryotes use oxygen as a major electron acceptor, as in the aerobic breakdown of pyruvic acid to carbon dioxide and water. In these aerobic processes, as compared with anaerobic reactions, much more energy is transferred to convert ADP to ATP. These are striking examples of the fact that many of the basic metabolic pathways are the same or similar in all cells. Most of the energy for life is the energy of sunlight captured by green plants in the reactions of photosynthesis and stored in carbohydrates. Chemosynthetic bacteria carry out processes very similar to those of other organisms, even though they power their reactions from reduced compounds such as methane, rather than from sunlight.

(10) Nowhere is the biochemical unity of life shown more dramatically than in experiments where genes of higher organisms, such as human beings, can be incorporated into the DNA of bacteria. There, using the bacterial synthetic machinery, the human gene can direct a human-type protein to be synthesized.

Although the sorts of data just listed are not proof of evolution, there is no other naturalistic explanation that can make as much sense of them. Similarly, there are no molecular data that falsify the concept of evolution. More than anything else, the molecular data point to a unity of life based on common descent.

Today the study of evolution with the procedures of molecular biology is a very active field of research. The elegance of its probes, the rigor of its methods, and the precision of its answers are making it possible to gain deeper understanding of the processes of evolution. The study of evolution at the molecular level is the study of the fundamental molecules of evolutionary change, DNA, and the molecules that are the products of DNA, the proteins. When one compares the bones of a horse and human being and then their hemoglobin molecules, the methodology is basically the same—comparing structures. The hemoglobin molecules, however, are immediate consequences of gene action, whereas the bones are the consequences of complex interactions of cells and their molecules extending over the long period of embryonic development. Proteins are basic; bones are derivative.

The procedures of classical evolutionary biology for estimating relatedness were crude. Hybridization was one. Among similar species, if A and B could be hybridized but neither could be crossed with C, it was assumed that A and B are more closely related to each other than either is to C. If hybridization was impossible, then one had to rely on comparisons of structure—the closer the structural similarity the closer the relatedness and, presumably, the shorter the time interval since a common ancestor.

Now it is possible to make detailed comparisons of the base sequences of different DNAs and the amino acids of proteins of different species. For DNA this does not measure what genes do, merely whether or not their base sequences are similar. When different species of frogs or of flies are compared, the sequences are the same for about 81–89 percent of the DNA. Two species of mice may have 95 percent of their sequences the same. A real surprise (shock?) is to learn that the base sequences of human beings and chimpanzees are 99 percent identical.

Similar experiments comparing amino acid sequences of many different vertebrate proteins reveal that human proteins resemble those of other primates closely, and this resemblance of proteins decreases in the following order: rodents, dogs, marsupials, monotremes, birds, and amphibians. This is the same sequence of relatedness that evolutionists had proposed, on other evidence, before molecular comparisons were possible.

The field of molecular evolution is now in a period of validation. Its discoveries are in accord with what previous methods have established as the major concepts of evolutionary biology—we would be in serious trouble if that had not been the case. The main contributions of molecular evolution to date have been to give more detailed, and often more precise, answers to important questions related to biological classification and the rates of evolution.

One of the most surprising discoveries of molecular biology that relates to evolution is that a minimum of 20–50 percent of loci that code for proteins are polymorphic. Before this was known it was assumed that, in general, natural selection would have given the "best" available allele for each locus. An alternative would be if two different alleles provided better adaptation when an individual was heterozygous for them. In that case both would be selected. But that alternate cannot explain cases where there are many alleles for a locus. Kimura (1983) and others have suggested that a large amount of the genetic variability in populations is not due to natural selection but to more or less random mutations that produce alleles that have little or no effect on survival —that is, they are neutral or essentially so. Selection would not act on neutral alleles, and so they could appear and change in frequency in an entirely random manner. This is a hotly debated subject, yet to be resolved. A distinct possibility is that the neutral alleles produce proteins that differ only in some nonessential part of the molecule and hence would not be subject to natural selection. So far the neutral theory is restricted to molecular evolution. Morphological evolution is seen as a consequence of natural selection.

This ends our formal analysis of Darwinism, which was accomplished mainly in a series of deductions stemming from the proposition, "If the hypothesis of evolution is true . . . " The first 10 deductions sought a direct test of the hypothesis. The last 5 emphasized the role of the hypothesis in bringing rational order to the data of modern biology and geology.

It is this second function of the concept of evolution that has so impressed scientists. To a considerable degree the two statements quoted earlier—Dobzhansky's that "Nothing in biology makes sense except in the light of evolution" and Medewar's that "The alternative to thinking in evolutionary terms is not to think at all" are fully accepted by scientists familiar with the data of biology and geology.

chapter 10

Life over Time

T he last edition of the *Origin* was published in 1872, and Darwin was interred in Westminister Abbey on April 26, 1882, a few feet from another of England's illustrious sons, Sir Isaac Newton. Together they had revolutionized the biological and physical sciences.

The amount of research performed in the century since Darwin's death that relates directly or indirectly to evolutionary biology is enormous in quantity but more modest in conceptual advances. For the most part the work has been "normal science," as Thomas Kuhn has designated the mopping-up activities that follow a scientific revolution.

Today very much more can be said about the history of life than was possible in Darwin's time. Generations of paleontologists have explored most of the earth's crust and filled the museums with fossils. The data of paleontology that were supportive of the concept of evolution a century ago have now reached the state where all with an open mind will regard them as complete proof. Nevertheless, there is much more to be learned. As is true with all science, each discovery provides more questions than answers.

Darwin spoke of "the imperfection of the geological record." It is still imperfect, but this is a consequence of more difficult questions being asked. In Darwin's day the principal information desired was whether or not organisms intermediate between major groups had ever existed, as the hypothesis of evolution demanded. *Archaeopteryx*, the fossil, said yes. But once that question was answered, it was only natural that scientists would wish to know about links between the archosaurian reptiles and *Archaeopteryx* and between *Archaeopteryx* and modern birds.

Thus for each missing link discovered there were two missing links to be sought.

The ultimate goal of paleontology is to document, to the greatest extent possible, the lineages of life. This has been accomplished but only in a very elementary manner—even for the chordates, which provide us with the best possibilities for discovering fossil lineages. For some of the very important questions—for example, the antecedents of the phyla—there is almost no solid information.

One problem relating to the interpretation of the data should be mentioned. Some paleontologists are reluctant, and wisely so, to say whether or not a specific fossil is on the *direct* ancestral line to a later fossil species or to a surviving species. To offer such an opinion with any high degree of confidence would require a huge sample of fossils from rocks of slightly differing ages covering the entire span from the putative ancestor to the putative descendant. Such a complete record has never been available; but for the basic question it is not essential. What is required are fossils that are intermediate in structure and occurring in strata of intermediate age.

In the sections that follow I will summarize what we know today about the history of life, particularly the Animal Kingdom.

The Origin of Life

The geological record provides almost no evidence for the origin and very early evolution of living organisms. We must rely, therefore, on hypotheses to guide the way we think about these early events. These hypotheses are based on what we know about the evolution of more recent organisms, on experiments replicating the conditions thought to have existed on earth at the time of the origin of life, and on detailed investigations of the least complex living organisms, mainly the archaebacteria. The following statements are commonly held hypotheses, which are constantly being tested and refined.

About 4 billion years ago, in an abiotic environment rich in organic molecules, the initial steps leading to life were taken. Short chain polymers of nucleotides and of amino acids were incorporated in vesicles surrounded by semipermeable phospholipid membranes. These vesicles could extract organic and inorganic molecules from the environment and synthesize more polynucleotides and polypeptides like themselves. This resulted in growth and, at some stage, the ability to replicate and reproduce individuals identical, or nearly so, with the original structure.

The control of both synthesis and replication resided with the poly-nucleotides (initially, probably ribonucleotides). They could replicate but not with complete precision—the resulting errors were mutations. The nucleotides directed the synthesis of polypeptides with specific arrangements of amino acids. Some of these polypeptides had catalytic abilities.

Initially all of the molecules required by the pre-cell vesicles for synthesis and energy were extracted from the environment. Since these vesicles could reproduce at the expense of the environmental pool of required molecules, a Malthusian catastrophe was inevitable, since the environmental pool was finite and the products of reproduction theo-retically infinite. Such a catastrophe, however, was avoided when some pre-cells mutated and acquired the ability to synthesize some of their molecules from simpler molecules previously unusable. Such pre-cells increased at the expense of those lacking these new synthetic abilities.

A basic level of structure and function was achieved with the ap-pearance of cells about 3.8 billion years ago. These may have resembled the prokaryotic cells, such as the archaebacteria, still abundant denizens of the earth. Cells became the structural and functional units of life and have remained so to this day. Two major levels of complexity were achieved: the simpler prokaryotic cells and the more complex eukaryotic cells, the latter probably appearing between 2 and 1.3 billion years ago.

The new synthetic abilities could not forever prevent a Malthusian catastrophe because once the supply of molecules that could be used for energy became exhausted, life could not continue. Some prokaryotic cells, however, perfected the ability to use the energy of sunlight for their intracellular syntheses. Photosynthesis then divided all life into two basic nutritional patterns: the *autotrophs*, which obtain energy by photosynthesis, and the *heterotrophs*, which depend on the energy-rich molecules produced by the autotrophs. One of the by-products of pho-tosynthesis, oxygen, was produced in increasing amounts and, in time, became a significant part of the atmosphere.

The twin phenomena of inheritable variability based on nucleic acids and natural selection by the environment tended to maximize each individual's ability to survive and reproduce. Survival depended on the ability to acquire resources, avoid destruction, and to leave offspring.

The constant pressures for scarce resources led to the evolution of new ways of life, which could allow exploitation of previously un-touched resources. The ability to exploit a new environment required adaptational changes to that environment. Living organisms increasingly became specialists, able to excel in one environment with a concomitant inability to succeed in others. This ever-increasing adaptation for a specific environment, in turn, required a more regulated (homeostatic) internal environment that would enable the life processes to be maximized.

Those individuals that were descended from the same ancestral cells and continued to have essentially the same structure and physiology could be recognized as species. With the onset of sexual reproduction, species became groups that could exchange genes among themselves but not usually with other species. Thus, species became the units of both genetic experimentation and evolutionary continuity.

The processes of mutation and natural selection perfected prokaryotic species for both moist terrestrial and aquatic environments. For about 2 billion years they were the sole living occupants of the earth. When prokaryotic cells evolved into eukaryotic cells between 2 and 1.3 billion years ago, symbiosis apparently played a major role. For example, a probable hypothesis for the origin of the mitochondria of eukaryotic cells is that they originated from archaebacteria that invaded other prokaryotes and remained as symbionts. Eukaryotic cells represented a new experiment in ways to seek resources.

The *Protozoans* can be thought of as variations on the theme of a generalized eukaryotic cell. Formerly classified as members of a single phylum of the Animal Kingdom, they are now parcelled out among many phyla of the Kingdom Protista or Protoctista. There are thousands of species that represent a great range of morphological and physiological types. Some species move by means of flagella, others by cilia or pseudopods, and some have no means of locomotion. Because they are unicellular, there are no organs, but some species have an extraordinarily complex structure. Free-living species are found in moist soil, fresh water, and the ocean. Some species are parasites. Several species of the genus *Plasmodium* are responsible for that most important human disease, malaria.

Every individual protozoan, since it is a single cell, is in close contact with the environment and so escapes the special physiological problems of multicellularity. Food is ingested or absorbed. The movements of the respiratory gases and excess molecules are largely by diffusion but sometimes assisted by specialized cell structures, such as contractile vacuoles. Some species, such as the foraminiferans and radiolarians, have beautifully intricate skeletons. Most species reproduce solely by asexual means but a few reproduce sexually, with meiosis.

The Rise of Multicelled Organisms

The earliest remains of complex metazoans so far discovered are in the Ediacarian strata of 670 million years ago (see figure 20). So if we estimate that it required about 30 million years from origin to earliest known fossils, we arrive at the prevailing hypothesis, which is that multicelled animals originated about 700 million years ago.

The Phanerozoic is generally regarded as the eon of metazoan life. The earth of the early Phanerozoic was a very different place from that familiar to us today. The world was then recovering from a period of about 300 million years of frigid conditions, extensive glaciation, and a marked reduction in the abundance and diversity of living organisms. Life had gone through a bottleneck, but during that seemingly endless period (almost as long as the time required for the entire evolution of terrestrial vertebrates) the first experiments in metazoan life must have begun. As the ice melted, low-lying portions of the continents were flooded, resulting in a vast increase in shallow seas, a most productive habitat for life.

A world map circa 600 million years ago, at the beginning of the Paleozoic Era, would have no resemblance to a map of the land masses of today. The antecedents of present-day continents consisted of large and small islands with distributions having little relation to their present locations. Throughout geological time the land masses were elevated and submerged, collided and separated. Our personal experience speaks to a near stability of the earth, but in the last few decades it has become clear beyond any reasonable doubt that the earth's crust is in a constant state of flux—as measured in the vastness of geological time. (The human life span does not give adequate time to detect most geological catastrophes. For example, we are not aware through our senses that the landmass of India is now crunching into the Asian continental plate, as it has been doing for millions of years, and pushing the Himalayas ever higher.)

The landmasses would have been of little interest to biologists in Ediacarian times since there could be no life. The land was yet to be colonized by autotrophic organisms so the heterotrophic metazoans, of necessity, were restricted to the sea where the photosynthesizers lived.

Even the atmosphere was different. Seemingly oxygen, so essential for nearly all modern heterotrophic life, was not present in the primitive atmosphere. It was produced as a by-product of photosynthesis carried out by prokaryotes and algae.

At first most of the oxygen was combined with iron and so not available for heterotrophs. But photosynthesis eventually produced oxygen that stayed in the atmosphere. It is believed that by 670 million years ago, the onset of the Ediacarian, the oxygen concentration in the atmosphere was about 1.6 percent instead of today's 21 percent (by volume). But that may have been enough to get things started.

It is only in recent years that metazoans of the Ediacarian Period have been discovered and described. Previously, the earliest fossil evidence of metazoan life was provided by rocks of the Cambrian Period, which began about 550 to 570 million years ago. The Ediacarian fauna are named from their site of discovery. The first important discovery of Precambrian rocks with metazoans came in 1946 in the Ediacara Hills north of Adelaide, Australia. All of the fossils discovered so far are of softbodied animals, which have left only an impression. There are considerable problems of interpretation of these fossils, but the most astonishing fact, if confirmed, is that many of the Ediacarian fossils can be assigned to extant phyla. That means that some living coelenterates, annelids, and arthropods have changed so little from Ediacarian times that both living species and species of the earliest known metazoans can be recognized as members of the same phylum.

What Is a Phylum?

The animal phyla represent major specializations in the ways organisms secure the molecules required by body cells and eliminate waste molecules. The species assigned to a major phylum are built on the same fundamental plan. For example, the fundamental plan for all species of the phylum Chordata can be imagined as an elongate, cylindrical animal with an alimentary canal, and at some stage of its life having gill slits in the pharynx, a notochord, and a dorsal nerve tube.

Yet within each phylum there is usually an extensive adaptive radiation that permits the different species to live in very different ways. These amazing adaptive radiations within phyla make it difficult to define the basic body plan of natural phyla. A "natural" phylum would include all species that share a fundamental body structure and a common ancestry, but delimiting natural phyla has not been easy.

Today many taxonomists believe that most multicelled animals fit into one of the following nine natural phyla. We will take a brief look at the characteristics of these major animal groups before continuing our walk through the history of life.

Phylum Porifera

The Porifera, or sponges, are not only multicellular but some are massive. Nevertheless, they exhibit a very low grade of structural differentiation. Many species are amorphous, and those with more definite structure are radially symmetrical. There are no organs or even well-defined tissues. All live in water. Their bodies are permeated by spaces and channels through which water is moved by the beating of flagella of special collar cells, or choanocytes. This water reaches the vicinity of every cell, bringing food particles and oxygen and removing the waste products of metabolism. Some species have symbiotic cyanobacteria, which are autotrophic and can meet some of the nutritional requirements of the sponge cells. Many species have skeletons composed of spicules of calcium or silicon compounds. Sponges, with their readily fossilized spicules, have been part of the fossil record since the early Paleozoic. Most of the living species are marine and different species occupy habitats from the edge of the sea to the abyss. A few species live in fresh water. Although the level of structural complexity of sponges may seem to be intermediate between the Protozoa and the Eumetazoa, and so may be a possible intermediate evolutionary stage, it is generally believed that they represent a dead end of evolution.

Phylum Coelenterata

The coelenterates are also largely marine species and of great geological antiquity. Their basic structural plan can be thought of as an organism with a radially symmetrical, saclike body. The body wall is composed of two layers of cells enclosing a central cavity, with a single opening to the outside. That single opening serves both as mouth and anus and is surrounded by a circle of tentacles that capture food. Coelenterates have a unique type of stinging cell in the epidermis that contains nematocysts. These explode, propelling a dart that serves for protection and the capture of food. There is a simple nervous system, and in the medusae (jellyfish) there is a simple vascular system. Marine colenterates show an enormous amount of diversity. Some are small plantlike creatures. Others are corals, sea fans, and sea anemones.

   Species able to reproduce sexually and asexually are common among the coelenterates and occur in many other phyla as well. Asexual reproduction—cloning successful individuals—rewards the genetically well adapted. But there is a disadvantage to this reproductive strategy.

A fixed genotype might not produce individuals that can succeed in slightly different environments or survive should there be substantial changes in the customary environment. Thus individuals with different genotypes will be an advantage to a species in some conditions.

This argument may seem at variance with the facts of life. The prokaryotes are the most abundant of all organisms, yet sexual reproduction among them is rare. Among the eukaryotes, sexual reproduction is unknown for *Amoeba* and its close relatives. Prokaryotes are of great antiquity, having been on earth far longer than the mammals. Thus they must be regarded as highly successful, yet we must note they have not evolved very spectacularly. It was only when some of the descendants of the prokaryotes hit upon the mechanism of sexual reproduction that an awesome evolutionary radiation became possible.

Phylum Platyhelminthes

The platyhelminth species ("flatworms") show many important advances over the coelenterates but are still comparatively simple in structure. They are bilateral rather than radial, as are all the remaining major phyla. Not only can the body be divided into a right and a left side but also into anterior and posterior as one end, the anterior end, almost always leads when the worm moves. Heads and tails, therefore, make their appearance in the platyhelminths.

One can imagine an archetypal platyhelminth as having a bilateral, flattened body without an internal cavity, a digestive cavity with a single opening, and simple nervous and excretory systems. The variations on this basic theme are enormous. Some species lack a digestive cavity and food passes directly to the cells. The most generalized species are aquatic, being found in salt and fresh water or in very moist terrestrial environments. Large numbers, such as the liver flukes (*Clonorchis*; figure 31), *blood flukes (Schistosoma),* and tapeworms *(Taenia)* have become parasites of other animals. Parasitism is often associated with reduction in some structures and functions, an increase in others, marked host specificity, and the possession of elaborate behavior patterns that enable the parasite to move from one host individual to another.

Phylum Nematoda

The nematodes, or round worms, differ from the platyhelminths in a number of morphological features, some of which are advances in structural complexity. One is the presence of an alimentary canal, in contrast to the saclike digestive system of the coelenterates and platyhelminthes. Food passes in one direction only in an alimentary canal—from mouth to anus. This means that parts of the canal can be specialized in different ways and digestion accomplished with assembly-line efficiency.

Another feature of the nematode worms is the presence of a special type of fluid-filled body cavity, a pseudocoel, which provides rigidity and serves in a sense as a skeletal system, and so is important in locomotion. The fluid of the pseudocoel also serves to a limited degree to expedite the movement of molecules to and from cells. In addition, since the pseudocoel is noncellular, it is a space that makes no metabolic demands. Organisms in more structurally complex phyla typically have a body cavity that is a true coelum.

31   A liver fluke (Clonorchis) *that parasitizes human beings. It is one of the flatworms, which represent an important level of organization among the invertebrates, having all of the organ systems present in more complex forms.*

The many species of nematodes are abundant in fresh water, the sea, soil, and as parasites of plants and animals. In fact, it is probable that there are more individual nematodes than individuals of any other metazoan phylum.

One might deduce from these facts that there would be a vast range of morphological types. Quite the opposite is true. The many species of nematodes have a remarkably constant morphology and seemingly have evolved an all-purpose body plan. That all-purpose body plan permits the free-living species to inhabit all aquatic and terrestrial environments as well as be parasites on many kinds of animals and plants.

This means, of course, that the reduction of structures observed in many parasitic flatworms does not occur in the parasitic nematodes. Thus we reach the important conclusion that the same habitat need not select similar types. Nematodes and tapeworms may find a home in the same human intestine yet remain very different creatures. Some parasitic nematodes have remarkable adaptations that enable them to get from one host to another, but many species produce prodigious numbers of eggs which by chance enter a new host in contaminated water or food. This is a very different means of dispersal from that perfected by the flukes and tapeworms, with their intermediate hosts and complex types of larvae. Once again, the same biological problem has been solved in different ways by different creatures.

Phylum Annelida

As we continue up the scale of increasing structural complexity, the next major phylum is Annelida (figure 32). All the major organ systems are present, which permits a level of activity not observed in the metazoans discussed so far. The annelid worms have three major structural advances over the previously considered phyla: a type of body cavity called a coelom, a closed circulatory system, and metamerism. It is a space in which many of the organ systems are situated. A closed blood system is one in which blood is always within the tubes of the system —within a heart, artery, capillary, or vein. Metamerism is the division of the body into a longitudinal sequence of similar segments. In the common earthworm, *Lumbricus*, most segments have a pair of excretory nephridia (excretory organs), an enlargement of the ventral nerve cord, in each of a few anterior segments a pair of "hearts," and similar blood vessels and peripheral nerves. Thus each segment contains portions of most of the major organ systems.

Different species of annelids are found in the major habitats: the ocean, where there is the greatest diversity of species, fresh water, and land. The terrestrial species are restricted to moist habitats, since they have no mechanisms for preventing the loss of water in very dry habitats—as the desiccating and dying earthworms on a pavement after a rain testify. One class of annelids, the leeches, includes many blood-sucking ectoparasites of vertebrates and invertebrates as well as free-living species that feed on other invertebrates.

32 *The marine annelid worm* Nereis. *In comparison with the flatworms, the annelid worms show important advances in having the body divided into segments (metamerism) and having appendages, in this case the flaplike structures on each segment.*

Phylum Arthropoda

There are more described species of arthropods than of all the other phyla combined. Both in diversity and numbers of individuals the phylum is awesome—crustaceans, spiders, mites, scorpions, insects—essentially all of the crawly creepy things that exist (figure 33). Arthropods are abundant in all major habitats: land, fresh water, and the ocean. They walk, fly, burrow, and swim. Clearly the arthropod way of doing things has been exceedingly successful in evolution.

Arthropods were the first animal phylum to successfully colonize dry land. Later there were two others, Mollusca and Chordata. In fact, some arthropod species have so liberated themselves from watery and moist environments that they can survive solely on the water obtained from their metabolism of even very dry food (for example, the flour beetle, *Tribolium*). Arthropods were also the first phylum with species that could fly.

A key to this ability to live on dry land is the presence of an outer covering that is impermeable to water and hence can prevent desiccation. This covering is an exoskeleton of chitin, which is secreted by epithelial cells, and which can be flexible when thin or rigid when thick or calcified. Growth can occur only by shedding the confining exoskeleton and then secreting a new and larger one. That may not seem like an efficient way of doing things but, considering their great success, one must be careful in claiming that anything the arthropods do is inefficient.

Other key features of the arthropods are paired, jointed appendages, segmented bodies, a greatly reduced coelom compared with the annelids (from which the arthropods probably evolved), and an open circulatory system. The last feature is in contrast with the annelids. Blood in the annelids is nearly always in vessels. In the arthropods blood leaves the heart in arteries and then passes into spaces, collectively called the hemocoel, that permeate the entire body and come near to every cell.

Phylum Mollusca

Among the invertebrate phyla the Arthropoda and Mollusca are the most complex structurally, and the two phyla are probably related—the best available hypothesis being that both evolved from a primitive annelid or preannelid stock. Nevertheless there is little superficial resemblance between arthropods and mollusks.

Once again we note that evolution has been profligate in producing incredible numbers of molluskan species, each obtaining in its own way the requirements for life and exhibiting its own peculiar behavioral patterns. Evolution has adapted molluskan species for life on dry land, in the sea, and in fresh water.

The mollusks are similar to annelids and arthropods in being bilateral with well-developed digestive, excretory, circulatory (usually open), and nervous systems—even though these systems differ from those of annelids and arthropods in fundamental ways. The important unique features of mollusks are their soft bodies, often covered by a shell, and their lack of metamerism. Compared to many of the arthropods, crustaceans and insects, most mollusks are rather sluggish creatures and the sense organs are poorly developed. The squid and octopus are notable exceptions. Both the lobster (an arthropod) and the squid (a mollusk) have achieved similar levels of complexity, one with and the other without a metameric body—similar ends are achieved by evolution in diverse ways.

Phylum Echinodermata

The last phylum consisting solely of invertebrates to be considered consists of the echinoderms, whose body plan is very different from that of any other phylum. Furthermore, what would seem at first glance to be the two most characteristic features of echinoderms, an exoskeleton and radial symmetry, turn out not to be so. The familiar sea stars and sea urchins appear to have exoskeletons, thus resembling the arthropods, but in fact this skeleton is internal, being formed beneath the outer epithelium. The apparent radial symmetry is secondarily derived from a true bilateral body plan that is obvious in the embryo and of which vestiges remain in the adult.

The echinoderm body plan, then, is that of a large bilateral animal with an overlying pentamerous symmetry, with an endoskeleton, and a water vascular system that is important in locomotion and internal transport. The internal organs reflect the pentamerous body form. There are no specialized excretory organs and the nervous system is not impressive, reflecting the rather sluggish behavior of echinoderms.

Echinoderms are marine, occupying all zones from tide levels to the great depths. A great diversity of species has evolved and the body form varies from plantlike crinoids, sluglike holothurians, to the muffin-shaped sea urchins and the obviously pentamerous sea stars.

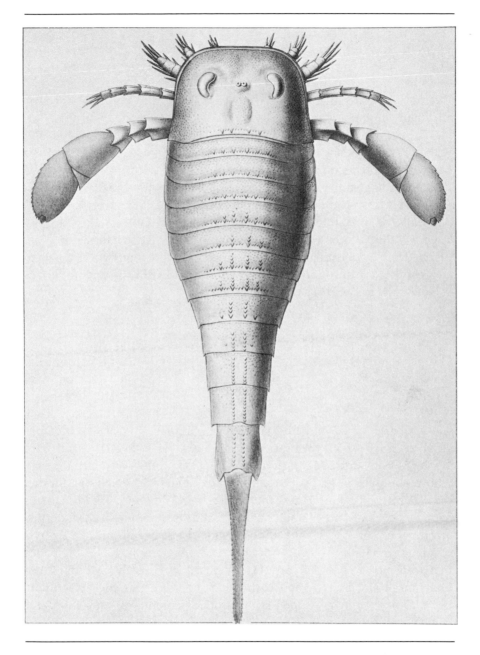

33 *The eurypterid* Eurypterus remipes, *a marine arthropod that lived during the Silurian in northern New York state. Eurypterids were fierce predators in those ancient seas, some reaching a length of 7 feet.*

Phylum Chordata

This, our own, phylum consists of the familiar vertebrates and more primitive forms such as amphioxus, the tunicates, and possibly some wormlike creatures, the hemichordates, none of which have a backbone. Again, adaptive radiation has been enormous, and chordates occupy all major habitats. The basic body plan is of an elongate cylindrical, metameric, bilateral animal with well-developed organ systems and with three diagnostic characteristics: a dorsal nerve tube, a notochord beneath it, and pharyngeal gill pouches. Chordate embryos have these three essential features, as does the adult stage of amphioxus. The adult vertebrates lack a notochord, the gill pouches are extensively modified, and only the dorsal nerve tube persists, as brain and spinal cord.

Very little is known of the relationship of the chordates with other phyla. The strongest clues come from comparative embryology, which suggest that the chordate's closest affinities are with the echinoderms.

Now back to the story of life over time.

Burgess Shale Metazoans

The passage of 150 million years and 9 thousand miles takes us from the Ediacara Hills in Australia to another exceptional site for fossil organisms, the Middle Cambrian Burgess Shale of British Columbia, which was formed about 530 million years ago. The Burgess Shale fossil site was discovered and first studied by Charles D. Walcott, a distinguished American invertebrate paleontologist. It has been worked since Walcott's time, and most recently Harry Blackmore Whittington (1985), who made very large collections, studied anew the material collected by Walcott and summarized the available information. He recognizes sponges, coelenterates, brachiopods, mollusks, annelids, echinoderms, arthropods, chordates, and minor groups for a minimum of 11 phyla of metazoans with about 129 species. And there are the 17 Problematica that were not assigned to any extant phylum. Thus this Middle Cambrian fossil site is extremely rich in major taxa, and the structurally more complex phyla—annelids, arthropods, mollusks, and chordates—are all present (figure 34).

The oldest known chordate, as distinct from the oldest known vertebrate, is *Pikaia gracilens*, from the Burgess Shale. It was described by Walcott as an annelid worm but, on the basis of more specimens and further study, Whittington believes it to be the earliest known chordate.

34   *A tiny sampling of animals from the Cambrian Burgess Shale of British Colum-*
bia. The extraordinary state of preservation of this enormous deposit has told us a
great deal about very ancient invertebrates.

The most abundant Cambrian fossils are trilobites, with their hard exoskeletons and numerous appendages, accounting for about 60 percent of fossil species. Trilobites were early experiments in how-to-be-an-arthropod. They were successful for about 350 million years but almost totally disappeared in the massive extinctions of the late Permian. Prominent but less abundant in the Cambrian were the brachiopods, comprising about 30 percent of all fossil species. They are shelled organisms with a superficial resemblance to the mollusks but are placed in their own phylum.

The Ediacarian and Cambrian periods together provided about 150 million years for an explosion of metazoan life. When the atmospheric oxygen concentration reached about 2 percent, organisms with shelly parts became abundant. The major metazoans represented were sponges, the archaeocyathines (spongelike and possibly the only generally recognized phylum to have become extinct), brachiopods, bryozoans, mollusks, annelid worms, and arthropods. The organisms related to us—echinoderms, graptolites, and possibly a chordate—were present. The graptolites, known only as fossils, are now thought to be related to the hemichordates, of which *Balanoglossus* is a living representative. Graptolites were colonial, sticklike, marine organisms. Thus all the major groups having species with hard parts are known except for the vertebrates.

All of the taxa of Cambrian metazoans continued into the Ordovician. Fossil remains suggest that the once-dominant trilobites decreased and that the brachiopods became the dominant group. Graptolites, crustaceans, sponges, and the cephalopods among the mollusks were well represented. The Ordovician strata have the earliest known reasonably good fossil vertebrates. Isolated scales, which can be identified because the same types of scales occur as parts of the better preserved ostracoderms of the Silurian period, are known from the middle Cambrian. All life was still in the sea during the Ordovician period. The land masses continued to move and were in a configuration totally different from the continents of today.

The event in the Silurian with the most far-reaching consequences was the beginning of the colonization of dry land by green plants. The Silurian seas were home for a rich variety of invertebrates, with the brachiopods, trilobites, and graptolites being abundant and diverse. The most fearsome Silurian sea creatures were the eurypterids, predatory arthropods with a strong exoskeleton.

Brachiopods and echinoderms were still abundant in the Devonian seas, and ammonites became more so. Spiders, mites, crustaceans, insects, and millipeds—all arthropods—were present on land in the Devonian. The earliest known insects were the collembolas, or spring-tails, and they are still with us. Land plants related to extant ferns, club mosses, and horsetails were abundant by the late Devonian.

Early Evolution of the Vertebrates

Although fragmentary remains of vertebrates are known from the late Cambrian, it is not until the late Silurian that well-preserved fossils are encountered.

There were no vertebrates on land during the Silurian but in the seas there were many kinds. The jawless fishes, such as ostracoderms, were abundant, and there were the beginnings of other major taxa of fishes that would reach their peak in the Devonian.

The ostracoderms are a group of jawless fishes, the Agnatha, that are ancestral to the few agnathans still living (lampreys and hagfishes) and probably to all other vertebrates. The ostracoderms flourished in the Silurian and Devonian but died out at the close of the latter.

The ostracoderms ("shell skin," so called because of the bony exoskeleton of some of them) are known from many parts of the world (figure 35). Fossils from Spitzbergen are among the most remarkable of all vertebrate fossils because of the amount of information gained from the internal anatomy of their heads. When the individuals died their bodies became buried in the bottom sediments. The soft parts decayed, leaving only the bony structure. Most of the head was bony, except for the spaces occupied by the nervous and circulatory systems. These spaces became filled with mud. Gradually both the bone and mud-filled spaces turned to stone, but stone of slightly different color. The Swedish paleontologist E. Stensiö dissected the fossils and revealed an astonishing amount of detail about the brain, cranial nerves, and blood vessels. In the end Stensiö made the anatomy of the ostracoderm head nearly as well known as that of the living agnathans.

Late in the Silurian and throughout the Devonian there was a wide variety of fishes with jaws such as the acanthodians, placoderms, and sharks. The first two groups were bony fishes, and those in the third group had a skeleton of cartilage. The abundance and diversity of Devonian fishes were so great that the period is called the Age of Fishes. A likely hypothesis is that these jawed fishes evolved from a very early agnathan stock. That, of course, is a common pattern of evolution: major diversifications occurring early in the evolution of the group.

The acanthodians show two important advances over the agnathans: jaws and paired appendages. The agnathans, lacking both, were not evolutionary successes over the long term. Jaws and paired appendages were crucial for the success of the higher vertebrates. The pectoral and pelvic fins are important organs of locomotion in their own right, but their evolutionary significance is far greater—they became the arms and legs of tetrapods.

Although only the general outlines of evolution in the early vertebrates are known, the picture becomes much more complete when we reach the fish–amphibian transition. One of the innumerable groups of fishes that are first recorded from the Devonian is a group of the lobe-finned fishes, the crossopterygians (figure 36). They were long thought to have vanished during the mass extinction at the end of the Creta-

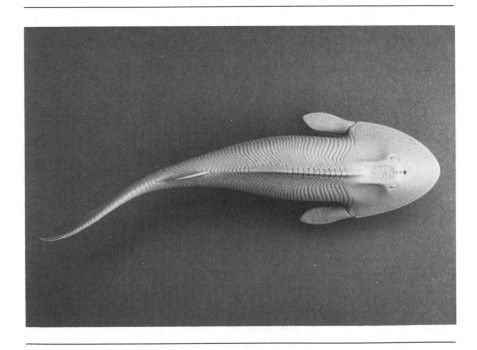

35 The ostracoderm Hemiclapsis, *from the late Silurian. The ostracoderms were the earliest jawless fishes and the most primitive vertebrates.*

ceous, but one was caught off the east coast of Africa in 1939, and since then many more have been obtained. They have been named *Latimeria*.

The crossopterygians are the direct ancestors of the amphibians. The details of skull structure are quite similar in these two major groups, and their nostrils communicate with the mouth, as in all higher vertebrates. The skeletons of the pectoral and pelvic fins are much like those of the arms and legs of the earliest amphibians. The paired fins of most fishes are wide at the base and then taper to a point. Lobe fins differ in being narrow where they attach to the body. That was essential in the development of arms and legs that could be turned, and hence become functional appendages for locomotion on land (figure 37).

By the end of the Devonian the first amphibians were present. One

36 Diplurus, *a crossopterygian fish from the Triassic. The structure of the pectoral and pelvic fins in this group of bony fishes is of a sort that could evolve into the arms and legs of amphibians and other terrestrial vertebrates. On the basis of this and other evidence the crossopterygians are thought to be the ancestors of the amphibians. Most of these fishes are extinct but one,* Latimeria *(a coelacanth), persists.*

of these, *Ichthyostega*, is an almost perfect blend of crossopterygian and amphibian characteristics. It might equally well be called an advanced crossopterygian or a primitive amphibian, but it is usually regarded as one of the first amphibians. It is a link between major taxa that is not missing.

In the early Carboniferous period primitive amphibians known as labyrinthodonts were common. They had little resemblance to the amphibians of today, their heads being covered with an armor of bony plates, which closely resemble the head covering of the crossopterygians. The most striking external features of the labyrinthodonts are the limbs, derived from the lobe fins of the crossopterygians. There is no evidence that the adults had gills but the embryonic stages did.

The replacement of water by air as the ambient environment of organisms resulted in changes in nearly all systems and many behavior patterns. The reasons start with cells themselves and their composition. Cells are composed mainly of water, their membranes are permeable to it, and water must be available to them. The evolution from aquatic to terrestrial life in air required extensive modifications of structure and physiology.

All fully terrestrial species that live in dry environments have outer coverings that are largely impermeable to water. However, no terrestrial organism, plant or animal, can be separated from the environment with a surface completely impermeable to water because oxygen (and carbon dioxide in plants) can enter the body only across moist cell membranes. Thus life in dry air requires most surfaces of the body to be highly impermeable to water but some surfaces to remain moist and serve a respiratory function.

In a sense the arthropods were preadapted for terrestrial life since they had a tough exoskeleton. Insects solved the problem by carrying oxygen directly to cells in tiny air tubes in which the humidity can be kept high. The terrestrial vertebrates solved this problem by having their respiratory surfaces within the body, in lungs, where the relative humidity can be kept at a high level.

Another difference between aquatic and terrestrial life is that a sedentary life is possible for aquatic heterotrophs but is difficult for heterotrophs living on land. Many kinds of invertebrates—hydroids, crinoids, barnacles, mussels, and tunicates, to name a few—rely on a constant flow of ocean water to bring them food and oxygen. A constant flow of air will supply terrestrial heterotrophs with oxygen but not with food. Although the green plant autotrophs can obtain carbon dioxide

37 Diplovertebron, *a primitive amphibian from the Carboniferous similar to, though more advanced than,* Ichthyostega.

from air, their roots must be in a moist environment if water and mineral salts are to be acquired.

No doubt the transition from life in a fully aquatic environment to one on dry land took a long time and first involved adaptations to living in moist terrestrial habitats. Many living species have adopted this strategy. Most amphibians lose so much water by evaporation from their skin that a frequent return to water is necessary. Many other terrestrial species—many species of soil organisms, terrestrial flatworms, earthworms, some crustaceans—can live only in very moist environments.

When organisms moved from aquatic environments, where water supported their bodies, to land, exoskeletons and endoskeletons were required for support. Those creatures that made the transition, mainly arthropods, mollusks, and vertebrates, were preadapted in the sense of already having skeletons.

Very special adjustments are required if the embryos of terrestrial metazoans are to survive in dry situations. This problem has been solved in two main ways. The first solution is to protect the egg and developing embryos with membranes that are almost completely impermeable to water. This solution was perfected in some reptiles during the Paleozoic and it maintains the embryo in an aquatic environment even though the eggs may be laid on dry land. Many living reptiles, all birds, and the monotremes retain this adaptation to dry land—the amniotic egg. The embryos of terrestrial invertebrates that develop in dry environments have evolved similar devices for supplying embryos with everything, except oxygen, required to bring them to a stage where they can live independently.

The second solution is the way taken by viviparous species. The embryos are retained within an aquatic or moist environment of the mother's body. Embryos of placental mammals float in the amniotic fluid and their blood systems exchange materials with the maternal capillaries in the placenta. It is only after birth that they begin to adjust to life on dry land.

During the late Carboniferous and especially during the Permian, which closed the Paleozoic, there was extensive glaciation. It was a stressful time for metazoans both on land and in the sea, and before its end there was massive destruction of life. About half of the families of marine metazoans became extinct. It was nip and tuck for our own lineage, which included a group known as the synapsids, that link the very primitive reptiles (figure 38) with the mammallike reptiles. Only a

38　The primitive reptile Seymouria.

few genera of mammallike reptiles survived the mass extinctions of the Permian, but these became more common in the Mesozoic. They eventually evolved into mammals.

During the Paleozoic there continued to be many changes in the relations of land and sea. At various times landmasses subsided and later were elevated. There were also extensive movements of the earth's crust. It now seems likely that by Carboniferous and Permian times the earth was astonishingly different from today. There was in reality a single supercontinent, Pangaea, which was composed of two main areas: Laurasia to the north, consisting of what was to break apart and become North America, Europe, and Asia (minus India), and Gondwana to the south, which was to become Africa, Antarctica, Australia, India, and South America. There was no Atlantic or Pacific ocean—just one huge unbroken expanse of water.

What was to become the Antarctic continent was situated in tropical and temperate paleolatitudes. (The earth's equator has moved as well.) Thus the puzzle of coal being found there, implying a warm, moist environment, is solved. Antarctica was warm and moist when it was part of Pangaea.

The Age of Dinosaurs

The Mesozoic ushered in the modern world. The Permian biocide cleared away the existing species and made way for a new proliferation of life.

On a global basis Pangaea started to fragment and the continental plates began their stately separation. The supercontinent first split into Laurasia and Gondwana, and then the landmasses began to move slowly toward the positions they hold today. This concept of continental drift (plate tectonics) has only recently been fully established. Obviously the movements and rearrangements of continents have had profound effects on terrestrial organisms and their distributions.

The Mesozoic began with the Triassic about 250 million years ago. Although evolution continued to be enormously innovative in producing minor taxa, only three major types of organisms were yet to appear—mammals, birds, and flowering plants (angiosperms). The Mesozoic witnessed the appearance, flourishing, and extinction of organisms, a phenomenon so typical of the Paleozoic. As was the case in the Paleozoic, the Mesozoic is divided into periods that coincide with radical changes in fauna. All taxa were involved in these changes, except at the level of phyla. With that single exception of the archaeocyathids, known only from the Cambrian, no other phylum has ever become extinct and there is no evidence that any new phylum has originated since the Cambrian, though that possibility remains for some of the phyla with few species and characterized by animals with soft bodies.

Vertebrate evolution was spectacular during the Mesozoic. The labyrinthodont amphibians vanished by the end of the Triassic. Reptiles continued the rapid diversification that began in the late Paleozoic but became much more pronounced in the early Triassic.

The earliest known fossil bird, *Archaeopteryx*, is from the Jurassic. Five avian families are known from the next period, the Cretaceous, but none of these persisted into the Tertiary, although others must have survived. The great radiation of birds was to occur in the Tertiary.

The transition from mammallike reptiles to mammals took place early in the Mesozoic (figure 39). The first known mammals occur near the Triassic-Jurassic boundary. Throughout the Mesozoic the mammals were small and, in view of their relative scarcity as fossils, probably uncommon. It is assumed that the Mesozoic mammals were no match for the reptiles.

The reptiles were the rulers of the Mesozoic world, that is, they evolved species that were successful in occupying all of the major habitats. The most spectacular species were the dinosaurs, which appear in late Triassic strata and peak in the Jurassic and Cretaceous. The first extensive information about them came from the American West, particularly Como Bluff, Wyoming.

The first searches for fossils at Como Bluff were made in 1877 by a local school teacher, Arthur Lakes, who sent several specimens to Professor Othniel Charles Marsh (1831–1899) of Yale and also reported his discoveries to Professor Edward Drinker Cope (1840–1897) of Philadelphia. Over the next decade Marsh employed many local individuals to collect for him. The fossils were cut out of the rock, packaged, and sent to Yale for final preparation and study. Cope obtained some material but not as much as Marsh.

By 1899, when Marsh ended his project, he had obtained 26 new species of dinosaurs and 45 new species of mammals.

The dinosaurs collected at Como Bluff and nearby sites were the first good specimens to be discovered anywhere. Fragments were known from Europe, but the Como Bluff specimens were the first that were complete enough to indicate the dinosaurs' size and general structure.

American paleontology during the late nineteenth century was notable not only for the wealth of important new fossils uncovered and described but for the famous "battle" between Marsh and Cope. Both were strong, proud, possessive, competitive, and even obsessive men whose goal in life was to describe the remarkable fossils from the American West. Since the laurels go to the first person to publish the description of a new organism, Cope and Marsh worked at a frenetic rate, which at times led to serious error. Their collectors in the field tried to be as secretive as possible so as not to reveal promising sites to the "enemy." Professor Cope's first visit to Como Bluff, where Marsh's men were working, caused consternation. There were even charges of specimens being stolen and of addresses being changed so that shipments of the crated fossils intended for one antagonist went to the other.

39 *The mammallike reptile* Lycaenops.

Birds, Mammals, and Flowering Plants

The Mesozoic era closed with the Cretaceous Period, and once again there was a major extinction of metazoans—the most impressive being that of the dinosaurs.

The Cenozoic era is variously divided but, for our purposes it will be considered to have two periods: the Tertiary, beginning about 65 million years ago and divided into five epochs, and the Quaternary, beginning about 1.6 million years ago and divided into the Pleistocene Epoch, which includes the most recent ice ages, and the Recent Epoch, which began with the close of the last ice age, about 10,000 years ago, and continues to this time.

The three main evolutionary events of the Cenozoic—the radiations of angiosperms, birds, and mammals—have much in common. All started slowly in the late Mesozoic and then exploded in the Tertiary.

The herbivorous Mesozoic vertebrates had to subsist on simply deplorable plant food. Epicurization is not a dominant force in evolution, and those herbivores had to eat what was there: primitive plants such as mosses, club mosses, ferns and their allies, horsetails, and gymnosperms—plants that at an earlier age formed coal. Horsetails with their cell walls impregnated with silica were used in Colonial America as scouring rushes, so eating them must have been like eating Brillo.

The angiosperms have played a key role in the evolution of animals, notably that of insects, birds, and mammals. One family of angiosperms, the grasses, has been especially important. Their fossil record begins in the early Tertiary, and they have become the basis of life for many birds and herbivorous mammals. Their seeds and fruits are a richer source of food than leaves. The seeds of three kinds of grasses —wheat, rice, and maize—have been the main sources of human nourishment since the beginnings of agriculture about 10,000 years ago. Other grass seeds important in agriculture are rye, millet, barley, sorghum, and oats. The seeds of other angiosperms, beans and peas, for example, provide protein in the human diet.

Birds are perhaps the most successful group of living vertebrates, and yet their fossil record is by far the poorest. The bones of birds are generally thin, easily broken, and infrequently fossilized. What is known of avian evolution, however, suggests a parallelism with the evolution of angiosperms and mammals: tentative experimentation in the Mesozoic and then rapid divergences in the early Tertiary. Most present-day avian families have no adequate fossil record, and those extant families with a fossil record are, in the main, known only from the Tertiary. A few do go back to the late Mesozoic.

Although the mammals had their beginnings in the Mesozoic, they formed an inconspicuous part of the biota until the Tertiary. There are three main types of the Class Mammalia living today—monotremes, marsupials, and placentals—which are assigned to taxa of different levels depending on the accepted authority. Colbert (1980) recognizes two subclasses, the Eotheria, with the monotremes, and the Theria, with marsupials and placental mammals.

The Eotheria are the very primitive Mesozoic mammals. All are extinct except for the platypus and two genera of echidnas (anteaters), which are restricted to the Australasian region. These relics from the past have a fascinating combination of reptilian and mammalian features. They have hair and suckle their young—both key mammalian characteristics. But those young are hatched from reptilian-type eggs, after which they suckle milk secreted by modified sweat glands on the mother's abdomen. The skeletal features are a mixture of reptilian and mammalian conditions. There is a fossil record of a Mesozoic monotreme that links the more primitive Mesozoic species to the living monotremes.

The marsupial embryos develop for only a short time in the uterus, have at best an incipient placenta, and are born alive—though as minute creatures that usually continue their development in the mother's pouch. The initial radiation of the marsupials occurred in the late Mesozoic and, although only some orders have become extinct, many species have.

Most living mammals are placentals and, as their name implies, one of their diagnostic characteristics is a placenta. This group accounts for about 95 percent of all Cenozoic mammals. Fossils of the most primitive order of placentals, the insectivores (including the living shrews, moles, and hedgehogs), are known from the Cretaceous. It is probable that all other placentals evolved from the insectivores, an event that probably occurred mainly in the early Tertiary.

The fossil record for primate evolution is not as complete as it is for some other placentals—the horses, for example—but the general pattern was the appearance first of lemurs, tarsiers, and the tree shrew (Tupaia) early in the Tertiary, and the splitting off later of the lines to monkeys, apes, and human beings.

Placental mammalian fossils have important features that make their study simpler and more accurate. For one thing their evolution has occurred in geologically recent times, which reduces the chance of their fossils being destroyed, and the strata with their remains tend to be near the surface and, hence, easier to work. But their most important feature is their teeth, which vary greatly, with crowns that have intricate patterns. The study of a single tooth is usually sufficient to determine the family and even the genus of a placental mammal.

There are 29 orders of placental mammals, and essentially all of them evolved very early in the placentals' evolutionary history. These 29 orders can be looked upon as experiments in how-to-be-a-placental-mammal. Thirteen orders were not long-term successes and became extinct. The 16 persisting orders show great adaptive diversification that has resulted in shrews, moles, bats, monkeys, human beings, anteaters, rabbits, whales, lions, aardvarks, horses, cattle, elephants, sea cows, and many other types.

The remarkable radiation of early Cenozoic mammals is correlated with the extinction of the dinosaurs at the end of the Tertiary. It is often thought that there must be a causal connection between these spectacular events but, if that is so, the connection remains unclear.

Still another great wave of extinctions, that of the Pleistocene Period, began about 1.6 million years ago. The Pleistocene was a time during which glaciers covered much of the northern hemisphere, and their traces remain to this day. The most dramatic event occurring among the vertebrates was the massive extinction of very large mammals. This took place on all continents, and, since it is the most recent of the mass extinctions, the details are fairly well known.

The Rancho La Brea Tar Pits

One of the finest sites for Pleistocene fossils is situated near the center of the city of Los Angeles. It consists of a series of solidified tar pits dating from the end of the Pleistocene, which began about 300,000 years ago and ended about 10,000 years ago. These tar pits contain a mixture of sand and asphalt, producing something with the consistency of quicksand, which entrapped a great variety of organisms and preserved them splendidly. They provide a record of local life for the interval of about 40,000 to 8,000 years ago.

Life in North America was very different at the end of the Pleistocene from what it is today. The most impressive difference is the number of huge mammals that simply vanished from the scene during and shortly after the Wisconsin (the last) glaciation. Among the dominant species that became extinct were many species of edentates such as giant sloths, glyptodonts, and armadillos. Some of these were huge. One glyptodont, for example, was about 3 meters in length and 1.5 meters in height. This group evolved in South America during most of the Tertiary when North America and South America were separated from one another. Late in the Tertiary, when the two continents were connected at the Isthmus of Panama, many species of edentates migrated north from South America. All were extinct by the end of the Pleistocene. (The armadillo now found in the southern part of the United States spread there in very recent times.)

There was a rich variety of carnivores with many species of dogs, wolves, hyenas, and many cats. The most common fossil at La Brea is the dire wolf, equal in size to a very large wolf living today but with massive teeth. The next most common fossils are those of the saber tooth cat ("tiger") with its huge upper canines. There was also a very large lion.

The New World was where the horse first appeared and underwent its major evolution. In the late Pliocene and Pleistocene there were about 40 species, including true horses, zebras, gazelle horses, asses, and onagers. All had died out in North America by about 8,000 years ago. Today's wild horses of the American West are descendants of horses that were brought from Europe and escaped—proving that today's environment is quite acceptable for wild horses. This disappearance of the horse tribe is absolutely astonishing, and no satisfactory explanation is available.

The New World was also the locality for most camel evolution. In fact, camels did not reach Asia, Europe, and Africa until late in the Tertiary. North America had six genera in the Pleistocene but all had vanished by its close. There was also a huge Pleistocene bison; it too, vanished, possibly due to competition with the smaller bison still present today.

The grandest animals of all, however, were the elephantlike mammoths and mastodons, which were widespread in North America, but they too vanished by about 10,000 years ago.

Although few small mammals became extinct at the end of the Pleistocene, many had done so in the late Pliocene and early Pleistocene. The overall effect of the Pleistocene, then, was to alter radically the mammalian fauna of North America. The mammalian fauna of the North American Pliocene resembled that of Africa today more than it resembled that of present-day North America. The massive extinction of large mammals in the Pleistocene was not restricted to North America. All continents experienced the same phenomenon.

In seeking an explanation, some have suggested extermination by human hunters, and there is evidence that human beings did hunt large mammals, including the mammoths. The extraordinary abundance of large mammals is offered as a counterargument; human beings could not have accomplished such a massive slaughter.

Human Evolution

Darwin avoided the question of questions in the *Origin* and said only that "light will be thrown on the origin of man and his history." Such light was most threatening for many Victorians, who fully accepted the Judeo-Christian dogma that saw human beings as the climax of special creation. But Darwin was suggesting that not only might human beings have evolved but that they might share a common ancestry with other mammals.

Which other mammals? Naturalists were fully aware of the close anatomical resemblances of human beings and the great apes, especially the gorilla and chimpanzee. Thus, if one accepted the Darwinian hypothesis, this close resemblance must be a consequence of sharing a common ancestor. Therefore two sorts of evidence were required to test this hypothesis: a sequence of human remains from increasingly remote times in the past and the discovery of a putative common ancestor of human beings and the great apes. Both types of evidence were slow in coming. In contrast with some mammals—horses, for example—fossil remains of human beings and great apes are exceedingly rare. Nevertheless, the great interest in these questions has led to very intensive searches for fossil evidence. New discoveries are being made constantly and the story is being revised, but the broad outlines are becoming apparent.

Between 2 and 4 million years ago in Africa there were at least several species, assigned to the genus *Australopithecus*, that can be accepted as the first human beings. They had only small brains, about 400 cm^3, many apelike features, and probably did not make fire or tools. Fossil footprints from Tanzania indicate that they were bipedal. Somewhat more than 2 million years ago another human type appeared in Africa —*Homo habilis* (figure 40a). In structure they had many apelike features, being about halfway between the australopithecines and modern human beings, and they used crude tools. By about 1.6 million years ago a third major human type appeared—*Homo erectus* (figure 40*b*). The brain size had increased to about 900 cm^3, and the skeleton was much like the modern *Homo sapiens*. One hypothesis holds that many different forms of *Homo sapiens* evolved from *Homo erectus* independently of one another. One form was the Neandertal man of Europe, which may have appeared about 500,000 years ago (figure 40*c*); another, which replaced the Neandertals, evolved about 200,000 years ago in Africa. About 34,000 years ago the Neandertals in Europe were rapidly replaced by modern *Homo sapiens*.

The evolutionary relationships among the various species of *Australopithecus* and *Homo* are far from settled, but the data show that, beginning with apelike forms, a sequence of human types beginning about 4 million years ago slowly approach modern human beings.

The details of evolutionary events during the evermore remote times before *Australopithecus* are poorly known. But again a general picture emerges. Primitive apes are known from the Oligocene of about 20 million years ago, and during the Miocene there were many kinds. *Proconsul* from the Early Miocene and *Ramapithecus* and *Sivapithecus* from the Middle Miocene are generalized enough to be ancestors of both human beings and the great apes. More adequate information must await the discovery of additional fossils covering the period between 12 and 4 million years ago.

The Role of Extinction in Evolution

The extinction of taxa has occurred throughout geological time, and it has given rise to one of the most striking features of the paleontological record, that is, the presence of different sorts of organisms in strata laid down at different times. Although extinction of one taxon or another must have been occurring constantly there are well-documented instances of mass extinctions during which a large portion of the total biota disappeared. Such a depopulation would have left many habitats empty, or only partially filled, and after every mass extinction the paleontological record shows a rapid radiation of that part of the biota that escaped destruction to fill those empty habitats. There never seems to be a return to the *status quo ante*; new species evolve for the new times.

The result of this phenomenon of origin and extinction is that each

40a   Drawings of early humans, based on fossils. Homo habilis.

geological period has it characteristic biota. In fact, the geological periods were first characterized by the unique species of animals and plants that were present for much of the period and that either became extinct or greatly reduced in numbers at the close. Some species are highly diagnostic and abundant at specific times and are used as "trace fossils" to identify specific strata. Although the geological periods may seem to end abruptly, "abrupt" is in terms of geological time, when many thousands of years might be involved.

Major extinctions occurred at the end of the Cambrian, Ordovician, Devonian, Permian, Triassic, Cretaceous, and as recently as 10,000 years ago at the end of the Wisconsin Ice Age. The last part of the Permian was especially difficult—about half of all the families of animals disappeared. The percentage was higher for certain groups. For example, 75 percent of amphibian families and 80 percent of reptilian families disappeared. Raup (1979) estimated the percentage of extinctions in the late Permian as 14 percent for Classes, 17 for Orders, 52 for Families, and possibly as many as 88 to 96 percent for species of marine organisms. Newell (1963) estimated that it took as much as 15 to 20 million years for the variety of animals to equal that before the massive Permian extinction.

The causes of extinction, whether at the small, constant, background level or as massive episodes, are slowly beginning to be understood. Extinction at a background rate occurs constantly. Presumably it is the consequence of competition among organisms, and, when taxa with a more successful way of life evolve, they replace the less well adapted taxa. Modest environmental changes could also favor one taxon over another.

It is assumed that drastic environmental changes must have been involved in the infrequent major episodes of extinction. It is probable that low temperatures were important. It is known that extensive glaciation occurred in the Permian at the close of the Paleozoic. In addition, the landmasses were moving and colliding. There was extensive volcanic activity and mountain building. The ambient world, home, was changing for all species. Of considerable importance were the periodic changes in sea level, which are correlated with the abundance of marine organisms. When the continents sink and the sea invades the low-lying areas, more space for marine life becomes available. When the land rises and the sea retreats, those areas are lost to marine life, and again become available for land life.

There was a massive extinction, with a subsequent changeover of

40b Homo erectus.

life, at the end of the Mesozoic—at the Cretaceous/Tertiary (K/T; the "K" is for Kreide, the German name for the Cretaceous) boundary. That event was especially noteworthy because it saw the end of the reptilian domination of the earth and the beginning of the dramatic evolutionary diversification of placental mammals, birds, and flowering plants. Equally dramatic changes occurred in the marine plankton.

There has been renewed interest in this K/T massive extinction event following the suggestion of Alvarez et al. (1980) that its cause was a huge meteor, about 10 km in diameter, striking the earth. Some have suggested that a meteor of 100 to 200 km in diameter would be required to produce the observed result. Such a missile would have raised a cloud of dust that might have darkened the sun for a long period of time and hence suspended that basic process necessary for nearly all life—photosynthesis. The reason such a hypothesis was suggested was the discovery of a thin layer of iridium at K/T boundary, varying from 30–160 times the amount normally found in the earth's crust. Iridium is very rare in the earth's crust but more abundant in meteorites, so it was suggested that the iridium at the K/T boundary was produced by the disintegration of a huge meteor that collided with the earth. The possibility of such a visitor from outer space has been hotly debated and remains a hypothesis.

Wolbach et al. (1985) noted a worldwide layer of graphitic carbon near the K/T boundary and suggested that a global wildfire was the origin of the carbon layer. Such a wildfire would have produced vast quantities of smoke that would darken the sky and interfere with photosynthesis and alter temperatures. A wildfire could have been started by a meteor, of course. Bourgeois et al. (1988) report evidence of a tsunami, with a wall of water possibly 50 to 100 meters in height that struck the Texas coast and washed far inland. The geological evidence for this great wave of water is at the iridium layer. They suggest that it was probably caused by a meteor.

There are good reasons for being cautious. Many species were not exterminated—the plants, for example, show no evidence of any drastic change at the K/T boundary but there were major changes a quarter of a million years earlier (Kerr, 1988). It must be borne in mind also that dramatic changes can result from what appear to be minor environmental changes. A few years ago there was an El Niño episode off the west coast of North America. The surface temperatures of the Pacific changed only a few degrees, but that was enough to cause drastic

40c *Neandertal man.*

changes—temporary "extinctions"—of some marine organisms that sea birds fed upon, resulting in breeding failures of the birds. A huge meteor does not literally have to blast organisms from the face of the earth. Its impact can result in relatively "minor" perturbations of the environment that can start a chain of events with major outcomes.

part three

CLASSICAL GENETICS

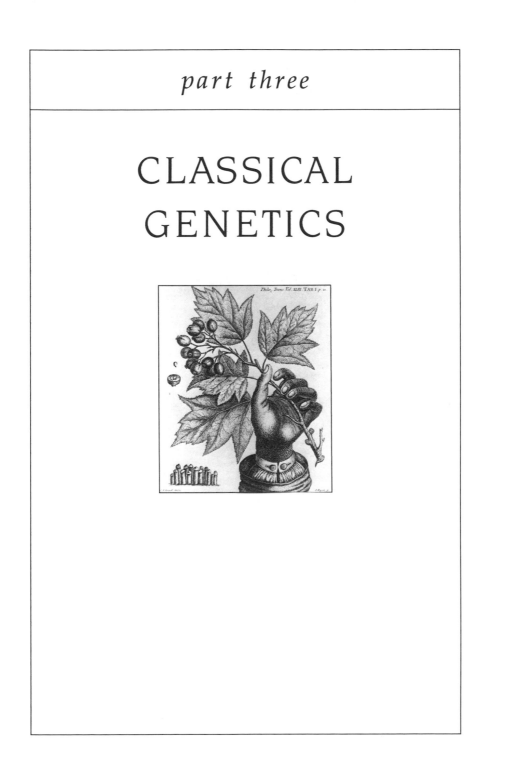

chapter 11

Pangenesis

The fundamental characteristic of life is its ability to produce more life almost exactly like itself by transforming the materials and energy of the nonliving world. *Genetics* is the field of investigation that seeks to understand this phenomenon of replication and hence must be considered basic to all biology. Replication and all other aspects of life are reflections of the structure and functioning of the genetic materials—the nucleic acids. *Morphology* and *physiology* are the structural and functional consequences of the activities of the genetic materials at all levels, from cell to the entire organism. *Developmental biology* deals with the growth and differentiation of the replicated individual. *Ecology* deals with the interactions of the environment with the genetically programmed individual and groups of individuals. *Evolutionary biology* is the field that investigates the long-term aspects of replication. *Systematic biology* (that is, classification or taxonomy) studies the diversity of life that is a consequence of replication being modulated by the environment over time.

Thus, there can be no more fundamental process in biology than genetics, since there is no more fundamental phenomenon of life than genetic replication. Genetics first and foremost, including its long-term manifestation—evolutionary biology—is the integrator of all biological concepts and data. True enough, but how does one initiate the inquiry?

What Is the Question?

Surprising as it may seem, one of the greatest obstacles to understanding the natural world is not knowing what questions to ask. This point can be brought home by looking at nature, a mountain for example. A person not familiar with geology would be hard pressed to suggest some of the scientific questions one might ask about that mountain. A professional geologist would be able to supply much information about the age, composition, and method of formation, but very few answers could be obtained by observations of that mountain alone. Instead, understanding would have to come from the synthesis of many observations and experiments in the fields of sedimentation, radioactive decay, erosion, vulcanism, chemistry, mineralogy, and plate tectonics.

This point can also be illustrated beautifully by a review of the difficulties scientists had in understanding what is involved in inheritance. Genetics, now the most rigorous and conceptually complete field of biology, has reached this stage only in our lifetime. For millennia human beings had no useful answers about inheritance because they were unable to formulate useful questions. In science, useful questions are those that are amenable to observation and experimentation and, hence, susceptible of being answered.

Thus, for most of human history, inheritance was no more than a vague principle having neither precise rules nor predictive value. Consider, for example, the sorts of data most readily available during the middle of the nineteenth century. The children of a human couple would, routinely, differ from one another in many ways. Some would be female, others male—a truly profound difference. What could possibly be the underlying reason? Furthermore, unless the children were identical twins, the siblings might differ strikingly in appearance and personality. One child might have little resemblance to the parents, yet another might have a strong familial resemblance. How could the same cause—reproduction by the same parents—lead to such diverse results?

Yet some regularities were recognized. The children of native Americans, African Americans, Asians, and Caucasians were observed to have the general characteristics of their race. But no precise rules were discovered that related the characteristics of offspring to those of the parents. Those vague answers were all that could be expected from the vague question, "What is the nature of inheritance?" Until our century there was no acceptable way to account for the observation that inheritance seems to consist of the transmission of similarities, differences, and even novelty.

Since the value of science lies not only in the information it provides but also in the manner of obtaining that information, there is value in exploring past attempts to understand the nature of inheritance. As with so many topics in biology, it is convenient to begin with the Greek philosophers. Not infrequently they defined the problem and suggested the main hypotheses that lasted into modern times. We will consider only two people: Hippocrates and Aristotle.

Hippocrates and Aristotle

Hippocrates, whose views on medicine were noted earlier, had some interesting things to say about inheritance. All who speculated on possible mechanisms for inheritance had this basic problem to solve: "Inheritance" means the transmission of "something" from parent(s) to offspring; what could this "something" be?

Writing about 410 BC, Hippocrates proposed *pangenesis* as an explanatory hypothesis for inheritance. Pangenesis assumes that inheritance is based on the production of specific particles ("seeds") by all parts of the body and on the transmission of these particles to the offspring at the time of conception. Darwin was to adopt a similar explanatory hypothesis long afterward, and pangenesis was to remain the only general theory of inheritance until the end of the nineteenth century.

One of the observations that led Hippocrates to this belief concerned a somewhat mythical race of people, the Macrocephali, who were characterized by very long heads. Long heads were thought to indicate nobility, so Macrocephali parents attempted to mold the soft skulls of their newborn to the desired shape.

> The characteristic was thus acquired at first by artificial means, but, as time passed, it became an inherited characteristic and the practice was no longer necessary. The seed comes from all parts of the body, healthy from the healthy parts and sickly from the sickly. If therefore bald parents usually have bald children, grey-eyed parents grey-eyed children, if squinting parents have squinting children, why should not long-headed parents have long-headed children. (Hippocrates 1950, p. 103)

Hippocrates was also proposing the concept of the inheritance of acquired characteristics—a point of view that was to be adopted by Jean Baptiste Lamarck in the eighteenth century as the mechanism of evolutionary change—a theory accepted by many well into the twentieth century.

Hippocrates' hypothesis for inheritance may not appear to be a monumental beginning—but it was. He identified a scientific problem (possibly the most difficult step of any), proposed an explanatory hypothesis, and wrote in a manner that we can understand. For such a scientific analysis to have occurred 2.5 millennia ago is quite exceptional. The roots of the way we think about scientific phenomena stem from the Greeks, even as much of our nonscientific mode of thought can be traced back to the ancient Hebrews (via the Hebrew and Christian Bibles).

Aristotle (384–322 BC) was active a century after Hippocrates. His *De Generatione Animalium* deals with problems of both genetics and development. This linking of two such seemingly disparate fields has a distinctly modern ring.

Aristotle assumed that there must be a physical basis of inheritance in the "semen." (He used this term for reproductive elements of both sexes; today we would say "gametes," referring to ova and sperm.) This point, so obvious to us today, was basic to all future work. Inheritance need no longer be thought of as caused by some vague spirit or emotion but by a *substance* transmitted by the parents.

The problem, then, becomes understanding the nature of semen. Aristotle discusses the prevailing hypothesis of pangenesis and lists four major observations and arguments in its support. First, noting that mating (in humans) gives pleasure to the whole body, one could argue that the whole body must contribute to the semen. Second, there were observations suggesting that mutilations may be inherited. One such case came from Chalcedon (on the Bosporus in present-day Turkey), where a man was branded on the arm. His child, born subsequently, had a defect on the arm. Third, it was commonly observed that the offspring resemble parents not only in general but often in strikingly specific ways. Hence the specific characteristics could be assumed to produce specific substances that become part of the semen. And fourth, if semen for the whole can be produced, there is no reason why the specific parts of the body could not contribute to the semen as well.

Yet Aristotle rejected the hypothesis of pangenesis for what to him were more compelling observations. Noting that children resemble parents not only in structure but also in such features as voice and gait, he questioned how nonstructural features could produce material for the semen. Then, too, babies of fathers with beards and gray hair are not similarly hirsute at birth. It had also been observed that children seem to inherit characteristics of more remote ancestors, who could hardly have contributed to the semen of their parents. Thus a woman of Elis (in the northwest part of the Greek Peloponnese) had intercourse with a black-amoor (a term applied then to any very dark-skinned individual). Her daughter was white, but her grandson was black.

Other evidence against the hypothesis of pangenesis was that parts could easily be removed from plants, yet these mutilated plants could produce offspring that were perfect and entire. And then there was the awesome argument that if, as in humans, two parents produce the semen with the gemmules for all parts of the body, wouldn't we expect offspring with two heads, four arms, and so on.

These and many other arguments and observations led Aristotle to reject pangenesis and to ask, "Why not admit straight away that the semen . . . is such that out of it blood and flesh can be formed, instead of maintaining that semen is itself both blood and flesh?" (Aristotle 1943, p. 65). An important distinction is being made: the *feature* itself is not inherited; instead, something is transmitted that will lead to the *development* of the feature in the offspring. As we say today, genetic information is transmitted. This tentative hypothesis was to be the conceptual limit for the next 2,000 years.

Interest in scientific questions almost ceased in the Western world throughout those long centuries when the Church held hegemony over the mind of man. It was not until well after the Renaissance that observation and experiment were applied in a systematic manner in an attempt to gain understanding of inheritance. Even then progress was exceedingly slow, again because it was impossible to find a productive question, this is, one that could be answered with the prevailing information and methods.

In the eighteenth and nineteenth centuries the standard way to seek information about inheritance was by crossbreeding. Individuals of the same species that differed from one another were mated and the offspring studied. The results were that the offspring were usually more or less intermediate or occasionally more like one parent than the other. But there seemed no clear way to use these centuries-old observations to obtain a deeper understanding of inheritance. In fact, so little progress was made before the closing decades of the nineteenth century that we may conclude that little of theoretical importance occurred between Aristotle and Darwin.

The Darwinian Answer

Darwin is an especially instructive example of a pre-Mendelian scientist attempting to explain inheritance, and in his lack of success we can see some of the reasons why so little progress was being made. Darwin is recognized, after all, as a person of tremendous ability, not only for his *Origin of Species* but for fundamental investigations on a broad range of biological subjects as varied as coral reefs, habits of earthworms, taxonomy of barnacles both living and fossil, animal behavior, the biology of plants, and the fertilization of orchids. His attempts to understand inheritance are described in a two-volume work. *The Variation of Animals and Plants under Domestication* (1868).

Darwin's problem was the same as for all who sought a rent in the veil of ignorance that enmeshed the subject of inheritance: how could one properly initiate the inquiry? It was exceedingly important that he do so. His momentous hypothesis for the origin of species by means of natural selection depended totally on a constant supply of new variants that persisted generation after generation upon which selection could act. He wrote, "It is obvious that a variation which is not inherited throws no light on the derivation of species, nor is of any service to man" (vol. 2, p. 1). Thus in the absence of inherited variations the population would remain the same—there could be no evolution.

Not everyone in the mid-nineteenth century believed that the inheritance of minute differences was either of much importance or subject to rigorous rules. There was support for the point of view that the main reason individuals differed from one another was the environment, not inheritance. Crops grown on poor soil differed from those on good soil; well-fed animals differed from those on short rations. The characteristics of parents were not transferred with complete fidelity to the offspring, and the offspring might have novel features unknown in their ancestors. Darwin reflected this point of view: "When a new character arises, whatever its nature may be, it generally tends to be inherited, at least in a temporary and sometimes in a most persistent manner" (vol. 2, p. 2).

Darwin sought data on inheritance from the experimental crossing of individuals of different kinds. He studied inheritance in the many varieties of pigeons, and through correspondence and reading he had an extensive knowledge of the results of other investigators. He was convinced that inheritance must be a phenomenon that is widespread, somewhat precise, and important.

His fine eye for the critical observation or test led him to lay great emphasis on some remarkable examples of inheritance that were so unusual that neither chance nor environmental influences seemed adequate explanations. One of the more spectacular examples was the "Porcupine Man." In 1733 Machin reported to the Royal Society on a puzzling condition of the skin of Edward Lambert, then in his teens. Edward was the son of a laborer who lived in Suffolk.

His skin (if it might be so called) seemed rather like a dusky coloured thick case, exactly fitting every part of his body, made of a rugged bark, or hide, with bristles in some places, which case covered the whole excepting the face, the palms of the hands, and the soles of the feet, caused an appearance as if those parts were naked, and the rest clothed. It did not bleed when cut or scarified, being callous and insensible. It is said he shed it once every year, about autumn, at which time it usually grows to a thickness of three quarters of an inch, and then is thrust off by the new skin which is coming up underneath.

Young Edward seemed entirely healthy and normal in all other respects. His father reported that Edward had normal skin at birth, but at the age of about two months it began to change. The baby had not been sick and there was no obvious cause. The mother had received no fright while with child. None of the many siblings exhibited the condition.

In 1756 Baker provided additional information. By then Edward Lambert was a married man. He had one surviving son with the defect. Five other sons had shown the defect but died. Baker reported that when a hand was drawn across the victim's skin it made a rustling noise. The surviving son married and subsequently two of his children showed the same defect. According to Darwin, a total of four generations were observed with the defect—always restricted to males. The modern medical term for this condition is ichthyosis hystrix gravior.

How is one to account for these exceedingly rare events, so atypical of what is usually observed? Was it merely a matter of chance (whatever that might mean), or was it the result of some undetected environmental influence? In *Variation* Darwin ruled out the probability, not possibility, of environmental influences being responsible and regarded this and similar instances as evidence that "something" is transmitted from parent to offspring:

> When we reflect that certain extraordinary peculiarities have thus appeared in a single individual out of many millions, all exposed in the same country to the same general conditions of life, and, again, that the same extraordinary peculiarity has sometimes appeared in individuals living under widely different conditions of life, we are driven to conclude that such peculiarities are not directly due to the action of surrounding conditions, but to unknown laws acting on the organisation or constitution of the individual;—that their production stands in hardly closer relation to the conditions than does life itself. If this be so, and the occurrence of the same unusual character in the child and parent cannot be attributed to

both having been exposed to the same unusual conditions, then the following problem is worth consideration, as showing that the result cannot be due, as some authors have supposed, to mere coincidence, but must be consequent on the members of the same family inheriting something in common in their constitution. Let it be assumed that, in a large population, a particular affection occurs on an average in one out of a million, so that the *a priori* chance that an individual taken at random will be so affected is only one in a million. Let the population consist of sixty million, composed, we will assume, of ten million families, each containing six members. On these data, Professor Stokes has calculated for me that the odds will be no less than 8333 millions to 1 that in the ten million families there will not be even a single family in which one parent and two children will be affected by the peculiarity in question. But numerous cases could be given, in which several children have been affected by the same rare peculiarity with one of their parents; and in this case, more especially if the grand-children be included in the calculation, the odds against mere coincidence become something prodigious, almost beyond enumeration. (vol. 2, pp. 4–5)

Even today it would be hard to supply better arguments that "something" had been transmitted from Edward Lambert to his sons and to their sons. It was most unlikely that the skin condition was a consequence of an environmental stimulus. If there was a physical basis for the inheritance of the porcupine-skin condition and similar variations, it should be possible to discover laws governing their transmission.

Assembling the Data

Darwin set about to discover these laws with the accepted procedures of his time, but, as we shall see, he was unsuccessful. Subsequently others, using entirely different approaches, were finally able to illuminate that black box of inheritance. He tells us in his autobiography how he began his great study:

After my return to England [in 1836 at the end of the voyage of the *Beagle*] it appeared to me that by following the example of Lyell in Geology, and by collecting all facts which bore in any way on the variation of animals and plants under domestication and nature, some light might perhaps be thrown on the whole subject. My first note-book was opened in July 1837. I worked on true Baconian principles, and without any theory collected facts on a wholesale scale, more especially with respect to domesticated productions, by printed enquiries, by conversations with skilful breeders and gardeners, and by extensive reading. (Barlow 1958, p. 119)

And he did record a prodigious amount of information related to "domesticated productions." Roughly half of *Variation* provides information on the presumed origin of domesticated plants and animals from wild ancestors. It was assumed that this had involved the selection by human beings of the hereditary variations that were thought desirable. Starting with domesticated dogs and cats, he went on to assemble the available data for horses, asses, pigs, cattle, sheep, goats, rabbits, pigeons, chickens, ducks, geese, peacocks, turkeys, canaries, goldfish, honeybees, silk moths, the common cereals, garden vegetables, and fruits. All made sense with the assumption that rapidly acting artificial selection by human beings was a counterpart of the excruciatingly slow natural selection that accounted for the origin of species. There were inherited variations, and careful breeding could perfect varieties of plants and animals especially desirable for human use.

Darwin assembled other sorts of observations that he assumed bore on the question of inheritance.

> Brothers and sisters of the same family are frequently affected, often at about the same age, by the same peculiar disease, not known to have previously occurred in the family. (vol. 2, p. 17)

> A rabbit produced in a litter a young animal having only one ear; and from this animal a breed was formed which steadily produced one-eared rabbits. (vol. 2, p. 12)

> I have been assured by breeders of the canary-bird that to get a good jonquil-coloured bird it does not answer to pair two jonquils, as the colour then comes out too strong, or is even brown. (vol. 2, p. 21–22)

> In one lot of eleven mixed eggs from the white Game and white Cochin by the black Spanish cock, seven of the chickens were white, and only four black: I mention this fact to show that whiteness of plumage is strongly inherited. (vol. 1, p. 240)

> I have been assured by three medical men of the Jewish faith that circumcision, which has been practiced for so many ages, has produced no inherited effect. (vol. 2, p. 23)

But then Darwin goes on to quote an authority who suggests that injuries can be inherited. "Nevertheless, Dr. Prosper Lucas has given, on good authorities, such a long list of inherited injuries, that it is difficult not to believe in them" (vol. 2, p. 23). Darwin recorded numerous examples of observations that, in the years after 1900, were to be critical to the advancement of our understanding of inheritance. The next two quotations give accurate descriptions of what came to be known as sex-linked inheritance.

> Colour-blindness, from some unknown cause, shows itself much oftener in males than in females; . . . but it is eminently liable to be transmitted through women. (vol. 2, p. 72)

> Generally with the haemorrhagic diathesis [hemophilia], and often with colour-blindness, and in some other cases, the sons never inherit the peculiarity directly from their fathers, but the daughters, and the daughters alone, transmit the latent tendency, so that the sons of the daughters alone exhibit it. Thus, the father, grandson, and the great-great-grandson will exhibit the peculiarity,—the grandmother, daughter, and great-granddaughter having transmitted it in a latent state. (vol. 2, p. 73)

Observations on the formation of hybrids, and their offspring if any, were to provide key data for the rules of inheritance. In Darwin's time the data were vague at best, but the following quotation is of extraordinary interest in describing phenomena that were to become the core of Mendelian inheritance.

> As a general rule, crossed offspring in the first generation are nearly intermediate between their parents, but the grandchildren and succeeding generations continually revert, in a greater or lesser degree, to one or both of their progenitors. Several authors have maintained that hybrids and mongrels included all the characters of both parents, not fused together, but merely mingled in different proportions in different parts of the body; or, as Naudin has expressed it, a hybrid is a living mosaic-work, in which the eye cannot distinguish the discordant elements, so completely are they intermingled. We can hardly doubt that, in a certain sense, this is true, as when we behold in a hybrid the elements of both species segregating themselves . . . Naudin further believes that the segregation of the two specific elements or essences is eminently liable to occur in the male and female reproductive matter; and he thus explains the almost universal tendency to reversion in successive hybrid generations. (vol. 2, pp. 48–49)

After this systematic and extensive survey of the data on inheritance, Darwin formulated an explanatory hypothesis. We can group the data in ten main classes of phenomena that he felt must be explained by a comprehensive hypothesis of inheritance.

(1) Some characteristics that are inherited. Most of these examples involved structures such as body size, color patterns, and an endless list of minor variations. Physiological characteristics were also inherited— such as color blindness and hemophilia. The inherited characteristic might be large or small, important or unimportant, and sometimes inherited and sometimes not. Any useful hypothesis would have to explain why features are inherited sometimes but not always. That was an awesome difficulty.

(2) The inheritance, or not, of mutilations. Some human societies habitually knock out teeth, perforate ears or nostrils, circumcise male babies, cut off a finger or two, yet their children do not show corresponding defects. There were other cases where it appeared that mutilations were inherited, and they were given on such good authority that Darwin found "it difficult not to believe them." Several times Darwin referred to the case of "a cow that had lost a horn from an accident with consequent suppuration, produced three calves which were hornless on the same side of the head" (vol. 2 p. 23). He concluded, "with respect to the inheritance of structures mutilated by injuries or altered by disease it is difficult to come to any definite conclusion" (vol. 2, pp. 22–23).

(3) Atavism. This is the occurrence in an individual of some characteristic not expressed in the immediate forebears but believed to have been present in remote ancestors. For example, it was believed that the wild ancestors of the domesticated sheep had been black. Thus, when a black lamb appeared in a flock of carefully bred white sheep, it was explained as the persistence of some long-dormant hereditary influence.

(4) Sex-linked inheritance. So far as the data went, it appeared that in most cases characters appeared to be inherited with equal facility from either parent. Nevertheless, Darwin knew of some cases, such as color-blindness and hemophilia where this was not the case. Darwin drew an interesting conclusion: "We thus learn, and the fact is an important one, that transmission and development are distinct powers" (vol. 2, p. 84).

(5) Inbreeding. If two organisms are crossed and their offspring bred with one another generation after generation, we speak of this as in-

breeding. The usual result is the production of a relatively homogeneous population:

> When two breeds are crossed their characteristics usually become intimately fused together; but some characters refuse to blend, and are transmitted in an unmodified state either from both parents or from one. When grey mice are paired, the young are not piebald nor of an intermediate tint, but are pure white or of the ordinary grey colour . . . In breeding Game fowls, a great authority, Mr. J. Douglas, remarks, "I may here state a strange fact: if you cross a black with a white game, you get both breeds of the clearest colours." Sir R. Heron crossed during many years white, black, brown, and fawn-coloured Angora rabbits, and never once got these colours mingled in the same animal, but often all four colours in the same litter. (vol. 2, p. 92)

Once again, the data of inheritance seemed to conform to no strict rules or regularities. Any comprehensive hypothesis to explain the data would have to be adjusted to this difficult fact.

(6) Artificial selection. Selection, either deliberate or unintentional, is a method that produces varieties of plants and animals of greater usefulness to human beings. It has been employed since the earliest days of agriculture. If a farmer desires to increase the size of chickens, and hence the quantity of food, he selects the largest hens and cocks to be the parents. In each generation he continues the same selection. With this procedure, within limits, it is usually possible to develop animals or plants with the desired characteristics in a few generations. One of the most puzzling aspects of selection was the ability to produce individuals with characteristics not present in the ancestral population. For example, Darwin's favorite material, pigeons, had been selected to produce the most unusual breeds—entirely different from the ancestral wild rock dove of Europe. Selection could create something new. It was clear that artificial selection could, in a few generations, produce varieties that differed as much from one another in structural details as did the various wild species of the same genus or even species of different genera.

(7) The causes of variability. "The subject is an obscure one; but it may be useful to probe our ignorance. Some authors . . . look at variability as a necessary contingent on reproduction, as much an aboriginal law, as growth or inheritance" (vol. 2, p. 250). Darwin believed that all domestic and wild species are variable. The differences are especially obvious in the domestic species, where many unique varieties have been selected. (Darwin found a report in the literature of a Dutch florist

who kept 1,200 varieties of hyacinth.) "Changes of any kind in the conditions of life, even extremely slight changes, often suffice to cause variability. Excess of nutriment is perhaps the most efficient single exciting cause" (vol. 2, p. 270). The kind of variation depends "in a far higher degree on the nature or constitution of the being, than on the nature of the changed conditions" (vol. 2, p. 250). (This last quote is one of numerous examples of the uncanny ability of Darwin to rise above the confusion of his time and see clearly what future research would establish.)

(8) Regeneration. When the tail or legs of salamanders are cut off, the lost structures are replaced. The ability to regenerate lost parts is common in many animals and plants. Darwin realized that both the original formation of structures in development and the replacement of lost parts must have a hereditary basis, since in both phenomena the final structures were characteristic of the species.

(9) Mode of reproduction. Some organisms, such as the invertebrate hydra, reproduce both by asexual and sexual means. The hydra that develops from a fertilized ovum is identical with that originating from an asexual bud. Thus, what is passed from parents to the offspring could not be restricted to eggs and sperm. The cells of the hydra's body wall that bud out to form the new individual must also transmit the hereditary information. It is noteworthy that Darwin realized that any comprehensive theory of inheritance would have to account both for regeneration and for development.

(10) Delayed-action inheritance. In the early stages of any scientific analysis, it may be difficult to distinguish between reliable and unreliable data. Lord Morton's Arabian chestnut mare is an example. This mare was crossed to a species of zebra, the now extinct quagga. The foal of this union was intermediate in form and color, a result that was not surprising. The mare was then sent to another farm where she was bred with a black Arabian stallion. There were two offspring.

> These colts were partially dun-coloured, and were striped on the legs more plainly than the real hybrid, or even than the quagga. One of the two colts had its neck and some other parts of its body plainly marked with stripes. Stripes on the body, not to mention those on the legs, and dun-colour, are extremely rare . . . But what makes the case still more striking is that the hair of the mane in these colts resembled that of the quagga, being short, stiff, and upright. Hence there could be no doubt that the quagga affected the character of the offspring subsequently begot by the black Arabian horse. (vol 1, p. 404).

That surely is hard to explain. Nevertheless many horse breeders believed that if a purebred mare was crossed to a less noble stallion she would be forever tainted and offspring of subsequent crosses could not be considered pure. Darwin regarded these examples of delayed-action inheritance as "of the highest theoretical importance," but they, as much as anything else, were to be the cause of his flawed hypothesis for inheritance.

Formulating the Hypothesis by Induction

Any useful hypothesis to explain inheritance would have to account for the ten classes of data just enumerated. No hypothesis comes immediately to mind to account for these very different sorts of data, all thought to have a bearing on inheritance. Therefore one must engage in an exercise not unlike Bacon's arranging his data relating to the nature of heat in tables, eliminating some of the data, and formulating a hypothesis from what remained. Here are examples of how one might reason.

(A) Since there are so many observations showing that offspring may resemble parents not only in general features but also in very specific characteristics, one must conclude that there is some physical basis for inheritance. This is suggested by class 1 above and is not negated by any of the other nine classes of data.

(B) Since the eggs and sperm are the only physical link between generations in those organisms that liberate eggs and sperm to unite outside of the body, all of the hereditary factors must be contained in them.

(C) But eggs and sperm cannot be the sole possessors of the hereditary factors, since in some organisms apparently identical offspring can be produced by sexual as well as asexual means (class 9).

(D) The observations on the regeneration of lost parts (class 8), together with point C, above, suggest that many (most? all?) cells of the body contain all the hereditary factors.

(E) The hereditary factors may be present but not expressed either on a short-term basis (parents not exhibiting the features while grandparents and grandchildren do) or on a long-term basis (atavism, class 3). This strongly suggests that the hereditary factors are relatively permanent and stable even when they are latent.

(F) The hereditary factors may change or entirely new ones may be formed, as in cases of the sudden appearance of new variations.

(G) Since the hereditary factors are present generation after generation, there must be some mechanism for their replication.

(H) The hereditary factors may act in a manner similar to infectious agents in that those of one individual may invade the cells of another—as with Lord Morton's mare (class 10).

Thus we may tentatively conclude that hereditary factors exist, are present in at least many of the cells of the body, that they may be transmitted via the gametes, may be expressed or remain dormant in a given generation, can persist unchanged for generations, may change under some unknown conditions, and are capable of increasing in number. All of this agrees with what we know today about genes.

Class H, however, is clearly not part of modern genetics—the hereditary factors, the genes, of higher plants and animals do not normally go wandering around the body, as seemed to be the case with Lord Morton's mare. We now know that the case was misinterpreted—there was no contamination of the mare by the semen of the quagga. Similar barring was found to occur in the offspring of Arabian and English race horses. Darwin was unaware of this, and his belief that the observation was correct was an important factor in his formulating a faulty hypothesis.

How could one unite these heterogeneous data into a single conceptual scheme? Darwin made the attempt.

> As Whewell, the historian of the inductive sciences, remarks:—"Hypotheses may often be of service to science, when they involve a certain portion of incompleteness, and even of error." Under this point of view I venture to advance the hypothesis of Pangenesis, which implies that the whole organization, in the sense of every separate atom or unit, reproduces itself. (vol. 2, pp. 357–358)

Darwin calls these minute units of reproduction the gemmules. Gemmules were assumed to possess the following characteristics: each and every part of an organism, and even parts of cells, were assumed to produce gemmules of specific types—the liver, liver gemmules and the eye, eye gemmules. Gemmules were assumed to be capable of moving throughout the body so that all parts of the body, including the eggs and sperm, would contain gemmules of all types, that is, they would contain a complete set of hereditary factors. During development the gemmules were assumed to unite with one another, or with partially formed cells, to produce new cells of the sort that had originally produced them. New gemmules were assumed to be produced continually. They were usually active in the offspring, but they might remain dormant for generations.

The hypothesis of pangenesis could account for each of the ten classes of phenomena. This is not surprising, of course, since hypotheses are formulated to account for the data.

(1) The transmission of characteristics from parent to offspring was explained as a consequence of the production of the specific gemmules in the parental body, their incorporation in gametes, and their development in the offspring. Edward Lambert's skin cells had produced porcupine-skin gemmules, and these were passed to his offspring via his sperm.

(2) Mutilations are usually not inherited since gemmules for the normal structure would have been produced before the mutilation. Thus, the regeneration of a salamander's leg is possible because the leg gemmules were already present throughout the body and after amputation could be assembled to produce a new leg. The few cases in which mutilations appeared to be inherited seemed to involve diseased parts. Darwin explained these cases as follows: "In this case it may be conjectured that the gemmules of the lost part were gradually all attracted by the partially diseased surface, and thus perished" (vol. 2, p. 398).

(3) Atavism was explained as a consequence of long-dormant gemmules becoming active after the passage of many generations. This was an especially gratuitous assumption. No more is being said than this: since some characteristics seem to reappear in a lineage after not having been present for many generations, and if characteristics are determined by gemmules, then the gemmules must have been dormant.

(4) Sex-linked inheritance is a consequence of gemmules being dormant in one sex. Thus a colorblind man transmits colorblind gemmules to his daughter, where they remain dormant. She transmits the colorblind gemmules to her sons, where they develop and result in colorblindness.

(5) The usually observed blending when two different forms are crossed is a consequence of the gemmules of each parent being mixed

in the offspring. Those cases in which the characteristics of one parent dominate is a consequence of that parent's gemmules "having some advantage in number, affinity, or vigour over those derived from the other parent."

(6) Artificial selection is possible because, by choosing individuals with desirable characteristics, one chooses as parents those individuals with the desirable gemmules. By continuous inbreeding of parents with the desired characteristics, one can slowly produce a variety of the sort required.

(7) The origins of variability were obscure but somehow the environment must be the cause—but not in a simplistic Lamarckian sense. But once new variations appear, they would produce new sorts of gemmules. This implies that somatic cells—cells other than reproductive cells—can influence the hereditary composition of eggs and sperm, a point of view that, much later, was to be regarded as a most serious genetic heresy.

(8) Regeneration could be accounted for since the gemmules for all structures are found throughout the body, so any part of the body would have the power to replace an adjacent lost part.

(9) The identical outcome of sexual and asexual reproduction finds explanation in that all parts of the body have gemmules for all parts. The gametes of hydra, as well as the cells of the body wall that are about to produce a bud, have the same library of gemmules.

(10) Gemmules from the quagga stallion passed to Lord Morton's mare via the semen and affected not only the first offspring but entered her ovary and expressed themselves when the mare was mated to the Arabian stallion and produced two colts.

What can we say? Darwin had performed a great service in assembling a huge mass of data and, in a real sense, had defined the field of heredity. His hypothesis of pangenesis was a notable advance over the hypothesis of pangenesis proposed by Hippocrates more than 2 millennia earlier. Darwin's most important contribution may have been his emphasis that inheritance has a physical basis and perhaps rules could be discovered for its mechanisms. He realized the weakness of his hypothesis of pangenesis, but he had tried to bring order where none existed. If his efforts served no other purpose, they at least gave other scientists a place to start, a catalog of the sorts of data to be explained by any comprehensive theory of inheritance, a discussion of the main problems, and a hypothesis that could be tested.

Galton's Rabbits

Charles Darwin's nephew Francis Galton (1822–1911), who had long been interested in his uncle's work, sought to test the hypothesis of pangenesis (1871a). His test was simple and direct. The hypothesis of pangenesis held that all parts of the body contain a full complement of gemmules. Thus all would be present in blood. Galton knew that blood could be transferred from one animal to another and that "it was not a cruel operation." He proposed to transfer blood between different strains of anesthetized rabbits and then study their offspring. One obvious deduction from the hypothesis is that if blood of black rabbits is injected into silver-grey rabbits and the silver-greys then bred with one another, one would predict that the blood of the black rabbits would have some effect. That is, the offspring of the two injected silver-greys would differ from those obtained in a cross of noninjected silver-greys. Galton found that his experimental rabbits all bred true. There was no evidence that the injected blood modified the offspring. This test of a deduction suggested, therefore, that the hypothesis was incorrect.

Darwin reacted promptly (1871) to this attack on his hypothesis, maintaining that Galton's experiments were no test at all since he (Darwin) "had not said one word about the blood" and "it is, indeed, obvious that the presence of gemmules in the blood can form no necessary part of my hypothesis," since gemmules were assumed to exist in creatures lacking a circulatory system.

This was, indeed, a strange rejoinder: if gemmules were present throughout the body, surely they would be present in blood. Possibly Darwin was having troubles with the Idols of the Cave. Galton replied (1871b) with mock contrition, saying how sorry he was to have misinterpreted what his uncle had said.

Darwin's hypothesis of pangenesis was based on gemmules, but no evidence was provided for their existence. They were invented to account for the observed phenomena of inheritance. This is legitimate scientific procedure. Atoms were invented to account for the data of chemistry; a planet later named Pluto was invented to account for irregularities in the orbits of the known planets. Atoms and Pluto were useful hypotheses before their reality was established, because they could be tested and eventually verified.

But the hypothesis of pangenesis was not very useful because it was so formulated that it could explain anything, and hence could not be tested. Darwin listed many diverse aspects of inheritance and said all were determined by gemmules. The hypothesis was not well regarded, even though there was not a better one to take its place. As Vorzimmer (1970, p. 257) was to write years later, the hypothesis of pangenesis was "so *ad hoc* as to withstand any criticism which sought to point up any fact inconsistent with it." But surely Galton's experiments transfusing blood should have been accepted as fatal to the hypothesis.

When Darwin wrote *Variations*, there was no possibility of anyone developing a concept to explain all of the data of inheritance. This was especially true when some of the "facts" that Darwin thought most important were later found to be erroneous. Biologists would have to reach the stage of genetic engineering before they could do to Lord Morton's mare what that quagga was thought to have done. Genetics became a rigorous science when an entirely different approach was taken. Success came when the attempt to explain everything about inheritance in one grand theory was abandoned and geneticists sought to explain seemingly minor phenomena—when Mendel concentrated on the colors and shapes of peas and when Morgan puzzled about the inheritance of white eyes in a tiny fly.

After 1900 genetics made great progress by first trying to explain very little and then, as confirmable hypotheses were developed, more and more puzzlements were studied, explained, and incorporated into the corpus of genetic theory. A remark of Hardin (1985, p. 4), made in another connection, is fully relevant here: "What began as knowledge about very little turns out to be wisdom about a great deal." Darwin began by trying to explain a great deal and ended by explaining very little.

chapter 12

The Cell Theory

We can look back and see that two main research approaches were of prime importance in gaining the understanding of inheritance that we now have. One, as noted, was cross-breeding—individuals that differ in some way are crossed and the offspring are compared with one another and with the parents. Hypotheses for the mechanisms of inheritance are then formulated from the data obtained. This was Darwin's approach, but neither he nor others in the last half of the nineteenth century were able to advance significantly our understanding using this approach.

The other line of research was based on this analysis: There is a structural bottleneck in the life cycle of both animals and plants. The two sexes usually produce small eggs and always very small sperm, and these combine to produce offspring that, in time, will closely resemble the parents. At least in some species, there is no further contact between parents and offspring after the liberation of eggs and sperm, and in many others there is no contact after fertilization, so the only physical link of parents and offspring are the eggs and sperm. Eggs and sperm must, therefore, contain the hereditary information that passes from one generation to another.

This last argument could not be extended to all species. In mammals and seed plants, for example, the early stages of development occur in close relation to the maternal tissues. There would be a possibility, therefore, of a maternal influence on inheritance during early development—in the case of mammals, for example, via the placenta. Nevertheless, one could follow Bacon and decide that, since there can be no maternal influence after fertilization in some species, a maternal influence cannot be universal. One could hypothesize further, with less confidence to be sure, that there may not be *any* maternal influence on inheritance in any species after fertilization.

Therefore, if our working hypothesis was that all the hereditary information must be contained in the eggs and sperm—the gametes—we might expect that a detailed study of gametes and fertilization could shed light on inheritance. Whether or not this proved to be a rewarding approach, we would have to accept that any comprehensive theory of inheritance must be compatible with whatever was discovered about the behavior of gametes.

One might simplify these two research approaches for studying inheritance by saying that breeding sought to discover the rules for inheritance of traits, and cell research (cytology) sought to discover the substance of inheritance. The two approaches were to be combined in 1902–1903 by W. S. Sutton, and thereafter the interplay of experimental breeding and cytology was to be the reason for the ensuing rapid and rewarding increase in our understanding of inheritance.

The Discovery of Cells: Robert Hooke

The birthday of cytology, the science and study of cells, can be fixed with considerable accuracy. It began with Robert Hooke, who was discussed in Part One for his work on microscopic structures and fossils. At the meeting of the Royal Society on April 15, 1663, he placed a piece of cork under his microscope and demonstrated its otherwise invisible structure (figure 41). That was the beginning of two centuries of observation and experimentation that were to establish the cell theory, which is that the bodies of animals and plants are composed solely of cells or cell products. Hooke imagined that cork consisted of a number of parallel tubes with cross partitions: "These pores, or cells, were not very deep, but consisted of a great many little Boxes, separated out of one continuous long pore, by certain *diaphragms*" (1665, p. 113). He observed similar structures in many other kinds of plants. It is generally thought that Hooke described those boxes as empty and let it go at that. Not at all: "Several of those Vegetables, whil'st green, I have with my *Microscope*, plainly enough discover'd these Cells or Pores fill'd with juices . . . as I have also observed in green Wood all these long *Microscopical* pores which appear in Charcoal perfectly empty of anything but Air" (p. 116).

This discovery of cells in cork and other plants could have been of general importance or it could have been a minor feature of a few kinds of organisms—cork oaks and those vegetables. Continued research was to show, however, that the bodies of plants consisted entirely, or almost entirely, of microscopic boxlike structures. Thus, Hooke had made an interesting observation that was not important at the time—it became an important discovery because of later research.

But what does all this have to do with inheritance? We can be certain that when Robert Hooke sat down to his microscope, he was not intending to unravel the mysteries of inheritance. There was no more reason to believe that cells had anything to do with inheritance than did, for example, the bristles he observed on the surface of a flea he described in such detail. Yet time and time again it turns out that the explanations in one field, in this case inheritance, emerge from discoveries made in an entirely different field, in this case cytology. But the major contributions of cytology had to await the perfection of the microscope two centuries into the future.

Cells became truly important only when the hypothesis that the bodies of *all* organisms are composed solely of cells or the products of cells was proposed, tested, and found to be highly probable. That hypothesis was formulated and tested early in the nineteenth century, and it is associated mainly with three nineteenth-century observers: R. J. H. Dutrochet, Matthias Jacob Schleiden, and Theodor Schwann.

But how could one possibly prove that "the bodies of *all* organisms are composed solely of cells or the products of cells?" The answer is, of course, that such a statement could not possibly be proven. How could one study all organisms? Most are long gone from this earth: Can we test the hypothesis, "The bodies of dinosaurs were composed of cells?" It would not even be practical to study the entire body of even one large living animal or plant. All that one can hope for in science is that a statement is "true beyond all reasonable doubt." Following Hooke's initial observations, it was found that cells were a common feature of plants. Small pieces of more and more individual plants and more and more species were studied, and all were found to have those cell-like structures. They were not all the box-shaped cells like those of cork—they came in various shapes and sizes. We must not forget that these early microscopists were not observing cells as we understand them today but cell walls.

Obſerv. XVIII. *Of the* Schematiſme *or* Texture *of* Cork, *and of the Cells and Pores of ſome other ſuch frothy Bodies.*

I Took a good clear piece of Cork, and with a Pen-knife ſharpen'd as keen as a Razor, I cut a piece of it off, and thereby left the ſurface of it exceeding ſmooth, then examining it very diligently with a *Microſcope*, me thought I could perceive it to appear a little porous; but I could not ſo plainly diſtinguiſh them, as to be ſure that they were pores, much leſs what Figure they were of: But judging from the lightneſs and yielding quality of the Cork, that certainly the texture could not be ſo curious, but that poſſibly, if I could uſe ſome further diligence, I might find it to be diſcernable with a *Microſcope*, I with the ſame ſharp Pen-knife, cut off from the former ſmooth ſurface an exceeding thin piece of it, and placing it on a black object Plate, becauſe it was it ſelf a white body, and caſting the light on it with a deep *plano-convex Glaſs*, I could exceeding plainly perceive it to be all perforated and porous, much like a Honey-comb, but that the pores of it were not regular; yet it was not unlike a Honey-comb in theſe particulars.

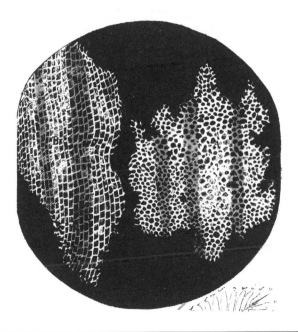

41 *Description and illustrations of cork cells from Robert Hooke's* Micrographia *of 1665.*

Schwann and Cells in Animals

With few exceptions, the bodies of animals contain no structures resembling the "cells"—that is, the cell walls—of plants. Thus it required a great deal of study and bold imagination before it became obvious that the concept of cells could be profitably applied to animals. This was first accomplished mainly by the German zoologist Theodor Schwann (1810–1882), in his monograph of 1839, published when he was 29 years old (figure 42). He emphasized the great difference between the cells of plants and the structures in animals but suggested that they are fundamentally the same.

> Though the variety in the external structure of plants is great, their internal structure is very simple. This extraordinary range in [external] form is due only to a variation in the fitting together of elementary structures which, indeed, are subject to modification but are essentially identical—that is, they are cells. The entire class of cellular plants is composed solely of cells which can readily be recognized as such; some of them are composed merely of a series of similar or even only of a single cell.
>
> Animals being subject to a much greater range of variation in their external form than is found in plants also show (especially in the higher species) a much greater range of structure in their different tissues. A muscle differs greatly from a nerve, the latter from a cellular tissue (which shares only its name with the cellular tissue of plants), or elastic tissue, or horny tissue, etc.

But if the components of muscles, nerves, and other parts of an animal body are so different from one another, how is it either possible or useful to call their component cells? When they are obviously so different, why maintain that they are fundamentally the same? And what is to be gained by claiming that these diverse structures in animals can be equated with the very different looking structures in plants? Schwann provides a partial answer:

> If, however, we go back to the development of these tissues, then it will appear that all of these many forms of tissue are constituted solely of cells that are quite analogous to plant cells . . . The purpose of the present treatise is to prove the foregoing by observations.

That is, in spite of the great diversity of structures that Schwann proposed to call cells, all develop from simpler structures that could be compared more readily with the cells of plants. The cells of an early embryo of animals look much alike, though in the course of development they become different from one another.

But Schwann proposed a much better criterion for defining cells—the presence of a nucleus. Only six years earlier, in 1833, Robert Brown (1773–1858) described a single circular (actually spherical) areola, or nucleus, in the living cells of orchids and many other kinds of plants. Previous observers had noted these structures, had illustrated them in

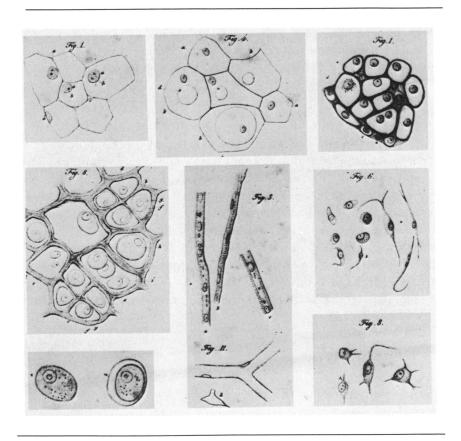

42 *Some illustrations from Schwann's monograph.* (Top row, left to right) *Onion cells, notochord of a fish, cartilage of a frog.* (Middle) *Cartilage of a tadpole, muscles of a fetal pig, areolar cells of a pig embryo.* (Bottom) *Ganglion cells of a frog, capillary in a tadpole's tail, cells of a pig embryo. Note that the nucleus and nucleoli are shown in nearly all cells.*

their publications, but had attached no importance to them. Brown found that many kinds of cells contain nuclei but did not speculate on their significance. Schwann changed the rules for defining cells by relying not on shape, which in plants meant the structure of the cell walls, but on the presence of a nucleus.

> The most frequent and important basis for recognizing the existence of a cell is the presence or absence of the nucleus. Its sharp outline and its darker color make it easily recognizable in most cases and its characteristic shape, especially if it contains nucleoli . . . identify the structure as a cell nucleus and make it analogous with the nucleus of the young cells contained in cartilage and plant cells . . . More than nine-tenths of the structures thought to be cells show such a nucleus and in many of these a distinct cell membrane can be made out and in most it is more or less distinct. Under these circumstances it is perhaps permissible to conclude that in those spheres where no cell membrane be distinguished, but where a nucleus characteristic of its position and form is encountered, that a cell membrane is actually present but invisible.

Although Schwann was a careful observer, his contribution was not primarily what he saw but how he interpreted the observations. Previous investigators had emphasized the boxes. Schwann emphasized what was *inside* the boxes. For him the animal cell became a bit of living substance containing a nucleus and bounded by a membrane and, in the case of plants, further encased in cell walls.

What does this new view of cells have to do with inheritance? Very little. Two other bits of information are required before cells can be considered to have an important relation to inheritance: the discovery that the gametes are cells and the realization that cells originate only from other cells.

Gametes as Cells

Schwann recognized that ova are cells since they exhibited the structure required by his definition—possession of a nucleus. The nature of spermatozoa was less clear. Even the name, meaning "sperm animals," indicates uncertainty about their true nature. In 1667 the Dutch naturalist Anton van Leeuwenhoek (1632–1723), or one of his students, had discovered and reported to the Royal Society of London that seminal fluid contain microscopic creatures that were imagined to enter the egg and achieve fertilization. That hypothesis was hotly contested, and some imagined that the spermatozoa were no more than parasites in the semen.

A little more than a century later, in 1784, the Italian biologist Lazzaro Spallanzani (1729–1799) conducted some remarkable experiments to ascertain the function of semen in the reproduction of frogs. During breeding the males clasp the females and, as we now know, deposit sperm on the eggs as they leave the female's cloacal opening. Another investigator with whom Spallanzani corresponded had attempted, without much success, to discover the role of male frogs in reproduction by putting trousers on them, believing that the batrachian trousers would prevent the semen from contacting the eggs. Now to Spallanzani:

> The idea of the breeches, however whimsical and ridiculous it may appear, did not displease me, and I resolved to put it in practice. The males, notwithstanding this incumbrance, seek the females with equal eagerness, and perform, as well as they can, the act of generation; but the event is such as may be expected: the eggs are never prolific [that is, they do not develop], for want of having been bedewed with semen, which sometimes may be seen in the breeches in the form of drops. That these drops are real seed, appeared clearly from the artificial fecundation that was obtained by means of them. (vol. 2, p. 12)

That is, he put some of the drops of semen on unfertilized eggs and observed that they developed. But which was the active agent—the "spermatic worms" or the fluid portion of semen? The answer was provided by George Newport, in 1854, who found that in frogs, sperm cells enter the egg at fertilization.

Here and elsewhere it is often difficult to give credit to the scientist who discovered an important biological phenomenon. After all, the discoverer of sperm, Leeuwenhoek, had thought sperm were the agents of fertilization but this was a hypothesis only—he had not proven it to be true beyond reasonable doubt. It remained for Newport to make the first convincing observations.

Sperm are so unusual in appearance that they were not recognized as cells by early observers. That they are was proven in 1841 by Rudolf Kölliker, who found that some of the testis cells are converted into sperm. Sperm, then, are highly modified cells.

In summary, the argument relating cells to inheritance has reached this stage.

(1) Gametes are the only physical link between generations in many organisms and possibly all.

(2) Therefore the gametes must contain all of the hereditary information.

(3) Since ova and sperm are cells, all of the hereditary information must be contained in these sex cells. Therefore, the physical basis of inheritance is the sex cells.

We still need that second bit of information: What is the origin of cells?

Omnis cellula e cellula?

The origin of new cells by the division of old cells was observed as early as 1835, but it was not at the time realized that this is a general phenomenon. The great authority, Schwann (1839), held a very different notion about the origin of cells.

> The general principles in the formation of cells may be given as follows. At first there is a structureless substance which may be either quite liquid or more or less gelatinous. This, depending on its chemical constitution and degree of vitality, has the inherent ability to bring about the formation of cells. It seems that usually the nucleus is formed first and then the cell around it. Cell formation is in the organic world what crystallization represents in the inorganic world. The cell, once formed, grows through its inherent energy, but in doing so it is guided by the organism as a whole in the way that conforms to the general organization. This is the phenomenon basic to all animal and plant growth. It is applicable to cases where the young cells originate in the mother cell, as well as those where they are formed outside of them. In both instances the origin of cells occurs in a liquid or in a structureless substance. We call this substance, in which cells are formed, a cell germinative substance or Cytoblastema. It can be compared figuratively, but only figuratively, with a solution from which crystals are precipitated.

This hypothesis for the origin of cells holds that they are episodic events in the life cycle of organisms. If true, the unit of inheritance must be the entire organism, not the cell. Schwann's hypothesis for the origin of cells was soon rejected by his contemporaries, since cell division was observed repeatedly in a variety of organisms and in different periods of development. More and more investigators began to suspect that the division of an existing cell was the sole mechanism for producing new cells.

This was an exceedingly difficult hypothesis to prove beyond all reasonable doubt. The microscopes and the techniques for studying cells in the early 1800s were most inadequate by later standards, and it took many observations on different sorts of organisms and tissues before the German pathologist Rudolf Virchow (1821–1902) was to express the view, in 1855, *omnis cellula e cellula* ("all cells from cells") and have it generally accepted. In a lecture given in 1858 he put it thus:

> A new cell can [never] build itself up out of any noncellular substance. Where a cell arises, there a cell must have previously existed (*omnis cellula e cellula*), just as an animal can spring only from an animal, a plant only from a plant. In this manner, although there are still a few spots in the body where absolute demonstration has not yet been afforded, the principle is nevertheless established, that in the whole series of living things, whether they be entire plants or animal organisms, or essential constituents of the same, an eternal law of *continuous development* prevails. (Virchow 1863, lecture 2)

Not everyone agreed with Virchow that all cells and all organisms come from preexisting cells, but as the nineteenth century progressed this hypothesis became ever more probable. In a few decades the experiments of the French scientist Louis Pasteur (1822–1895) would establish the improbability of spontaneous generation of living organisms and the cells within them. Thus by the end of the nineteenth century it had been established beyond all reasonable doubt that:

Omnis vivo e vivo
Omnis cellula e cellula

There was no question, then, that inheritance is based on cell continuity, and we may now work with the hypothesis that all the hereditary information is contained in the germ cells. Also possible was the hypothesis that all cells contain the hereditary information necessary for the development of the individual and for its transmission, via the sex cells, to the next generation.

The Technology of Cell Research

For most of human history we have relied almost entirely on our sense organs for information about the environment. Each sense organ detects only a narrow window in the range of possible stimuli. Our eyes, for example, can respond only to that portion of the electromagnetic spectrum between violet and red, so we see wavelengths only between these two colors, which from the shortest to the longest waves are: violet, indigo, blue, green, yellow, orange, and red. The human eye cannot detect the shorter ultraviolet, X-rays, and gamma rays or the longer infrared and radio waves. Special instruments must be used to study them.

Our unaided eyes fail also to tell us about objects that move very rapidly. The individual gray blades of a rapidly moving fan merge into a continuous circle that is less gray. A bullet leaving a rifle barrel is wholly invisible. Nor can we see objects that are very small. The apparent uniformity of a half-tone illustration is a result of the individual dots of ink being too close for the human eye to resolve. The twin headlights of an automobile appear as a single source of light when far away. As the automobile approaches, the single light source divides into two.

Human eyes vary in their ability to resolve two objects, that is, to determine whether an object is single or multiple. The limit of resolution is about 100 microns at reading distance and about one millimeter at a distance of 10 meters. A more general statement is that the human eye can resolve objects separated by an arc of 1 minute. That value was determined by Robert Hooke (1674), who wondered how far apart double stars had to be before they could be seen as two. When they were closer than 1 minute of arc, most people see only a single point of light. Some people can do better, and the maximum resolving power for their unaided eyes is about 26 seconds of arc.

The small size of cells is not the only problem that makes their study difficult. Most animals and their tissues are opaque, and examination with a microscope would reveal little. It was discovered, however, that if very thin slices of tissues are cut, light can pass through and some structure becomes visible. But then another problem arises. Animal tissues consist mostly of water, and thin slices of them quickly dry and become a shriveled mess. This is a special problem with animal cells, which lack the supporting walls of plant cells.

Thus special methods had to be developed by microscopists in the early nineteenth century if they were to learn about the cellular nature of organisms and, later on, the internal structure of cells themselves. It was necessary, therefore, to preserve tissues in such a manner that the cellular structure would remain intact and thin slices of them could be made.

The first step is fixation. It was found that many chemicals such as alcohol, formaldehyde, or solutions of picric acid, potassium dichromate, mercuric chloride, or osmium tetroxide would kill and harden cells, often by coagulating their proteins. It was hoped, of course, that this could be done in such a way that the parts of the cells would resemble the living state to an acceptable degree. The fixed tissue could then be embedded in paraffin wax and slices made with a sharp razor or an instrument devised for this specific purpose—the microtome. Slices 10 micrometers in thickness became possible but, for the most part, they revealed very little because the tissues were nearly uniform in appearance. But the inventive microscopists tried everything and found that some dyes would stain some structures in cells but not others. A start was made in 1858 when it was found that a dilute solution of carmine would stain the nucleus more intensely than the cytoplasm. Carmine was available commercially as a stain for fabrics. It is derived from the dried bodies of female cochineal insects (*Coccus cacti*), which live on cactus plants in Central America and southwestern United States. Another important biological stain, hematoxylin, was introduced in 1865. It is an extract of the logwood tree (*Haematoxylon campechianum*) of Central America. It also has more affinity for the nucleus than the cytoplasm.

A decade later a large number of aniline dyes were developed for the textile industry, and many of these were found to be useful for differentiating cells and their internal structures. One was eosin, the first aniline dye, which had been synthesized in 1856. It proved to have a great affinity for the cytoplasmic proteins. Thus by using both hematoxylin and eosin one could stain the nucleus blue and the cytoplasm pink and so reveal structure where, before staining, all looked nearly uniform.

Great improvements in microscopes were made in the last part of the nineteenth century. Many were due to Ernst Abbe (1840–1905), who was employed by the Zeiss Optical Works in Jena, Germany. For most of his life Abbe was both professor of physics at the University in Jena and the principal lens designer for Zeiss. Later he became its owner. In 1878 he developed an oil-immersion objective and, in 1886, the apochromatic objective. In the hands of a skilled microscopist magnifications of 2,500 diameters became possible, and this was near the theoretical limit to which the compound microscope could be perfected. The limitation is due to the nature of light itself—two objects can be resolved only if the distance between them is at least equal to half of the wavelength of the type of light being used.

Nevertheless, cytologists of the last third of the nineteenth century were able to use the available technology to establish as highly probable the hypothesis that the physical basis of inheritance is the cell nucleus and, more specifically, the chromosomes within it. (It was not until the twentieth century that cytologists could see far more by using the phase-contrast and electron microscopes.)

But a constant difficulty faced the nineteenth-century cytologists. Living tissues are most fragile, and how could one be sure whether a given structure in a prepared slide had been there when the cells were alive or was an artifact resulting from the very drastic treatment to which cells had been subjected? Consider the saga of a cell subjected to the following procedure by a cytologist of the late 1800s:

> Sections of vegetable tissues present a beautiful appearance under the microscope when doubly stained. They should first be soaked in alcohol, if green, to deprive them of chlorophyll, then subjected to a solution of chloride of lime (1/4 ounce to a pint of water) until thoroughly bleached. Soak then in a solution of hyposulphite of soda (1 drachm to 4 ounces of water) for one hour, and after thoroughly washing in several changes of water transfer them to alcohol. Prepare some red staining fluid by dissolving 1/2 a grain of magenta crystals in 1 ounce of alcohol. Soak the specimen in this for thirty minutes, then rapidly rinse it in alcohol and place in a blue fluid made by dissolving 1/2 grain of anilin blue in 1 drachm of distilled water, adding 10 minims of dilute nitric acid and alcohol enough to make 2 ounces. Let the specimen remain only two or three minutes in this, rapidly rinse in alcohol, put in oil of cajeput, thence to turpentine, and mount in balsam. (Wythe 1880, p. 348)

Apart from giving thanks to the Muses for the metric system, one may wonder how accurately the final preparation reflected the structure of living cells.

But what can a study of cellular structure tell us about inheritance? Possibly some cytologists working after the publication of Darwin's *Variation* in 1868 must have wondered if they could see the gemmules that formed the basis of the hypothesis of pangenesis. When they examined cells they could see all sorts of spheres, granules, and fibers. Some might be the gemmules, but how could one establish whether or not any of these organelles had a role in inheritance? Or, in fact, how might one establish the function of *any* intracellular structure?

Despite these difficulties, there was plenty to be learned by studying fixed and stained cells. This was a necessary stage in the development of cytology—the study of as many different species as possible to find what were the structures common to all cells and to see the range of variation.

chapter 13

The Hypothesis of Chromosomal Continuity

D uring the last half of the nineteenth century, the hypothesis that the bodies of animals and plants are composed solely of cells and cell products was established as true beyond all reasonable doubt in the minds of most competent microscopists. We can speak, therefore, of the cell theory, using the term "theory" to apply to an entire body of data, hypotheses, and concepts relating to an important natural phenomenon. In this particular case it is the synthesis of all data about the structure and functioning of cells, which are the smallest units that can maintain the living state—that is, they are able to use substances acquired from the environment to maintain themselves, grow, and reproduce.

There was another important reason for studying cells: analysis at a simpler level of organization contributes to understanding at more complex levels. For example, the interactions of chemical substances are better understood when we know their molecular structure. The movements of the human body can be studied at many levels. One may observe and describe the complex movements of a ballet dancer or baseball pitcher, each beautiful and important in its own way. Understanding of this movement is increased when we obtain information about the many muscles and their attachments that make the motions possible. Other sorts of understanding come when we study muscles at the cellular level, where we learn about the activity of myosin, actin, and the other molecules involved in the movement of muscles. Knowledge obtained at each level of organization contributes to an understanding of the total phenomenon, while each level retains its own validity. One cannot completely understand a Waslaw Nijinsky or a Fernando Valenzuela merely by knowing about the actin and myosin in their muscle cells, any more than one can predict the properties of water from knowing about hydrogen and oxygen.

The Ephemeral Nucleus

As noted before, the difficulty in studying living cells made fixed and stained preparations the favorite material. It was only in those that one could clearly see intracellular structures. In stained cells the most prominent structure is the nucleus. Many dyes, especially basic dyes such as carmine or hematoxylin, stain the nucleus heavily, and this, together with its apparently universal presence, suggested that it must be important. It is only human to suspect that what is visible is more important than what is invisible. Or more accurately, one cannot make observations or do experiments on what has neither been seen nor imagined.

But what is the origin of the nucleus? It took nearly a half century of observation and experiment by numerous cytologists to find out. As late as the 1860s and 1870s some prominent cytologists continued to believe that at least some nuclei can have a nonnuclear origin. Concurrently other equally competent cytologists were claiming that all nuclei originate from existing nuclei. Various methods were suggested—usually some form of pinching in two or fragmentation, a process that later came to be known as amitosis. For example, in what was considered favorable material—the early embryos of sea urchins—the nucleus appeared to pinch in half immediately before the cell as a whole divided (figure 43).

There was no necessary reason, of course, why there should be a single mechanism for the origin of nuclei. Considering the large amount of variability of natural phenomena, one should not be surprised if there were a variety of modes of origin. Nevertheless, scientists seek to discover the regularities in nature, and it would be more intellectually satisfying if the concept of a constant mechanism of nuclear origin proved to be correct.

Another seemingly universal phenomenon that was difficult to explain for those maintaining a continuity of the nucleus was the disappearance of the nucleus just before the cell divides. That is, the spherical body that Brown and Schwann held to be a constant cell structure vanishes. In stained preparations it could be seen that, as the spherical nucleus vanishes, rod-shaped bodies not previously present made their appearance. But possibly these rod-shaped bodies were artifacts. Intensive study of these rods (later called chromosomes) led to the next major advances in cytology.

Schneider, Flemming, and Cell Division

In 1873 Friedrich Anton Schneider (1831–1890) published what can be accepted as the first reasonable account of the complex nuclear changes, now called mitosis, that occur when the cell divides.

Schneider's account was part of a general study of the morphology of *Mesostoma*, one of the platyhelminths. Nearly all of his paper is devoted to the anatomy of this small flatworm, but, being a careful observer, he described everything that he saw. Fertilization is internal in *Mesostoma*, and early development takes place in the uterus.

Schneider noticed that shortly before the living cell divided, the outline of the nucleus became indistinct. He found, however, that by adding a little acetic acid it became visible, though folded and wrinkled. Later the nucleolus (at the very center of the nucleus) disappeared, and all that remained of the nucleus was a clear area in the cell. However, acetic acid treatment revealed a mass of delicate, curved fibers. These strands, the chromosomes (a term not to be introduced until 1888), lined up in the center of the cell and seemed to become more numerous. When the cell divided, they passed to the daughter cells (figure 44).

What was one to make of these observations? The answer was far from clear. If one could not see the strands in the living cells and if they appeared suddenly when acetic acid was added, would it not be cautious to assume they were artifacts? Nevertheless, the fact that strands were observed repeatedly, and that they seemed to undertake these strange movements, argued that they might be present, though invisible, in the living state.

Schneider did not realize he was giving a reasonably accurate first description of a process—mitosis—that was soon recognized to be of tremendous importance in replication. His primary concern was to study the morphology of *Mesostoma* and to use the data obtained to ascertain the relation of these little animals to other invertebrates. In these post-Darwinian years one of the main preoccupations of biologists was to use morphology to try to sketch the broad outlines of organic relationships and evolution.

The events in cell division, and especially the changes in the nucleus, were recognized as important phenomena to be studied. In fact, they seemed to be about the only constant changes that occurred in cells.

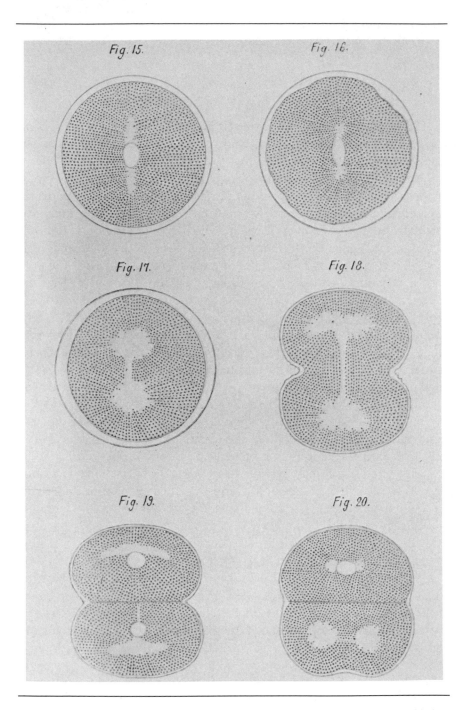

43 Cell division in a sea urchin embryo. Oskar Hertwig's figures 15–20 published
in 1876 show what can be seen in living embryos. Figures 21–26 show what can be
seen after the embryos have been fixed in osmic acid and stained with carmine. Now
the chromosomes, spindles, centrosomes, and asters have become visible.

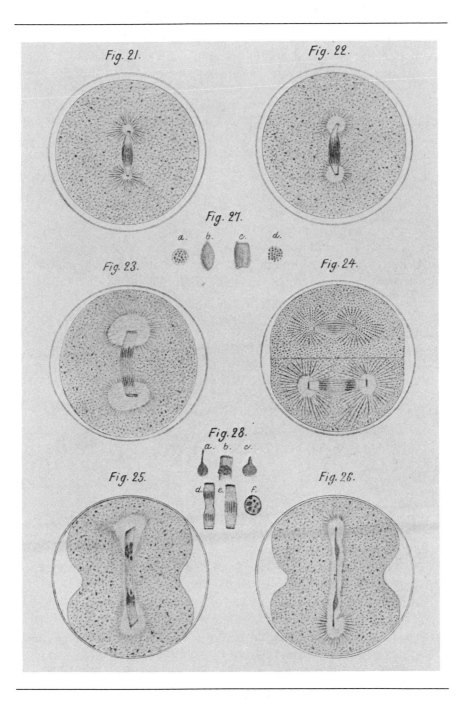

Fig. 21.

Fig. 22.

Fig. 27.

a. b. c. d.

Fig. 23.

Fig. 24.

Fig. 28.

a. b. c.

d. e. f.

Fig. 25.

Fig. 26.

Again, one studies what is available. But a basic difficulty remained—chromosomes *are* artifacts in most materials, that is, they could not be seen in living cells and became obvious only after the most drastic treatment. Their very name, meaning "colored body," indicates their artifactual nature—no one had discovered that the living cell possess any colored rodlike structures. What was needed to settle the matter was a species of plant or animal in which chromosomes could be observed while the cells were alive. These cells could then be fixed and stained. A comparison could then be made, and if the structures seen in living cells were much like those in stained cells, one could conclude that fixation and staining were not producing artifacts. That was accomplished by Walther Flemming.

Flemming was successful in determining that the nuclear events observed in cell division in fixed and stained material have their counterparts in living cells. Although he did not discover mitosis, we owe to him more than to any one else the concept of mitosis that we hold today. Only details were added after Flemming. His success was due to his selection of material for study, his verification in living cells of the things observed in fixed and stained cells, and his access to microscopes that were very much better than any available previously (figures 45 and 46).

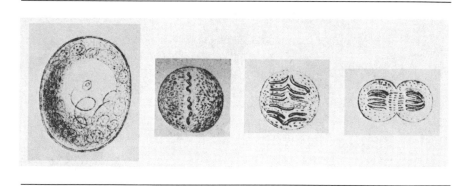

44 *Friedrich Anton Schneider's (1873) illustrations of the nuclear changes during cleavage in embryos of the flatworm* Mesotoma. *The left figure is of an uncleaved ovum (clear area with a nucleus and nucleolus) surrounded by follicle cells. The spiral structures are sperm. The other figures show the "strands," now chromosomes, and their movements during cell division.*

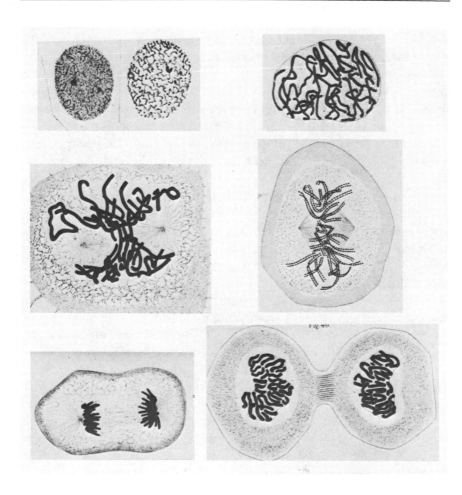

45   *Walther Flemming's (1882) illustrations of mitosis in fixed and stained cells of
a salamander embryo. The two cells at the left in the first row are in the resting
stage. There are no chromosomes recognizable as such, but there are two nucleoli.
The right figure is of prophase. The nucleoli have disappeared, but the nuclear
membrane is still intact. The cytoplasm is not shown. The left figure in row two is
of early metaphase. The nuclear membrane has disappeared and the centrosomes
have moved apart. The right figure is an especially fine preparation showing that
the metaphase chromosomes are double, that is, each composed of two chromatids.
The chromatids separate and move to the poles of the spindle, as in the bottom left
figure. In the lower right figure the cell has divided and the chromosomes of the
daughter nuclei are being surrounded by a nuclear membrane.*

Flemming examined many sorts of cells and found that those in the epidermis of salamander embryos were worth detailed study. The chromosomes are huge by microscopic standards, but, of much greater importance, with careful observation they can be seen in living cells. The sequence of events was as follows: A nucleus not undergoing mitosis is said to be in the *resting stage*. This is an unfortunate term since it implies inactivity, and it is now realized that great physiological activity is occurring during this time. Flemming saw no chromosomes in the resting-stage nuclei of living cells. In fact, the nucleus appeared to lack all internal structure. When such cells were fixed and stained, the nucleus was seen to contain a dense and deeply staining network together with one or two large spherical granules, the nucleoli.

Changes in the nucleus are the first indication that mitosis is under way. In the apparently structureless living nucleus long, delicate threads make their appearance. When they can first be seen, that is the start of *prophase*. (Mitosis is a continuous process that, for descriptive purposes, was divided into discrete stages by cytologists.) These threads condense into chromosomes that assemble in the middle of the cell at *metaphase*, at which time the nuclear membrane disappears. In stained cells the chromosomes were seen to be in an elongate fibrous structure—the spindle. Stained cells also revealed the presence of tiny granules, the centrioles, at the ends of the spindle. They also revealed another set of fibers, the astral rays, that radiate from the centrioles. During *anaphase* the chromosomes separate into two groups and move to opposite poles of the spindle. When the chromosomes have reached the ends of the spindle, that is *telophase*. At this time the chromosomes in living cells become less and less distinct and the nuclear membrane reforms. The nucleus is, once again, in the *resting stage*.

What is one to conclude about this process? It is obvious that *all* cell structures must be reproduced if the daughter cells are to be essentially identical with the parent cell. Flemming was able to explain how this is accomplished for chromosomes. If the chromosomes of a single cell are to be divided equally between the daughter cells, they must double in number at some stage in the cell cycle. Flemming observed that when the chromosomes first appear in early prophase they are double, so sometime between their disappearance in the previous telophase and their reappearance in prophase, each chromosome must have doubled.

Early embryos provided another way to determine the stages in mitosis—even in preserved cells. In fertilized echinoderm eggs, for example, cell divisions occur every half hour or so (depending on the temperature); and, of greater importance, all the embryos develop synchronously. Thus, if small samples are preserved every few minutes and stained and studied later, one can ascertain the sequence of events. Embryos have rapid cell divisions—clearly a great advantage if one wishes to study the process. In most adult tissues a dividing cell is seen infrequently—except where there is wound regeneration or growth.

It is now known that chromosomes are permanent cell structures even though they are readily visible only during mitosis. We also recognize the individuality of chromosomes, that is, they usually exist in homologous pairs with each pair containing a specific set of genes on

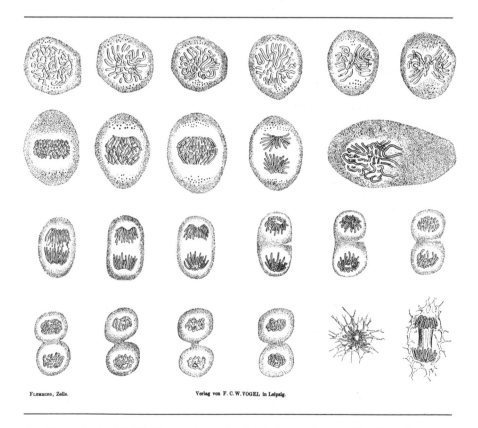

FLEMMING, Zelle.   Verlag von F. C. W. VOGEL in Leipzig.

46 Flemming's (1882) illustrations of mitosis in the living cells of a salamander larva. The drawings are arranged in sequence beginning with prophase at the upper left and ending with two cells in the lower row. The last two figures show a polar view of chromosomes and a side view of telophase. The rightmost figure in row two shows that the chromosomes are double.

each member of the pair. Could any of this be concluded from the observations of Flemming? Not really. In fact, the following hypothesis could not be denied. We will assume, with Darwin, that inheritance and the functioning of cells are due to specific gemmules. The gemmules are widely distributed during the resting stage when they are, presumably, directing the activities of the cells. Before the start of mitosis the gemmules congregate in the nucleus and join one another, like beads on a string, to form long strands—the chromosomes. The gemmule-bearing chromosomes are then divided in mitosis and each daughter cell receives an allotment. The chromosomes then break down and the gemmules are dispersed throughout the cell, where they carry out their activities. Flemming's data can neither support nor deny this hypothesis. Later research, however, invalidated it.

To some cytologists, Flemming's observations suggested that chromosomes might be involved in inheritance. The argument was something like this: since the mitotic process ensures that each daughter cell receives its allotment of chromosomes, this must indicate, beyond much doubt, that such an elaborate and precise mechanism for duplication and distribution of chromosomes is of fundamental importance. And what can be more important than ensuring that the elements controlling inheritance are allocated to each cell? This is a vague notion to be sure, but notions are generally vague before the answer is known.

But one might respond that, since the daughter cells grow to be essentially identical with the parent cell, *all* cell products must be reproduced and passed to the daughter cells. It could be merely an accident that the process of reproduction and distribution is more readily visible for the chromosomes. There is no reason, therefore, not to assume that chromosomes, cell membranes, and all those granules and globules in the cytoplasm have an equal chance of being involved in inheritance.

Flemming and many other contemporary cytologists had made a strong case for mitotic divisions of the nucleus being a concomitant of cell division. But if the hypothesis that the nucleus always divides by mitosis is true, that is, each daughter cell has the same chromosomes as the parent cell, then the number of chromosomes should double in each generation. This is inevitable, for if the nuclei of egg cells and sperm cells have been formed by mitosis, and if they unite in fertilization, the zygotes must have twice the number of chromosomes as the parents. Yet they do not—Flemming and other cytologists were aware that the number of chromosomes seems to be about the same in all individuals and in all available generations of a species.

Obviously there is a problem with this hypothesis. There must be some mechanism for reducing the number of chromosomes at or before fertilization. One might imagine that, when egg and sperm nuclei fuse at fertilization, the chromosomes fuse to one another or that half are destroyed. Or possibly there were some changes in chromosome number when the eggs and sperms were formed in the gonads.

The Chromosomes and Inheritance

In the last half of the nineteenth century, many biologists began to suspect that only the nucleus, and especially its chromosomes, might be the vehicle of inheritance. As early as 1866 Ernst Haeckel (1834–1919) suggested that the nucleus is responsible for the transmission of what we call genetic information. This was really a long shot. It was still seven years before Schneider's description of mitosis in *Mesostoma*, and Haeckel had no data to support his hypothesis. Nevertheless, he was one of the most prominent biologists of the day, and any idea of his, no matter how slight the factual basis, would be noticed. Therefore, Haeckel's hypothesis of nuclear control of heredity would stimulate others to think along such lines.

Darwin's *Variation* was published two years later and, again, something was being said about inheritance by a very important scientist. The great difference between the hypotheses of Haeckel and Darwin indicated that the field was still wide open.

In 1884–1885 four German biologists independently concluded that the physical basis of inheritance must be the chromosomes of the nucleus. They were Oskar Hertwig, Edouard Strasburger, Rudolf Kölliker, and August Weismann. The first three were active cytologists. Weismann, in part because of poor eyesight, was restricted mainly to theoretical considerations.

The essential problem for the student of the cell who sought to discover the physical basis of inheritance was to find some cellular phenomenon that could account for what was known about inheritance. That is, how could one find in cells something that is consistent with the results of breeding experiments? Expressed still another way, one had to discover some cytological phenomenon that would parallel genetic phenomena—an especially difficult task when, at that time, there were no precise genetic rules.

The sorts of data and arguments available that suggested the hypothesis of chromosomal involvement in inheritance to the four biologists mentioned above were as follows:

First, it was usually observed that both parents seemed to have an equal share in transmitting their characteristics to the offspring. A study by Joseph Gottlieb Kölreuter, the great plant hybridizer, was an excellent example. More than a century earlier, he had crossed two very different species of the tobacco genus, *Nicotiana paniculata* and *Nicotiana rustica*. It was important to use species that differed considerably so that he could test the influence of each parent on the progeny. So far as Kölreuter could tell, the hybrids were the same whether the cross was *Nicotiana paniculata* pollen × *Nicotiana rustica* ovules or the reverse. Neither parent had a dominant role.

What could be the basis of this equality found by Kölreuter and many others? It could not be the transmission of an equal amount of material, for it was well known that in animals there is a great difference in the quantity of material in ova and sperm. If the results of inheritance depended solely on the quantity of material transmitted by the female and by the male, and if their gametes were the sole physical link between the generations, then one would expect the female's influence on the offspring to be far greater than the male's. Since this appears not to be so, another candidate must be sought for the physical basis of inheritance.

Is there, therefore, *any* cellular component of sperm and ovum that is equivalent? If so, it might be a candidate for a central role in inheritance. In the late 1880s a possible candidate was being suggested by the then most recent research. Hertwig and many others were finding that shortly after fertilization there are two nuclei in the zygote—the female pronucleus and the male pronucleus. These appeared to be identical. Could this equivalence in structure be the basis of the equivalent importance of the two gametes in inheritance?

Second, there seemed to be both a stable and an unstable component to inheritance. In almost all ways offspring were observed to closely resemble their parents in general body structure. Thus, whatever was transmitted from parent to offspring via the gametes must have a high degree of stability. Yet this stability could not be complete since offspring were rarely exactly like their parents and, moreover, the offspring of the same parents might differ from one another.

There did seem to be a possible cellular basis for the stability—the chromosomes of the nucleus. During cell division the cytoplasm and its formed structures appear to be divided passively and by chance. That is, if a granule or globule happens to be at one end of the spindle, it ends up in the daughter cell at that end. The chromosomes, on the other hand, go through a complicated mitosis that results in each daughter cell apparently receiving an identical set of chromosomes. It seemed to Hertwig and the others that such precision in the distribution of the nuclear material during cell division could mean that the chromosomes were involved in transmitting the genetic information. There was no other likely candidate, except possibly the centrioles. But centrioles were so small and difficult to see that cytologists were not even sure that they were present in all cells. They did not appear to be in ova just before fertilization, and higher plants seemed to lack them.

Third, there was at least some experimental data that suggested the involvement of the nucleus in inheritance. It proved possible to cut some species of protozoans, which are only a single cell, into two parts—one part with the nucleus and the other without. Both parts may heal. The part with the nucleus was observed to regenerate any missing structures and to live as a normal, reproducing individual. The part without the nucleus did not regenerate to form a whole animal and it never reproduced. Its fate was death.

Gamete Formation

In 1887 August Weismann proposed the following hypothesis:

> At least one certain result follows, viz. that there is an hereditary substance, a material bearer of hereditary tendencies, and that this substance is contained in the nucleus of the germ-cells, and in that part of it which forms the nuclear thread, which at certain periods appears in the form of loops or rods. We may further maintain that fertilization consists in the fact that an equal number of loops from either parent are placed side by side, and that the segmentation nucleus [the embryo's nucleus] is composed in this way. It is of no importance, as far as this question is concerned, whether the loops of the two parents coalese sooner or later, or whether they remain separate. The only essential conclusion demanded by our hypothesis is that there should be complete or approximate equality between the quantities of hereditary substance derived from either parent.

One deduction from this hypothesis would be:

> If then the germ-cells of the offspring contain the united germ-plasms of both parents, it follows that such cells can only contain half as much paternal germ-plasm as was contained in the germ-cells of the father, and half as much maternal germ-plasm as was contained in the germ cells of the mother. (1889, pp. 355–356)

Weismann's hypothesis for this halving of the hereditary material soon found support.

In the 1880s Edouard van Beneden (1846–1912), Theodor Boveri (1862–1915), and particularly Oskar Hertwig (1849–1922) discovered that there are two unusual cell divisions during the formation of gametes that result in the chromosome number being reduced by half—as Weismann said must happen. These two divisions are highly modified mitotic divisions and were given the name meiotic divisions—the names being so similar that they remain confusing to this day.

In the female *Ascaris*, a roundworm, the ovary begins to form early in development, and the huge increase in the number of its cells is a consequence of mitotic divisions. Each nucleus has four chromosomes, the diploid number, and before each division of the cell every chromosome is seen to have doubled by early prophase, forming two chromatids. The eight chromatids are divided between the two daughter cells, and the result is four chromosomes in each—a chromatid becoming a chromosome once it has become separated. As the *Ascaris* female matures, her ovary has many enlarged cells, the ova, still with the diploid number of chromosomes. The ovum remains diploid until it has been released from the ovary and entered by a sperm. It is only then that meiosis begins.

At the onset of meiosis each of the four long chromosomes shortens to form a tiny sphere. These four chromosomes then come together in pairs, a process known as synapsis. Then each chromosome is duplicated. Thus the cell will have two groups of four chromosomes each, each group of four being known as a tetrad. The tetrads are divided in a highly unequal cell division that results in the formation of a tiny sphere called "polar body" and a large egg cell. Each contains the diploid number of four chromosomes. These four chromosomes are not separate—they are in pairs. Thus, each tetrad has been divided into two dyads.

At the second meiotic division one observes a key feature of meiosis: The chromosomes are not duplicated. Thus each dyad enters the spindle and, at anaphase, its two chromosomes go to opposite poles. The result is a tiny second polar body, with two chromosomes, and a large ovum, also with two chromosomes. Thus meiosis in the female has, in two divisions, reduced the diploid number of four chromosomes to the monoploid number of two chromosomes. These two very unequal meiotic cell divisions result in essentially all of the yolk, the substance that will permit early development, remaining intact.

Weismann's hypothesis demanded a similar change in the male. When the testis was studied it was found that during early development its cells increase in number by mitotic divisions, that is, each daughter cell has the diploid number of four chromosomes. However, in the mature testis, the last two divisions before the cells differentiate into sperm are meiotic divisions. The four chromosomes synapse, duplicate, and form two tetrads. At the first meiotic division each tetrad is divided, a dyad going to each pole. In contrast with the first meiotic division in the female, two cells of equal size are produced. At the second meiotic division there is no chromosomal replication and the dyads are divided, each daughter cell ending with two chromosomes.

Thus one original diploid cell in the testis, with four chromosomes, will form four cells after the two meiotic divisions—each with two chromosomes, the monoploid number. There are no further divisions of these cells, and each differentiates into a sperm cell. The two meiotic divisions in the male produce four sperm, whereas in the female only a single ovum results. (Sometimes in the female the first polar body also divides; when it does, the two meiotic divisions produce a single ovum and three, rather than two, polar bodies.)

An essential difference between meiosis and mitosis is this: in mitosis there is one duplication of each chromosome for each cell division; in meiosis there is only one duplication of each chromosome for the two subsequent divisions. Mitosis is a mechanism for maintaining constancy of chromosomal number in cell division, whereas meiosis is a mechanism for halving that number in the gametes.

Fertilization

The basic fact of fertilization—namely, that a sperm rather than the seminal fluid is required to initiate development of the ovum—was discovered by J. L. Prévost and J. B. Dumas in 1824. However, the actual role of the sperm was not established by their work. As noted before, George Newport (1854) proved that the sperm penetrate the ova of frogs. It remained for Oskar Hertwig (1876) to show what happens within the egg. He noted, as had others before him, that shortly after fertilization the ova of sea urchins appear to have two nuclei. One of these is situated just under the cell surface. Hertwig suggested that this one was derived from the sperm. The other nucleus was near the center of the ovum. Hertwig suggested that it was the female nucleus. Five minutes after fertilization the putative sperm nucleus moved inward toward the center of the cell. By 10 minutes after fertilization the two nuclei were side by side in the center of the ovum. By 15 minutes there was a single nucleus.

Hertwig believed that he was observing the essential feature of fertilization: the union of a paternal pronucleus formed from the sperm with a maternal pronucleus in the ovum. That union produces the embryo's nucleus that would, by mitotic cell divisions, produce the cells of the new individual.

The roundworm *Ascaris* provided much better material for studying the details of fertilization, again because of its few large chromosomes. Boveri (1888) described the process in detail.

The first illustration, **A,** is of an entire ovum shortly after the sperm has entered. The paternal pronucleus is in the lower right-hand quadrant. The two darkly stained irregular masses are the two chromosomes—two is the monoploid number. The structure forming the wrinkled cap immediately above the paternal pronucleus is the acrosome, which is the portion of the sperm head composed of Golgi material. The dark granular mass in the center of the ovum is the centrosome. It, too, originated from the sperm. There are four black bodies near 12 o'clock. The upper two are the chromosomes of the second polar body. The lower two are the monoploid number of chromosomes of the maternal pronucleus. The second polar body is shown in the sectioned embryos of **B, C,** and **E** as well.

In **B** the maternal and paternal pronuclei have moved closer to one another and their chromosomes have become indistinct. In **C** the chromosomes have elongated greatly and, although we now know there are only two in each pronucleus, this cannot be seen in the illustration (this is a vivid example of the great difficulty cytologists had in coming to the realization that the chromosomes of any one species are constant in number and are individually unique—much of the time they seemed to be as confusing as spaghetti). One can distinguish two dark granules in the centrosome. These are the centrioles.

In **D** the chromosomes have become distinct once again (from **B** through **C** they were going through a modified resting stage) and each pronucleus shows two. The centrosome has divided in two, each with a centriole in the center. This process continues through **E**. In **F** four chromosomes, two from each pronucleus, are lined up on the spindle and shortly thereafter each is seen to be double, that is, be composed of two chromatids. The chromatids will separate to form independent chromosomes, and one will go to each pole.

The form of the mitotic apparatus is shown well in **F**. At each end of the spindle one finds the tiny centriole, surrounded by a dark granular area—the centrosome. In fixed and stained cells fibers are seen to radiate from each centrosome, forming an aster. Other fibers extend from one centrosome to the other, forming the spindle. In **F** the cell is in metaphase of the first embryonic division and the chromosomes are lined up in an equatorial plate.

It became clear from the work of van Beneden, Boveri, and others that each parent transmits the same number of chromosomes to the zygote. Furthermore, the chromosomes in maternal and paternal nuclei appeared to be structurally identical. These observations could help explain the long-held belief that the hereditary contribution of each parent is roughly the same.

This was exciting and important research, and soon many investigators studying a large variety of plants and animals found with very few exceptions that what had been found for the little roundworm Ascaris

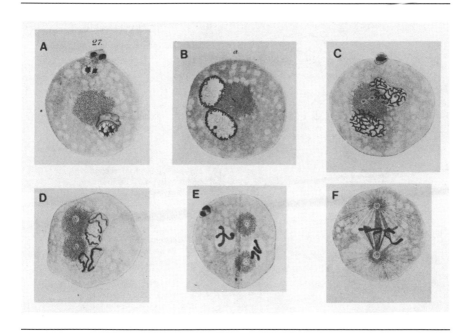

47 *Theodor Boveri's (1888) illustrations of fertilization in the roundworm* Ascaris. *See text for details.*

was true for all other organisms. To be sure there were some minor variations, but an intensive study of these served only to increase the depth of our understanding of the entire process. A concept of universal application had been discovered.

E. B. Wilson in a most astonishing statement, foresaw the future:

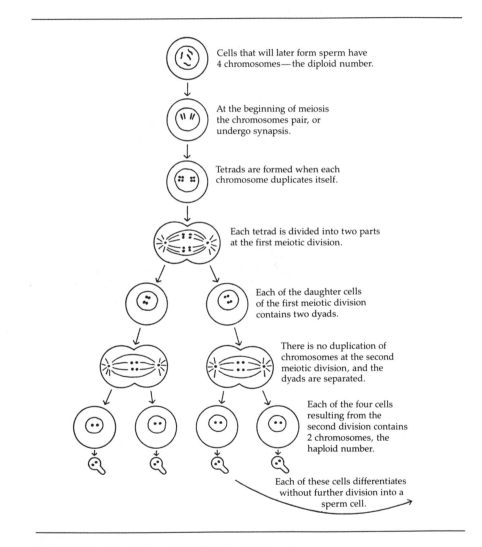

Cells that will later form sperm have 4 chromosomes—the diploid number.

At the beginning of meiosis the chromosomes pair, or undergo synapsis.

Tetrads are formed when each chromosome duplicates itself.

Each tetrad is divided into two parts at the first meiotic division.

Each of the daughter cells of the first meiotic division contains two dyads.

There is no duplication of chromosomes at the second meiotic division, and the dyads are separated.

Each of the four cells resulting from the second division contains 2 chromosomes, the haploid number.

Each of these cells differentiates without further division into a sperm cell.

48 *Outline of meiosis and fertilization in* Ascaris.

These facts justify the conclusion that the nuclei of the two germ-cells are in a morphological sense precisely equivalent, and they lend strong support to Hertwig's identification of the nucleus as the bearer of hereditary qualities. The precise equivalence of the chromosomes contributed by the two sexes is a physical correlative of the fact that the two sexes play, on

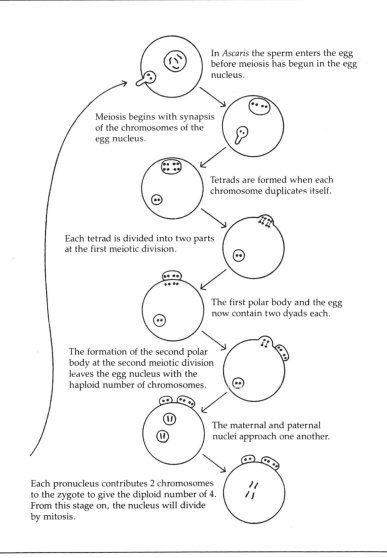

In *Ascaris* the sperm enters the egg before meiosis has begun in the egg nucleus.

Meiosis begins with synapsis of the chromosomes of the egg nucleus.

Tetrads are formed when each chromosome duplicates itself.

Each tetrad is divided into two parts at the first meiotic division.

The first polar body and the egg now contain two dyads each.

The formation of the second polar body at the second meiotic division leaves the egg nucleus with the haploid number of chromosomes.

The maternal and paternal nuclei approach one another.

Each pronucleus contributes 2 chromosomes to the zygote to give the diploid number of 4. From this stage on, the nucleus will divide by mitosis.

the whole, equal parts in hereditary transmission, and it seems to show that the chromosomal substance, the *chromatin*, is to be regarded as the physical basis of inheritance. Now, chromatin is known to be closely similar to, if not identical with, a substance known as *nuclein* (C_{29} H_{49} N_9 P_3 O_{22}, according to Miescher), which analysis shows to be a tolerably definite chemical composed of nucleic acid (a complex organic acid rich in phosphorus) and albumin. And thus we reach the remarkable conclusion that inheritance may, perhaps, be effected by the physical transmission of a particular chemical compound from parent to offspring. (1895, p. 4)

Prophetic yes, but at the time a hypothesis that could not be tested. In reality, attempts to use the data of cytology alone to understand inheritance appeared to have come to a dead end by the close of the nineteenth century. How was one to establish a causal link between the behavior of chromosomes and the data on inheritance derived from breeding experiments? For that matter, the study of inheritance by breeding experiments had come to a dead end as well. Both cytology and what we would now call genetics were awaiting the arrival of a new paradigm. That was to occur, in a most dramatic fashion, in the year 1900.

chapter 14

Mendel and the Birth of Genetics

The year 1900 marks the onset of modern genetics. Then, a modest, unappreciated, and almost forgotten paper written in 1865 by a long-dead Augustinian monk became known to the biological community at large and started a scientific revolution.

The story is familiar to many. Two scientists, Hugo de Vries (1900) and Carl Correns (1900), are credited as being the first to understand the importance of what Mendel had accomplished in his experiments with garden peas. De Vries had crossed numerous "species" and varieties of plants during the 1890s. In those days the term "species" was sometimes applied to varieties of different domesticated plants that we would now consider to be a single species but differing by one or a few alleles with large effects. De Vries adopted the point of view that these different "species" should be considered "as a composite of independent factors" or units, and that,

> The units of species-specific traits are to be seen in this connection as sharply separate entities and should be studied as such. They should be treated as independent of each other everywhere, as long as there is no basis for doing otherwise. In every crossing experiment only a single character or a definite number of them is to be taken into consideration. (Stern and Sherwood 1966, p. 108)

This was an important way of thinking about the physical basis of inheritance. Instead of inheritance being based on some vague substance or force, it was hypothesized to be due to discrete, nonblending factors. De Vries spoke of these specific traits that affected the same structure, and were inherited, as antagonistic characters. He noted that only one was expressed in the hybrid offspring, or, as we would now say, in the F_1 or first filial generation. Nevertheless, when the pollen and ovules were formed, "the two antagonistic characteristics separate, following for the most part simple laws of probability" (Stern and Sherwood, p. 110).

De Vries stated that his basic conclusions had been reached before he knew of Mendel's work. The story of how de Vries came to know of Mendel's paper is of considerable interest. He did not uncover it through a "literature search" but by one of those extraordinary chance events that seem to be of such importance in scientific discovery. A fellow Dutch scientist, Professor Beyerinck of Delft, knew that de Vries had been hybridizing plants and wrote asking if he would be interested in an old reprint dealing with the same subject. It was Mendel's "Versuche über Pflanzen Hybriden," which reached de Vries in 1900, just as he was preparing to publish his own experiments. He was able to do so knowing that he was confirming Mendel's earlier, and more extensive, experiments.

The story about Correns is equally interesting. He also had been performing genetic experiments with plants and was trying to develop a hypothesis to account for the data. In the autumn of 1899 the solution came to him in a "blind flash," which, more often than not, seems to be the origin of the truly important breakthroughs in science. A short time later he found a reference to Mendel's paper and looked it up. He published his own data and showed how it confirmed what Mendel had found.

Perhaps it is time for us, also, to see what Mendel had done. Gregor Mendel's famous paper is not a scientific report in the usual sense, but instead consists of the lectures that he presented to the Natural History Society of Brünn in 1865. The full data were never published, but the portion that he did include, coupled with his extraordinary analysis of the data, puts his contribution in the same class as *On the Origin of Species*.

Mendel was fully aware that experiments in plant breeding, usually called hybridization, had been conducted for years by many famous scientists. No general rules had emerged, as we have already seen from Darwin's lack of success in *Variation*, published only two years after Mendel's paper. Mendel had started his experiments trying to understand inheritance shortly after the publication of Darwin's *Origin*, and one of the reasons for so doing was the need for "reaching the solution to a question whose significance for the evolutionary history of organic forms must not be underestimated." Thus, Mendel's work started out as normal science within the paradigm of the theory of evolution. Only later was it to become the beginning of a new paradigm—Mendelian genetics. This is an important point: how a discovery in one field of science may be of great importance in another.

Plant hybridizers of the mid-nineteenth century had a wealth of readily available material. Numerous varieties of the same species of both food and ornamental plants had been selected. Many of the varieties were very different from one another—so different that they might be given their own scientific names. Once varieties had been developed, continued selection was practiced so they would "breed true."

Mendel decided to work with garden peas and started with 34 varieties. He grew the separate varieties for two seasons to make sure that they bred true. Finally he reduced the number to 22 varieties.

Peas had important advantages. Not only were many varieties available, but they were easy to grow and had short generation times. The offspring obtained by crossing the varieties were fertile. The structure of the flower was also important. The stamens and pistils are enclosed by the sepals and petals and, if the flowers are covered to prevent insects from reaching them, they self-fertilize, that is, pollen falls on the stigma of the same flower.

Nevertheless, experimental crosses could be made. This was done by removing the anthers before they matured and, later, placing pollen from another plant on the stigma. Thus, Mendel could cross any of his varieties; if he covered the flowers alone, the next generation would be a consequence of self-fertilization.

Some of his varieties had round seeds and in others the seeds were wrinkled; some of his varieties had yellow seeds and in others they were green. In all he used 7 pairs of contrasting characters as follows:

Character affected	Varieties
Seed shape	*round* or *wrinkled*
Seed color	*yellow* or *green*
Seed coat color	*colored* or *white*
Pod shape	*inflated* or *wrinkled*
Pod color	*green* or *yellow*
Flower position	*axial* or *terminal*
Stem length	*long* or *short*

These different sorts of peas were crossed by removing the immature anthers from the flowers of one variety and placing pollen from the other variety on the stigma. In the resulting first filial generation, the F_1, all of the offspring of any cross were the same and all resembled one of the parents. If *round* and *wrinkled* were crossed, all of the F_1 were *round*. There was no blending of the contrasting characteristics. Mendel spoke of the characteristics that appeared in the F_1, in this case *round*, as being *dominant;* the characteristic that did not appear, in this case *wrinkled,* he spoke of as *recessive.*

These results, so familiar to us today, were rather unexpected in the 1860s. The general rule was that the F_1 individuals tend to be intermediate. And in most cases they are, for the simple reason that, if varieties differ in many ways, the F_1 will usually be more or less intermediate. But Mendel concentrated on the inheritance of details, not all of the differences. In a sense he forgot the whole plant and asked only if the peas had *round* or *wrinkled* seeds.

But a puzzle remained. How could one account for the fact that if a cross is made between a *round* and *wrinkled* parent, all of the F_1 individuals were *round*? Did *round* in the parent and *round* in the offspring have the same genetic basis? There was an experimental way to answer the question. Mendel knew that the *round* parent bred true, that is, always gave *round* offspring when allowed to self-fertilize. They were genetically pure for *round.*

What would happen if the *round* F_1 hybrids were allowed to self-fertilize in order to ascertain the genetic basis of their roundness? So the flowers of the F_1 plants were covered to protect them from being cross-pollinated by insects and allowed to "self." The results were surprising. Both *round* and *wrinkled* appeared in the second filial generation, the F_2.

Most plant breeders would have reported only that both varieties appeared in the F_2 and that *wrinkled* had "skipped a generation." But Mendel did a simple and revolutionary thing. He *counted* the numbers of individuals with each characteristic. There were 5,474 *round* and 1,850 *wrinkled* in the many crosses of this sort.

Mendel made similar crosses with each of the seven pairs of contrasting characters, and the general results were the same: both dominant and recessive individuals, but never intermediates, appeared in the F_2, and there was a great excess of dominants. For example the F_2 of *yellow* × *green* crosses produced 6,022 *yellow* and 2,001 *green.*

Mendel the mathematician noticed the similarity of the ratios of dominant to recessive in the F_2 of these monohybrid crosses (that is, crosses in which only a single pair of contrasting characteristics is followed). In the *round* × *wrinkled* crosses it was 2.96 to 1 and in the *yellow* × *green* crosses 3.01 to 1. Mendel had reason to suspect that the theoretical ratio was 3 to 1 and not 2.96 to 1 or 3.01 to 1. Or we might say 3/4 (75 percent) showed the dominant characteristic and 1/4 (25 percent) the recessive characteristic.

When Mendel followed the inheritance of two pairs of contrasting characteristics, the dihybrid cross, uniform results were obtained once more. The F_1 exhibited the two dominant characteristics only, and the F_2 exhibited all four characteristics in a 9:3:3:1 ratio. That is, 9/16 of the F_2 showed both dominant characteristics, 3/16 showed one dominant and one recessive, 3/16 showed the other dominant and other recessive, and 1/16 had both recessive characteristics.

Thus, if the original P (parental) generation cross had been *round-yellow* × *wrinkled-green*, all of the F_1 would be *round-yellow*. In the F_2 he obtained 315 *round-yellow*, 108 *round-green*, 101 *wrinkled-yellow*, and 32 *wrinkled-green*. For this total of 556, the ratios of different kinds are 9.8:3.4:3.2:1. Those ratios are derived from the data, but Mendel proposed a hypothesis that suggested that in a theoretically ideal experiment the ratios would be 9:3:3:1.

Again the mathematician Mendel saw how a 3:1 ratio is related to the 9:3:3:1 ratio. When two, or even more, pairs of contrasting characteristics are involved, one must recognize that the 3 to 1 ratio still holds for the individual characteristics. In the cross already discussed that gave 9:3:3:1 ratio for two pairs of characteristics, the ratio is still 3:1 for the single characteristics. Consider the cross that gives 9/16 *round-yellow*, 3/16 *round-green*, 3/16 *wrinkled-yellow*, and 1/16 *wrinkled-green*. Looking at *round* and *wrinkled* separately, we find 9/16 + 3/16 = 12/16 that are *round* and 3/16 + 1/16 = 4/16 that are *wrinkled*. Since 12/16 = 3/4 and 4/16 = 1/4, we observe a 3:1 ratio for the single pair. The same holds true for the *yellow* and *green* pair.

If we then ask what are the frequencies of the different combinations of characteristics in the F_2 that result from a dihybrid cross, the answer comes from a simple multiplication of the fractions for the separate characters. Thus, of the 3/4 that will be *round*, 3/4 of them will also be *yellow* and 1/4 will also be *green*. Therefore 3/4 × 3/4, or 9/16, will be both *round* and *yellow*; 3/4 × 1/4, or 3/16, will be both *round* and *green*. Of the 1/4 that are *wrinkled*, 3/4 will also be *yellow* and 1/4 will also be *green*. Therefore 1/4 × 3/4, or 3/16, will be *wrinkled-yellow* and 1/4 × 1/4, or 1/16, will be *wrinkled-green*. That is the derivation of the 9:3:3:1 ratio.

These striking regularities were observed by Mendel in all of the crosses. He assumed that there must be some underlying principle.

Model for Monohybrid Crosses

There was. Figure 49 is a model for the explanatory hypothesis that Mendel proposed to account for monohybrid crosses. Both the scheme and the terminology were to become standard half a century later in the early 1900s.

Several useful terms will be used to describe genetic crosses. *Genotype* designates the genetic makeup, in symbols, of an individual or

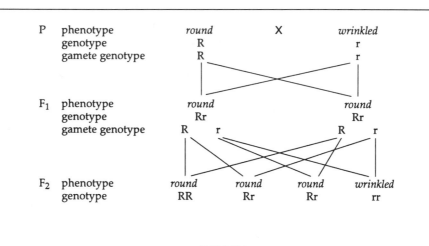

49 *Model for a Mendelian monohybrid cross. The genotypes of the P (parental) generation are as Mendel would have shown them. The genetic checkerboard at the bottom shows the origin of the F₂ (genotypes as we would show them today) from the F₁ pollen and ovules.*

gamete—for example, **R** or **r**. *Phenotype* refers to the expression of the genotype in the individual or part—*round* or *wrinkled* seeds. The hereditary factors are now called *genes*. Thus there is a gene for seed shape, and in Mendel's example it existed in two states, **R** or **r**. The different states of a gene are spoken of as *alleles*. Thus **R** and **r** are alleles of the gene for seed shape. In practice, however, the term gene is often used as a synonym for allele. Thus we speak, with some loss of precision, of the **r** or **R** genes. Upper case is used for dominant alleles; lower case is used for recessive alleles.

The first thing that a knowledgable reader is likely to note is the "error" in the genotypes of the P generation. They are shown as monoploid, that is, as **R** and **r**, instead of diploid, **RR** or **rr**. The reason is that Mendel used the symbols for genotype to indicate *kinds* of hereditary factors, not *number* per cell.

There can be only one type of offspring, **Rr**, since there is only one type of pollen and one type of ovule in this cross. When these F_1 plants mature, each flower will produce ovules and pollen. Now comes one of the most important features of Mendel's model: he assumed that a gamete could have hereditary factors of only one kind, that is, in this cross a gamete would have **R** or **r** but not both. This "purity of the gametes" presented a very difficult problem for geneticists in the early 1900s. Their minds had been influenced by the concept of innumerable gemmules. How could a gamete be "pure," that is, have only **R** or **r** gemmules?

Mendel next assumed that ovules and pollen grains would combine at random and, consequently, the kinds of offspring would be in frequencies determined by the frequencies of the different kinds of gametes.

It must be emphasized that the apparent simplicity of the scheme shown in figure 49 works only because each F_1 produces 50 percent **R** pollen or ovules and 50 percent **r** pollen or ovules. The lines are so drawn that all possible combinations occur in equal frequency. The **R** gamete on the left, for example (let's pretend that it is a pollen grain), has an equal chance of combining with either an **R** or an **r** ovule. The same is true for the **r** pollen.

The genetic checkerboard at the bottom of figure 49 is another conventional scheme to show Mendel's hypothesis for the 3:1 ratio in the F_2 of a monohybrid cross. The model applies to *all* of the crosses involving a single pair of contrasting characters, yet it will hold true only if the following conditions are met:

(1) In each pair of contrasting hereditary units, one member of the pair is dominant and the other recessive. Dominance and recessiveness are operational definitions—determined experimentally by examining the phenotype of an individual that has both types of hereditary units.

(2) The dominant and recessive hereditary characteristics do not modify one another in any detectable way when they exist together. Thus, in the F_1 the r alleles from the *wrinkled* parent in the cross of figure 49 are combined with R alleles from the *round* parent. There is no expression of the r allele in the F_1, but in the F_2 one quarter of the individuals are *wrinkled*—and just as wrinkled as the *wrinkled* grandparent.

(3) By some mechanism unknown to Mendel the two sorts of alleles in the F_1 segregate in such a manner that each gamete contains only one kind of allele—they are "pure." In the example being discussed, the gametes will contain R or r.

(4) By still another unknown mechanism the R gametes and the r gametes are produced in equal numbers.

(5) Combinations between the pollen and ovules are entirely at random, and the phenotypic frequencies of the offspring will depend on the frequencies of the different classes of gametes.

It should be emphasized that the congruence of data and model is not fortuitous. Although items 1 and 2 above could be accepted as true, items 3–5 were entirely hypothetical—they were invented to explain the data. This is a perfectly acceptable scientific procedure. The model is to be regarded as a hypothesis that will stand or fall on the basis of tests of deductions made from it.

One critical test was easy to make. The F_2 individuals in the figure 49 cross consist of 3 plants showing the dominant *round* characteristic for every 1 showing the recessive *wrinkled* characteristic. However, if the hypothesis is true, the *round* seeds must be of two sorts and in a ratio predictable from the hypothesis. Thus, for every seed that has the R genotype, still using the original Mendelian scheme, there will be two that are Rr.

There is no way of distinguishing visually between an R and an Rr seed, but if the seeds are planted and the flowers allowed to self-fertilize, the offspring will give the answer. Thus the R genotype should breed true and the Rr genotype should give a 3:1 ratio of *round* to *wrinkled*. Mendel planted the seeds and found this to be true.

Model for Dihybrid Crosses

Figure 50 shows the model for the dihybrid cross discussed before. A pure-breeding *round-yellow* plant is crossed with a pure-breeding *wrinkled-green*. The F₁ individuals are uniform—showing the dominant phenotype of both pairs of contrasting characters.

In the formation of gametes by the F₁, Mendel assumed, as for the monohybrid cross, that each gamete would receive only one type of the two contrasting units—either **R** or **r**. The same was assumed to be true

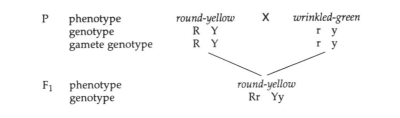

P	phenotype	*round-yellow*	X	*wrinkled-green*
	genotype	R Y		r y
	gamete genotype	R Y		r y

| F₁ | phenotype | *round-yellow* |
| | genotype | Rr Yy |

<table>
<tr><td colspan="5" align="center">POLLEN</td></tr>
<tr><td></td><td>RY</td><td>Ry</td><td>rY</td><td>ry</td></tr>
<tr><td>RY</td><td>RR YY
round-yellow</td><td>RR Yy
round-yellow</td><td>Rr YY
round-yellow</td><td>Rr Yy
round-yellow</td></tr>
<tr><td>Ry</td><td>RR Yy
round-yellow</td><td>RR yy
round-green</td><td>Rr Yy
round-yellow</td><td>Rr yy
round-green</td></tr>
<tr><td>rY</td><td>Rr YY
round-yellow</td><td>Rr Yy
round-yellow</td><td>rr YY
wrinkled-yellow</td><td>rr Yy
wrinkled-yellow</td></tr>
<tr><td>ry</td><td>Rr Yy
round-yellow</td><td>Rr yy
round-green</td><td>rr Yy
wrinkled-yellow</td><td>rr yy
wrinkled-green</td></tr>
</table>

OVULES (labeled to the left of the checkerboard)

50 *Model for a Mendelian dihybrid cross. The genotypes of the P generation are as Mendel would have shown them, not the current practice, which is* **RR YY** *and* **rr yy**. *The genetic checkerboard shows the origin of the F₂ (genotypes as we would show them today) from the F₁ pollen and ovules.*

for **Y** and **y**. At this point, still another assumption had to be made: there would be an independent assortment of both pairs. Thus each gamete would have either **R** or **r** and, in addition, either **Y** or **y**. There would then be four classes of both pollen and ovules: **RY, Ry, rY,** and **ry**. The model also demanded that these classes be in equal frequency —25 percent for each.

A simple game with two different coins—a penny and nickel perhaps—may help to understand the origin of the four classes of gametes. Let each coin represent a gene and the "head" represent the dominant allele and the "tail" the recessive allele. Both coins are tossed and the results recorded. If enough throws are made, one expects 1/4 to be heads for both coins (the **RY** category above); 1/4 will be tails for both coins (the **ry** category above); 1/4 will be heads for the penny and tails for the nickel (= **Ry**); and 1/4 will be tails for the penny and heads for the nickel (= **rY**).

When there are four classes of gametes, it is not practical to use lines to show the possible combinations, as in figure 49. It is more convenient to use a checkerboard as in the lower part of figure 50. This shows all possible combinations of pollen and ovules that produce the F_2. All genotypes are now shown as diploid. There are 16 boxes in the checkerboard and, if the phenotypes are combined, we find that 9 of the 16 are *round-yellow*, 3 are *round-green*, 3 are *wrinkled-yellow*, and 1 of the 16 is *wrinkled-green*. That is a 9:3:3:1 ratio.

Notice that only *round-yellow* and *wrinkled-green* phenotypes were present in the P and F_1 generations. The model demands, however, that two *new* types of seeds appear: *round-green* and *wrinkled-yellow*— and they do.

Here are the data that Mendel reported for the F_2.

	Actual	*Expected*
round-yellow	315	313
round-green	108	104
wrinkled-yellow	101	104
wrinkled-green	32	35

The "actual" numbers are the counts of the seeds. The "expected" are what the numbers would be in a perfect 9:3:3:1 ratio. The reason for the close correspondence of the actual and expected numbers is that, for the purposes of illustrating his lectures, Mendel selected the experiments that gave results close to those predicted by his model.

The model for the dihybrid cross allowed even more elegant testing of the hypothesis. The hypothesis predicted that, except for the 32 *wrinkled-green* seeds, all other classes although phenotypically uniform, would consist of genetically different individuals. This could be tested by planting the F_2 seeds, allowing the plants to self-fertilize, and then counting the F_3 seeds.

Consider first the 32 *wrinkled-green* seeds. The model predicts that these will breed true if allowed to self-fertilize. The seeds were planted and 30 grew. All proved to be *wrinkled-green*.

The 101 *wrinkled-yellow* seeds were identical so far as the eye could tell. We can see from the model in figure 50 that 3/16 are in this category but two genotypes are represented: 1 of the three is **rrYY** and the other 2 are **rrYy**. Thus 1 of every 3 seeds of the **rrYY** class would be expected to breed true and produce only *wrinkled-yellow*. Two of the 3 in the **rrYy** class would be expected to produce offspring in a ratio of 3 *wrinkled-yellow* to 1 *wrinkled-green*. The 101 seeds were planted and 96 grew. Of these 28 (32 expected) produced all *wrinkled-yellow* and 68 (64 expected) produced *wrinkled-yellow* and *wrinkled-green* in a ratio of 3 to 1. Thus the deduction from the hypothesis was found to be true.

The same analysis was done for *round-green*. The 3/16 belonging to this class were predicted to consist of 1 **RRyy** for 2 **Rryy**. The 1/3 that were **RRyy** should breed true. The 2/3 that were **Rryy** should produce seeds in a ratio of 3 *round-green* and 1 *wrinkled-green*. The 108 seeds were planted and 102 grew. One would have expected 34 to breed true and 68 to give the 3:1 ratio. The actual numbers were 35 and 67.

The most complex test of the hypothesis was based on the 9/16 of the F_2 that were *round-yellow*. The checkerboard shows that 1 of the 9 is **RRYY**, 2 are **RRYy**, 2 are **RrYY**, and 4 are **RrYy**. Thus, only one of the 9, the **RRYY** class, should breed true. The **RRYy** should give offspring in a ratio of 3 *round-yellow* to 1 *round-green*. The **RrYY** should produce 3 *round-yellow* to 1 *wrinkled-yellow*. And finally the **RrYy**, which are the same as the F_1 in figure 50, should give a 9:3:3:1 ratio. The 315 seeds were planted and 301 produced a crop. The model predicts that the actual numbers in each class (in the order just listed) should be 33, 67, 67, and 134. Mendel found the actual numbers to be 38, 65, 60, and 138.

In every case the actual numbers are very close to the expected numbers. The expected numbers are based on the probability of the gametes behaving according to strict rules of the model. The expected and actual numbers were never identical. We should not expect them to be so any more than we should always expect to get 5 heads and 5 tails for every 10 tosses of a coin.

The fact that the F_2 gave an F_3 that did not differ significantly from the hypothesis in these rather demanding tests of the deductions is strong support for the validity of the hypothesis.

Mendel's Laws

Mendel's experiments on crossing varieties of peas and his remarkable analysis of the data permit the following eight major conclusions. In 1865 the conclusions were for peas and peas only. To be sure, Mendel had made some preliminary crosses with beans, but the results were confusing.

(1) The most important conclusion is that inheritance appears to follow definite and rather simple rules. Mendel proposed a model that would account for the data of all his crosses. Furthermore, the model had great predictive value—a goal of all hypotheses and theories of science.

(2) When plants of two different types are crossed, there is no blending of the individual characteristics. Of the seven pairs of contrasting characteristics, one type was dominant and the other recessive. That is, in a hybrid formed by crossing a pure-breeding plant having the dominant characteristic with a pure-breeding plant having the recessive characteristic, the offspring are all the same and phenotypically identical with the dominant parent.

(3) Since the hybrid described above is identical in appearance with the pure-breeding dominant parent, we can conclude that there is not an exact relation between genotype and phenotype. Thus, the *round* phenotype can be based on either an **RR** or **Ry** genotype.

(4) The hereditary factors responsible for the dominant and recessive condition are not modified by their occurrence together in a hybrid. If two such hybrids are crossed—or, in the case of peas, an F_1 hybrid is allowed to self-fertilize—both dominant and recessive-appearing offspring will be produced. These offspring show no evidence that the hereditary factors responsible for their appearance have been modified by their association in the same individual.

(5) When hybrids such as **Rr** are crossed, the two types of hereditary units—**R** and **r**—segregate from one another and at fertilization recombine at random. The offspring will be in a phenotypic ratio of 3:1 and

genotypically there will be, using the modern convention for genotypes, 1 **RR**, 2 **Rr**, and 1 **rr**. Segregation is called *Mendel's first law.*

(6) This ratio can occur only if each gamete receives only one type of hereditary factor—in the example, either **R** or **r**.

(7) When crosses involve two pairs of contrasting hereditary units, such as **Rr Yy** crossed with **Rr Yy**, each pair behaves independently. That is, the different types of hereditary units assort independently of one another, so the gametes can be only **RY, Ry, rY,** or **rr**. Thus all possible combinations will be obtained, with the strict rule that each gamete can have only one kind of each of the pairs of hereditary units. The different classes of gametes will be in equal frequency. This phenomenon of independent assortment is known as *Mendel's second law.*

(8) The Mendelian hypothesis, and its formulation in a model, was so specific that deductions could be made and these could be tested by observation and experiment.

The relation of Mendel's model to his hypothesis is subtle. The model is not "fact" but an attempt to explain the data obtained by observation and experiment. He observed regularities, for example, 3:1 and 9:3:3:1 ratios. The regularities implied that specific underlying natural processes must be at work. He tried to imagine how such ratios could be achieved and developed a hypothesis for what the rules governing inheritance might be. One of these was that a pollen grain or an ovule would have only one allele, **R** or **r**, but never both. Another regularity was that the gametic genotypes were in definite proportions and that fertilization was random. He had no independent evidence that any of these hypothetical rules were true. The rules were concocted to account for the data. He was able to test the validity of the rules by a wide variety of experimental crosses. These tests established that the rules were true beyond all reasonable doubt for the varieties of peas that he used, but they provided no basis for understanding the cellular mechanisms involved.

No other field of experimental biology had reached an equivalent stage of development in 1865. But, as we have seen, no biologist at that time seemed to realize that such was the case. To be sure, Mendel's work would not have been important if it had applied only to garden peas—any more than Hooke's discovery of cells would have been important had cells been observed only in cork. The field of plant breeding was full of data from which no general conclusions could be drawn. Mendel wrote to a foremost scholar in the field, Karl Wilhelm von Nägeli, and explained his results. Nägeli was not impressed and must have regarded the data for peas as just one more example of the tremendous variation in the results obtained in hybridization experiments.

Nägeli suggested that Mendel try another plant—*Hieracium*, the hawkweed. Mendel did and failed to find consistent rules for inheritance. The problem was with *Hieracium*, not Mendel. It was exceedingly difficult to make experimental hybrids in *Hieracium* with its tiny flowers. Nevertheless Mendel thought he had done so in many instances and was surprised at the lack of uniformity in the results. Long after Mendel's death it was discovered that a type of parthenogenetic (asexual) development occurs in *Hieracium*. No uniform ratios are to be expected if some of the offspring are the result of fertilization and others of parthenogenesis.

So even Mendel came to believe that his results had a restricted application, and, in any event, his model was ignored throughout the last third of the nineteenth century. During those decades the leading students of heredity had abandoned the paradigm of experimental breeding and concerned themselves mainly with the behavior of chromosomes in meiosis, mitosis, and fertilization. They believed that they were laying a physical basis for inheritance, and further research was to prove them correct.

Much is usually made of the fact that Mendel's seminal work had been published in an obscure journal of an obscure society, so that it was either forgotten or unknown for 35 years—a 35 years that saw the flowering of cytology and intense interest in its possible relation to inheritance. A more accurate statement would be, I suspect, that the paper was unappreciated rather than unknown. It was known to Focke (1881), who discussed it briefly in his standard treatment of plant hybridization, and it was mentioned later by Bailey (1895), another prominent plant hybridizer. As already noted, Mendel had corresponded with Nägeli, one of the most prominent students of heredity at the time, who was unimpressed with Mendel's data and analysis.

Mendel's work on peas is not an isolated example of an important discovery being made but not understood by the scientific community at the time it was announced. New paradigms are not readily identified and adopted. The difficulty in changing what one does with hand and mind promotes resistance to new ideas and to the undertaking of new research programs.

This was not a problem for de Vries and Correns in 1900. They understood the importance of Mendel's conclusions because they had done similar work and had developed a similar explanatory hypothesis before they read Mendel's paper. They were working in the new paradigm before they knew of their paradigmatic progenitor.

The same point can be made for William Bateson (1861–1926). He had been studying variation and hybridization for years, and although he had not observed the regularities of the Mendelian model, he knew what sorts of experiments needed to be done. Consider the following:

On Tuesday and Wednesday, July 11 and 12, 1899, the Royal Horticultural Society held an International Conference on Hybridisation (the Cross-Breeding of Species) and on the Cross-Breeding of Varieties at Chiswick and London. Volume 24 of the Society's journal reports on the conference. Thus we have the opinions of many of the world's outstanding plant hybridizers immediately before Mendel changed their science. Most of the articles in the journal describe the results of crosses, but Bateson (1900) gave a more theoretical talk. This is part of what he had to say:

> What we first require is to know what happens when a variety is crossed with its *nearest allies*. If the result is to have a scientific value, it is almost absolutely necessary that the offspring of such crossing should then be examined *statistically*. It must be recorded how many of the offspring resembled each parent and how many showed the characters intermediate between those of the parents. If the parents differ in several characters, the offspring must be examined statistically, and marshalled, as it is called in respect of each of those characters separately.

It is almost as though Bateson is advising a graduate student, by the name of Mendel, how to plan his Ph.D. research program!

There are many interesting aspects of the story about Mendel. One is the almost universal attention that is given to the scientist who makes the discovery. Until recently scientists, especially biologists, could hardly expect to "make their fortune" as scientists. The rewards to a scientist come from the joy of probing nature for its regularities and the approval of one's peers for research well done and for formulating bold and imaginative hypotheses. To this day scientists look at the Mendelian paper in awe. How could he have gone so far beyond the existing paradigm and made observations that were, well after his death, to revolutionize the biological sciences?

Another interesting point for students is that, time and time again, it seems that when the field is "ready" the discovery will be made. If Mendel had never lived, the history of genetics probably would not have been greatly different. About the year 1900 someone or another would have reached similar conclusions. It just happened that it was de Vries and Correns. Tschermak was so close that he is usually included with de Vries and Correns as a codiscoverer. In a year or so Bateson might have independently discovered the Mendelian rules for inheritance. Sometimes (but certainly not always) there seems to be an element of inevitability in the progress of science.

Initial Opposition to Mendelism

This telling of the Mendelian story may imply that, in 1900, with the publication of the papers of de Vries and Correns, "pure science" had finally triumphed. Not at all. There was vigorous, at times vitriolic, opposition to Mendel's conclusions. This scientific donnybrook mainly involved four Englishmen—William Bateson, a firm believer in Mendelism versus Karl Pearson, Francis Galton, and W. F. R. Weldon, who were biometricians and anti-Mendelians. The two schools were fundamentally different in their approaches. Bateson sought information about inheritance from experimental crosses. Weldon, Galton, and Pearson sought to apply mathematical, and especially statistical, methods to biological problems. The opposition of these biometricians is all the more surprising when we remember that Mendel had relied so heavily on mathematics.

The consequence was that Bateson and the breeders continued to perform experiments that showed to what extent Mendel's principles could be extended to other species, and Weldon and others continued to point out that not all could be explained by the original Mendelian hypothesis.

Weldon (1902) summarized Mendel's conclusions and subjected his ratios to statistical tests. He concluded that, "if the experiments were repeated a hundred times, we should expect to get a worse result about 95 times, or the odds against a result as good as this or better are 20 to 1" (p. 235). Years later still another mathematically inclined scientist, R. A. Fisher (1936), was to deal with the problem of Mendel's data being "too good." It seemed impossible to him that the expected and actual data could be so similar. Nevertheless, others repeated the crosses and found the same ratios. Six other plant breeders between 1900 and 1909 attempted to check Mendel's results (Sinnott and Dunn 1925, p. 47). In the case of the *yellow* × *green*, for example, the total number of seeds scored was 179,399. Of these, 134,707 were *yellow* (75.09 percent) and 44,692 (24.91 percent) were *green*. Mendel had reported 75.05

percent *vs* 24.95 percent. Apparently it was not all that difficult to obtain data that were "too good." It bears repeating that Mendel never published his full data. His 1865 paper was based on lectures and it would have seemed reasonable for him to select the data from those crosses that best illustrated the hypothesis he was proposing.

The debates between those who accepted Mendelism from the start and those who did not are instructive in showing that even a discovery as important as Mendel's and as well grounded in data and brilliant analysis need not be easily accepted by prominent (and aging) scientists. William Bateson, the defender, was able to make this shift quickly, but a generation later he too would resist a new paradigm—one suggesting that the Mendelian model was understandable in terms of chromosomal mechanics. In protecting and advancing Mendelism in its infancy, Bateson played a role similar to that of his countryman Thomas Henry Huxley, who, a half-century earlier had been vigorous and effective in the defense of Darwinism.

We will return to the development of Mendelian genetics shortly; but, also in 1902, a paper was published that was to unite the fields of animal and plant breeding with cytology. Thus the twin approaches to the study of inheritance were to become linked and mutually supportive—and require concordance.

chapter 15

Genetics + Cytology: 1900–1910

Two of the most important, and debated, questions in cytology in 1900 were whether or not chromosomes are permanent cell structures and whether or not those in a single cell are all the same or different from one another. It is clear to us today that these questions had to be answered before there could be notable progress in our understanding of inheritance.

In 1900 these questions had not been decided to the satisfaction of all. The "disappearance" of the chromosomes when the nucleus of a just-divided cell entered the resting stage presented a serious problem for those who believed in the permanence and individuality of chromosomes. The most obvious interpretation was that chromosomes were temporary structures, forming *de novo* at the onset of mitosis and disintegrating at its end. Others believed chromosomes were permanent and that as they entered the resting stage they joined, end to end, to form a continuous spireme. The spireme was thought to fracture into chromosomes at the onset of the next mitotic division. But did it fracture at the same place in each mitotic division and thereby maintain the individuality of chromosomes?

In the second edition of E. B. Wilson's *The Cell* (1900, pp. 294–304) evidence was given that supported the hypothesis of some type of permanence and individuality of the chromosomes. He notes that Rabl's observations, made in 1885, were evidence that "the chromosomes do not lose their individuality at the close of division, but persist in the chromatic reticulum of the resting nucleus." Wilson cites the studies of Boveri, van Beneden, and others on *Ascaris* as demonstrating that "whatever be the number of chromosomes entering into the formation of a reticular nucleus [a resting nucleus], the same number afterward issues from it." The best evidence for this was provided by *Ascaris* where, at the end of telophase, the nuclear membrane forms lobes that surround the ends of the chromosomes. These lobes persist and

at the succeeding division the chromosomes reappear exactly in the same position, *their ends lying in the nuclear lobes as before* . . . On the strength of these facts Boveri concluded that the chromosomes must be regarded as "individuals" or "elementary organisms," that have an independent existence in the cell. Boveri expressed his belief that "we may identify every chromatic element arising from a resting nucleus with a definite element that entered into the formation of that nucleus, from which the remarkable conclusion follows *that in all cells derived in the regular course of division from the fertilized egg, one-half of the chromosomes are of strictly paternal origin, the other half maternal.*"

The last statement was a reasonable conclusion to make from the available data. Recall that in *Ascaris* each pronucleus contributes two chromosomes to the diploid complement of four. If each replicates exactly for each mitotic division during development, every body cell must have chromosomes identical with those in the female and male pronucleus. But there were awesome "ifs," and this explanation remained a hypothesis awaiting further testing.

Wilson was assembling evidence to support this hypothesis, but many cytologists were not convinced that chromosomes had permanence or individuality. Their skepticism was appropriate. It is interesting to note how slim the evidence was—those lobes on the nuclear membrane of *Ascaris* being about the best. One of the most influential cytological studies dealing with these questions at the turn of the century was that of T. H. Montgomery (1901). He made a detailed investigation of sperm formation and egg formation in numerous species of *Hemiptera* (true bugs). The conclusions were not only important in their own right but, building on them, others were able to make important conceptual advances in our understanding of inheritance.

The species of *Hemiptera* are ideal from several points of view. The chromosomes are not overly numerous, they often differ from one another structurally, and the species are easily collected. One of the most important features, however, is the organization of the testes. The immature cells are at one end and, as each one passes along the organ, the various stages in spermatogenesis occur in sequence, ending with mature sperm. In a single testis, therefore, one can study the entire process of meiosis and sperm formation.

Montgomery amassed data and interpreted them to show that chromosomes are permanent cell structures; that they exist in homologous pairs consisting of one originally inherited from the mother and the other from the father; that synapsis during meiosis consists of the coming together of these homologous chromosomes; that in meiosis each spermatid receives one chromosome of each type. He described accessory chromosomes that later investigators were to associate with sex determination.

Montgomery was 28 when his classic paper was published—almost the same age as W. S. Sutton, who made the next major advance in relating the data of cytology and genetics. Both died before they were 40.

Sutton's Model

Looking retrospectively at the conceptual development of the science of genetics, we can recognize 1902 as a year of momentous events. The young Walter Stanborough Sutton (1877–1916) was to demonstrate in a paper that year, and in a second one in 1903, that there is an exact parallel in the behavior of the Mendelian hereditary units and of the chromosomes in meiosis and fertilization. The most economical hypothesis (applying Occam's razor), therefore, is that the hereditary units are parts of the chromosomes. Alternatively, the hereditary units might be parts of unknown cell structures that behave exactly like chromosomes in meiosis and fertilization. With that choice of hypotheses, it would be more practical to study the known, chromosomes, rather than the unknown.

Two of the premises of Sutton's hypothesis were those supported so strongly by Montgomery and Wilson, in whose laboratory at Columbia he was a student, namely, that chromosomes persist in some form during the nuclear cycle, that is, they are permanent structures and, furthermore, that chromosomes have individuality (that is, as we now realize, each pair of homologous chromosomes has a unique cluster of genes).

Sutton's 1902 paper was a study of the chromosomes in the testis of a grasshopper of the genus *Brachystola,* which provided additional support for the hypotheses that chromosomes are permanent cell structures and that there are differences among them. But how could one ascertain these things by looking at slides? Sutton was working long before it became possible to study the fine structure of chromosomes. When in mitosis or meiosis they were seen as deeply stained, opaque structures. The only practical means of identification was chromosomal size and shape. Even this was beset with problems, since the chromosomes change in size during mitosis, beginning as long, delicate threads in prophase and becoming short and thick by metaphase. His solution was to use relative sizes, since the chromosomes seemed to change their sizes synchronously.

The spermatogonial cells, which are those that will eventually differentiate as sperm, undergo a series of mitotic divisions in the testis before the onset of meiosis. The youngest spermatogonia have 23 chromosomes. One of these is the so-called "accessory chromosome" that had been observed by Montgomery and others in some species and was a puzzle. Neglecting the accessory chromosome for a moment, we note that there were 22 other chromosomes of various sizes and shapes. When Sutton measured these carefully, he found that there were not 22 different sizes but only 11. In other words there are 2 of each size—11 pairs of chromosomes (figure 51).

Although it was not easy to identify individual chromosomes, it was possible to recognize that the 11 pairs consist of 8 large pairs and 3 small pairs. Careful study showed that the spermatogonia go through eight mitotic divisions and, in the metaphase of each, there are 8 large and 3 small pairs of chromosomes. This was the evidence that Sutton accepted as indicating that the 22 chromosomes of *Brachystola* are of 11 kinds.

The spermatogonia then begin to form sperm by first undergoing meiosis. The chromosomes of the same size synapse in pairs, forming 11 tetrads—8 large and 3 small. After the second meiotic division each of the 4 cells will form sperm without further divisions. Each sperm cell will receive 1 each of the 8 long and 1 each of the 3 short chromosomes.

The cells of the female were not as easy to study. Sutton reported, however, that the female has 22 chromosomes—again consisting of 8 pairs of long and 3 pairs of short chromosomes. The fact that both male and female nuclei have the same 8 pairs of long and 3 pairs of short chromosomes was additional evidence for the specificity of chromosomes. Sutton was proposing that the size differences were real and not that "as usually has been taken for granted, these differences are merely a matter of chance."

Thus it seemed that the diploid number for the male is 11 pairs plus the accessory and that the female has only the 11 pairs. (Sutton made an error. Later workers found that the female has 24 chromosomes consisting of the 8 long pairs, 3 short pairs, and a pair of accessory chromosomes.) The previous year, McClung (1901) had suggested that the accessory or "X" chromosome (so-called because its function was unknown) might be involved in the determination of maleness, a subject to which we shall return.

According to Sutton's observations, the mature ova of *Brachystola* would, therefore, have a monoploid number of 11 chromosomes. The sperm would be of two sorts: half would have the 11 chromosomes only and half would have 11 plus the accessory. Fertilization would result, therefore, in two sorts of offspring. Some would have 22 chromosomes, and be females, and others would have 22 chromosomes plus the accessory and, if McClung were correct, be males.

The Cytological Basis of Mendel's Laws

In 1903, in an even more remarkable paper, "The Chromosomes in Heredity," Sutton summarized what he had learned or surmised about the chromosomes in *Brachystola*:

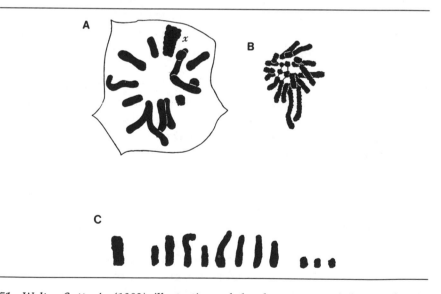

51 *Walter Sutton's (1902) illustrations of the chromosomes of the grasshopper* Brachystola. *A shows the monoploid set of the male,* B *the diploid set of the female, and* C *are the chromosomes of* A *arranged by size.*

(1) The diploid chromosome group consists of two morphologically similar chromosome sets. Every chromosome type is represented twice or, as we say today, chromosomes are in homologous pairs. Strong grounds exist for the belief that one set is derived from the father and one set from the mother at the time of fertilization.

(2) Synapsis is the pairing of homologous chromosomes.

(3) Meiosis results in a gamete receiving only one chromosome from each homologous pair.

(4) The chromosomes retain their individuality throughout mitosis and meiosis in spite of great changes in appearance.

(5) The distribution in meiosis of the members of each homologous pair of chromosomes is independent of that of each other pair. While each gamete receives one of each pair, *which one* is a matter of chance.

If one accepts this interpretation of the nature of chromosomes and their behavior in meiosis and fertilization, Sutton pointed out, there is a striking resemblance between the behavior of chromosomes and the behavior of the Mendelian units. He proposed the hypothesis that Mendel's results could be explained if the hereditary units were parts of chromosomes. Let us assume that Mendel's *round* and *wrinkled* alleles are on one pair of homologous chromosomes, as shown in figure 52. Let us further assume that *yellow* and *green* are on a different pair of homologous chromosomes. A cross of *round-yellow* × *wrinkled-green* will be made. When meiosis occurs, the gametes of the *round-yellow* parent will receive one of each homologous chromosomes and have the genotype **RY**. The *wrinkled-green* parent will form **ry** gametes. All individuals in the F₁ will have the same genotype, namely, **Rr Yy.**

Meiosis in the F₁ will result in the segregation and independent assortment of the four chromosomes, each gamete receiving one or the other member of each pair. Thus one would expect four types of gametes, **RY, Ry, rY,** and **ry.** Furthermore, the meiotic divisions would have resulted in equal proportions, 25 percent, of each. Since the four genotypic classes of gametes are produced in equal proportions, we can use a genetic checkerboard to derive the F₂ generation. The result is 9:3:3:1 ratio.

Thus the strict parallel between the genetic and the cytological data supported Sutton's hypothesis that Mendel's units of inheritance are parts of chromosomes. The discovery of parallelism between apparently dissimilar natural phenomena is one of the most important scientific methods for discovering causal relationships.

Sutton's model provided a formal explanation of the major Mendelian assumptions. For example, the problem of the "purity of the gametes" is solved if the hereditary units are parts of chromosomes. Thus, when gametes are formed in the F_1, normal meiotic divisions would prevent two homologous chromosomes from going to the same gamete. There could be no F_1 gametes with **R** and **r** or **Y** and **y,** for example.

The chromosomal movements in meiosis also account for segregation. Since the homologous chromosomes with their alleles go to opposite poles of the spindle, **R** goes to one gamete and **r** to another. The chromosomal movements can account, as well, for independent assortment. If the pairs of homologous chromosomes move to the poles of the spindle *independently of each other*, the chromosomes with their alleles will be distributed as shown in figure 52. Sutton did not know whether or not this was so, since he could not see any difference between two homologous chromosomes. In this case the data of genetics helped the cytological analysis: if the hereditary units are parts of chromosomes and if the hereditary units assort independently, the chromosomes must also assort independently.

Sutton had formulated a hypothesis that was useful, that is, it was specific enough to permit testable deductions. If we are to use the genes-are-parts-of-chromosomes hypothesis, it will be necessary to find a parallel between all types of genetic behaviors and chromosomal behaviors. Any variation in chromosomal phenomena from the usual condition must be reflected in the genetic results. Similarly, if genetic ratios are obtained that cannot be explained in Mendelian terms, one must find a chromosomal basis for the deviation.

Some of the deductions have been mentioned earlier. Here is a summary. We will first assume the correctness of what Sutton had to say about chromosomes, including that each chromosome can have only one allele of a contrasting pair, and what Mendel had to say about inheritance. Thus the segregation of the different alleles, **Aa** for example, must mean that there is a segregation of the meiotic chromosomes as well. Sutton interpreted his observations to show that there is. Furthermore, that seemingly inexplicable fact that the gametes are "pure," that is, can have only one allele of a contrasting pair, means that only one member of a pair of homologous chromosomes can enter a gamete. Cytological observations strongly suggested that this is so. In a similar manner, the independent assortment of alleles could be accounted for

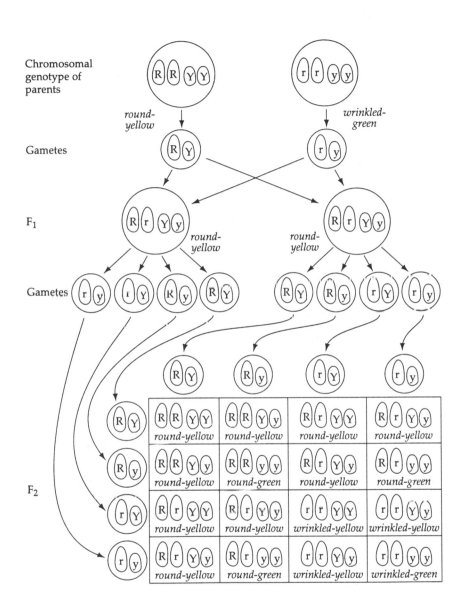

52 *The distribution of hereditary factors if they are parts of chromosomes. Compare with figure 50.*

by the independent assortment of chromosomes at anaphase of the second meiotic division. This, however, was only probable and would remain so until it became possible to distinguish between the members of a homologous pair of chromosomes.

The deductions so far mentioned could be tested because both the cytological and the genetic data were available. Sutton went on to deduce that non-Mendelian results must be expected to occur if his hypothesis was correct:

> We have seen reason, in the foregoing considerations, to believe that there is a definite relation between chromosomes and allelomorphs [now called alleles] or unit characters but we have not before inquired whether an entire chromosome or only part of one is to be regarded as the basis of a single allelomorph. The answer must unquestionably be in favor of the latter possibility, for otherwise the number of distinct characters possessed by an individual could not exceed the number of chromosomes in the germ-products; which is undoubtedly contrary to fact. We must, therefore, assume that some chromosomes at least are related to a number of different allelomorphs. If then, the chromosomes permanently retain their individuality, it follows that all the allelomorphs represented by any one chromosome must be inherited together. (p. 240)

Thus, Sutton is deducing that there must be many genes on the same chromosome and, if there are, they must be inherited as a unit. If so, there would be no possibility of independent assortment. We can predict, therefore, that an exception to the original Mendelian ratios must occur if we find more pairs of alleles than there are pairs of homologous chromosomes.

The elegance of Sutton's analysis must not make us forget that it was but another step in the long and difficult research which established that the nucleus, or some part of it, is the physical basis of inheritance. Nearly 40 years had passed since Haeckel's "lucky guess" and nearly 20 years since his guess gained the support of Hertwig, Strasburger, Kölliker, and Weismann.

Sutton was 25 and a student with E. B. Wilson in the Zoological Laboratory of Columbia University when he published the 1902 paper. Wilson had long been interested in the possibility that the chromosomes were the physical basis of inheritance. Furthermore, he had a magisterial grasp of cytology and embryology, having already published the first two editions of *The Cell*. One of his closest friends was Th. Boveri, whose brilliant research had added so much to the knowledge of chromosomes and their possible participation in heredity. Wilson had come to Columbia from Bryn Mawr in 1891, and in 1900 Thomas Hunt Morgan followed him from the same institution. The complex and synergistic interrelations of Wilson, Sutton, and Morgan were to climax during the following decade in the work with *Drosophila*. But in the beginning it took a young and brilliant scientist whose mind was not saturated with a tremendous mass of competing hypotheses and confusing facts to see conceptual order where the giants could not.

E. B. Wilson told the story of how Sutton first explained his hypothesis to him in a memorial volume prepared after Sutton's death:

> I well remember when, in the early spring of 1902, Sutton first brought his main conclusions to my attention, by saying that he believed he had really discovered "why the yellow dog is yellow." I also clearly recall that at that time I did not at once fully comprehend his conception or realize its entire weight.
>
> We passed the following summer [1902] together in zoological study at the sea side, first at Beaufort, N.C., later at South Harpswell, Me., and it was only then, in the course of our many discussions, that I first saw the full sweep and the fundamental significance of his discovery. Today the cytological basis of Mendel's law, as worked out by him, forms the basis of our interpretation of many of the most intricate phenomena of heredity, including the splitting up and recombination of characters in successive generations of hybrids, the phenomena of correlation and linkage, of sex and sex-linked heredity and a vast series of kindred processes that were wholly mysterious before. (Sutton 1917)

After a distinguished career as a physician, Sutton died at the age of 39. In his brief life in biological research he had produced two papers that probably can stand with those of Mendel and Watson and Crick, the discoverers of the structure of DNA, in fundamental importance and in the brilliance of the analysis. Yet according to Darlington (1960), as late as the mid 1920s in England, "Seven men might have been willing to assert their belief in the chromosome theory [of heredity] and give their reasons for it. But against this view there were seven hundred who held a contrary opinion." The interval between the time some important concept in science becomes true beyond all reasonable doubt to the discoverer and a few cognoscenti and its acceptance by a majority in the scientific community tended to be long in the years before World War II. It is often much shorter now that there are so many more scientists working on the same problems and progress is so rapid.

Cytology at the turn of the century was largely a descriptive science. To be sure, one could treat cells with various chemical reagents and differentially stain some of the cell structures, but little else. It was not practical at that time for those testing the hypothesis that the physical basis of inheritance resides in the chromosomes to proceed as follows: If the hypothesis is true, the removal of individual chromosomes should result in some change in the organism.

Nevertheless, Boveri (1902 and especially 1907) found a way to accomplish this feat, by creating sea urchin embryos with abnormal chromosome sets. The diploid number of chromosomes is 36 in the species of sea urchin that he used. The chromosomes are small and apparently uniform. There was no *a priori* reason to assume that the individual chromosomes might differ from one another. (Recall that Weismann had suggested that each chromosome has all of the hereditary information.) Nevertheless Boveri sought to test the hypothesis that the chromosomes differed from one another and that a normal set of 36 would be necessary for normal development.

In a normal fertilized egg the 36 chromosomes replicate before first cleavage to form 72 chromosomes, and these are divided equally at the mitotic first division, with 36 going to each daughter cell. Mitotic divisions throughout development would maintain this number.

Boveri knew that if concentrated sperm are used to fertilize sea urchin eggs, two sperm may enter the same egg. Since the monoploid number of chromosomes is 18, the dispermic embryo would have 54, that is, 18 each from the two sperm pronuclei and 18 from the egg pronucleus. Each chromosome would replicate before first cleavage to produce 108. The embryo would next undergo an atypical first division by cleaving into four cells (since each sperm brings in a "division center"—centrioles and centrosome). There is no way that each of these four cells could receive the normal complement of 36 chromosomes. If the 108 are divided equally among the four cells, each cell would receive 27 chromosomes. In fact, examination of fixed and stained cells showed that the distribution of chromosomes among the four cells was most uneven. Thus, if each cell must have the normal complement of 36 chromosomes for development to be normal, these dispermic eggs would be expected to develop abnormally. They did—out of 1,500 embryos, 1,498 were abnormal. There was reason to believe that the normal embryos could have been due to experimental error.

In another experimental procedure Boveri found that if the dispermic eggs are shaken, one of the division centers might not divide. The result would be three division centers, arranged in a triangle with spindles connecting them. Such an embryo would divide into three cells at first division. Again the chromosomes were divided irregularly but, in this case, there would at least be a *chance* that each cell could receive a normal set of 36 chromosomes—if the total of 108 is divided by 3, the result is 36. Of 719 embryos of this sort, 58 developed normally. According to Boveri, these data correspond fairly well with the chance expectations that the cells of some embryos will receive the normal set of chromosomes and so can develop normally. The conclusion was, therefore, that every cell in the embryo must have the normal set of 36 chromosomes if development is to be normal. This must mean that each chromosome in the set is endowed with a specific quality in spite of their appearing to be identical.

Sutton and Boveri had used entirely different methods to reach a similar conclusion: chromosomes are the physical basis of inheritance. They had not shown, of course, that chromosomes are the *only* bearers of hereditary information.

Sutton's hypothesis relating genes and chromosomes was made and tested without his ever seeing a gene, let alone seeing a gene as part of a chromosome. He related gene and chromosome because they behaved in a strictly parallel manner in meiosis and fertilization. To be sure this was indirect evidence, but the discovery of causal relations in science is often based on the parallel behavior of phenomena. For example, long ago, it was noted that the daily cycle of tides was associated with the relative position of the moon and to a lesser degree to the relative position of the sun. The relationship of moon and tides can be checked in several ways and the hypothesis so firmly established that one can predict, with a high degree of accuracy, tides in the future. Parallel behavior is the only practical way to study the relation of moon and tide. One cannot perform the more critical experiment of excising the moon from the solar system and noting the consequences. Correlations need not, however, always denote a causal relation. The 28-day lunar cycle and the 28-day menstrual cycle of women were long suspected to be causally related, but thus far we have no convincing evidence that this is so.

Which method is superior, the experiments of Boveri or the correlations found by Sutton? So far as supporting the hypothesis is concerned, the two are about equal. Beyond that there is a large and important difference. What would be the next step in Boveri's approach? It is hard to see how deeper insights into the nature of inheritance could have been obtained with the methodology available to Boveri. He was able to perform a more direct test of the relation between chromosomes and inheritance by altering the chromosomes and studying the consequences. One might think next of removing individual chromosomes, but that could not be done and also there was no way of distinguishing one chromosome from another.

Sutton's approach, on the other hand, was far more elegant than Boveri's. He was able to link Mendelism and cytology, which of course Boveri could not, and even to suggest testable deductions. Sutton had set the stage for the culmination of classical genetics in the work of Morgan's *Drosophila* group a decade later. And, it is interesting to note, eventually the Morgan group was able to manipulate individual chromosomes by genetic methods.

Variations in Mendelian Ratios

In the year 1902 still another publication of fundamental importance appeared—the beginning of a series of *Reports to the Evolution Committee of the Royal Society*. The first one was by Bateson and Edith R. Saunders (1902). In 1897, before they knew of Mendel, they had begun a series of crosses of different varieties of plants and animals. Once they read Mendel's paper they realized that "the whole problem of heredity has undergone a complete revolution" (p. 4). The Mendelian paradigm could account for their results of their crosses.

They also provided us with some of the basic terminology for Mendelian genetics:

> This purity of the germ-cells, and their inability to transmit both of the antagonistic characters, is the central fact proved by Mendel's work. We thus reach the conception of unit-characters existing in antagonistic pairs. Such characters we propose to call *allelomorphs*, and the zygote formed by the union of a pair of opposite allelomorphic gametes, we shall call a *heterozygote*. Similarly, the zygote formed by the union of gametes having similar allelomorphs, may be spoken of as a *homozygote*. (p. 126)

In time *allelomorph* was shortened to *allele*, which is closely related to two other terms—*gene* and *locus*. In the years before molecular genetics those were differentiated as follows. A *gene* was a portion of a chromosome that produces an indivisible effect (the atoms of heredity!), which must be detectable, of course (or we would never know of its existence). The position that the gene occupies on a chromosome was its *locus*. An *allele* was a detectable variation in the expression of the gene, caused by mutation. Every gene whose existence we know of has at least two alleles—otherwise we would not know of its existence. Mendel knew only of those genes in peas that had two alleles that produced different phenotypes. Peas have many genes that are unknown to this day—because they never changed in such a way that a new variant could be recognized. A gene reveals its existence when it *mutates* in such a manner that the new mutant allele has a detectable effect.

The results of crossing varieties of peas—dominance and recessiveness, segregation, independent assortment with the consequences that one observes a 3:1 ratio in the F_2 of a monohybrid cross and a 9:3:3:1 ratio in the F_2 of a dihybrid cross—exhibited a high degree of uniformity and thus raised the question of the universality of these findings.

Although the typical Mendelian monohybrid and dihybrid ratios have, to this day, proved to be the most common patterns, they are not the only ones. When a wide variety of species were studied, examples were encountered where heterozygotes were intermediate or where two different genes might interact to produce an entirely new phenotype.

In the first report to the Evolution Committee, Bateson and Saunders (1902) described numerous crosses, many of them begun before they knew of Mendel's work. Saunders described her experiments with wild species of the genus *Lychnis*, the campion. Some of the species are *hairy* and others are *glabrous*, that is, without hairs. The results must have been absolutely bewildering at first but readily explained once she knew of Mendel's model.

(1) Crosses of *hairy* × *glabrous* produced an F_1 consisting of 1,006 *hairy* and 0 *glabrous*. This was interpreted to be a simple monohybrid cross with *hairy* as the dominant allele. If we use a general scheme of **A** for the dominant allele and **a** for the recessive, the cross is **AA** *(hairy)* × **aa** *(glabrous)*. All the F_1 will be **Aa**, and *hairy* in appearance.

(2) When the F_1 individuals, **Aa**, were crossed, they produced an F_2 consisting of 408 *hairy* and 126 *glabrous*. This is a ratio of 3.2:1, which for small samples such as this can be taken to be a 3:1 ratio. It must be remembered that the theoretical ratio is

achieved only if certain conditions are met—one being that all genotypes, in this case **AA, Aa,** and **aa,** are equally viable. If any one was not as hardy, the ratio obtained would depart from the expected value.

(3) When an F_1 individual was crossed with a pure-breeding *hairy,* the offspring consisted of 41 *hairy* and 0 *glabrous.* Mendel's model would predict such results. The F_1 would be **Aa** and the *hairy* would be a homozygous dominant, **AA.** Since all the offspring would receive an **A** allele from the dominant parent, all the offspring would be *hairy,* in this case being either **AA** or **Aa** in equal frequencies.

(4) When an F_1 individual was crossed with a pure-breeding *glabrous,* the offspring were 447 *hairy* and 433 *glabrous.* This is a cross between a heterozygote, the F_1 **Aa,** and a homozygous recessive, **aa.** One would expect 50 percent **Aa** and 50 percent **aa.** This type of cross is valuable because it determines the genotype of individuals that are phenotypically identical. It is known as a "backcross" or, better, a "testcross." Thus all of the F_1 are *hairy,* but one cannot tell by observation whether they are homozygous, **AA,** or heterozygous **Aa.** If they were homozygous, all the offspring would be *hairy* with a genotype **Aa.** If heterozygous, as in this case, there would be equal numbers of *hairy* and *glaborous.*

Thus the crosses involving *hairy* and *glabrous* found their formal explanation in Mendel's model for a monohybrid cross.

A variation in the expected results emerged from some other experiments made by Saunders, for example, with plants of the mint genus *Salvia* (Bateson et al. 1905). True breeding strains with *pink* and *white* flowers were used.

(1) When *pink* is crossed with *white,* all of the F_1 are *violet.* This is a case of blended inheritance in which there was no clear-cut dominant allele. The hypothesis would be that the phenotype *violet* occurs when both the *pink* and *white* alleles are together in the heterozygote. Even though there is no clear-cut dominant allele, let us use the symbol **A** for *pink* and **a** for *white. Violet* would then be **Aa.**

(2) We can test this hypothesis by crossing two *violet* F_1 plants. Each would be **Aa** so the F_2 should be 1/4 **AA,** 2/4 **Aa,** and 1/4

aa or 1/2 *pink*, 2/4 *violet*, and 1/4 *white*. Saunders reported that one cross of the F₁ plants gave 59 *violet*, 25 *pink*, and 34 *white* and in another 225 *violet*, 92 *pink*, and 114 *white*. Thus the deduction was confirmed.

In cases such as this, where there is blended inheritance, the F₂ ratio is 1:2:1. If we had classified the plants as *colored* and *white*, the ratio would have been the familiar 3:1. It is always the case that the "3" in monohybrid crosses of this kind consist of two genotypes, 1/3 being homozygous and 2/3 being heterozygous.

In the same publications Bateson reported his early experiments with chickens, which were to show an entirely new sort of inheritance that was very difficult to explain. One of the characteristics studied was the shape of the combs typical of the various breeds (figure 53). One type was called *pea* and another *single*.

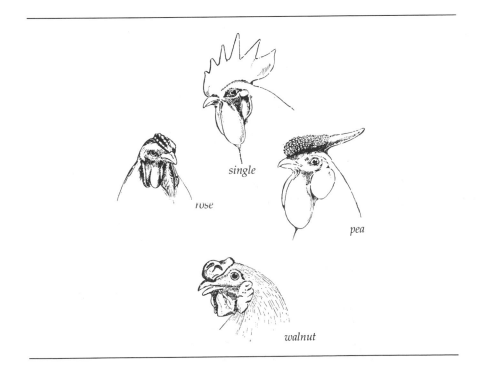

single

rose

pea

walnut

53 *Comb shapes in chickens.*

(1) When *pea* was crossed with *single*, all of the F_1 had *pea* combs (figure 54).

(2) When the F_1 were crossed, the offspring were 332 with *pea* combs and 110 were *single*. Nothing strange about that—the same old F_2 ratio of 3:1.

(3) Equally familiar was the cross *rose* comb with *single*. All were *rose* in the F_1.

(4) When these F_1s were crossed with one another, the F_2 gave 221 *rose* to 83 *single*. This, again, is a 3:1 ratio.

(5) When an F_1 was crossed with a *single*, the offspring were 449 *rose* and 469 *single*. This is the expected 1:1 ratio when a heterozygote is crossed with a homozygous recessive. The model would be **Aa** × **aa** with the offspring being 1/2 **Aa** and 1/2 **aa**.

    Figure 54 provides an explanation. Or does it? When the crosses *rose* × *single* and *pea* × *single* are viewed separately they appear to be simple cases of a monohybrid cross giving a 3:1 ratio in the F_2. Both *rose* and *pea* are dominant to *single*. But notice that the genotype of *single* is given as **rr** in the first cross and **pp** in the second. That would seem to imply that there are three alleles at the locus. That was unexpected at a time when, as Mendel had found and Bateson had emphasized, there were only pairs of antagonistic characters.

(6) The plot really thickened with a cross of pure-breeding *pea* with pure-breeding *rose*. The F_1 generation was uniform, but all had a type of comb not seen in either parent. It was *walnut*, a comb shape that was known in other breeds (figure 53).

(7) When those F_1s were crossed, the F_2 consisted of 99 *walnut*, 26 *rose*, 38 *pea*, and 16 *single*. These raw data suggested that a 9:3:3:1 ratio might be involved. If so the expected numbers would be 99 *walnut*, 33 *rose*, 33 *pea*, and 11 *single*. The expected and actual numbers were close enough for this to be accepted as a dihybrid cross with a 9:3:3:1 ratio in the F_2. This would mean that two different genes are involved, which we can designate as **P** and **R**.

(8) If this hypothesis is correct, namely, that the F_1 is **Pp Rr,** then the results of a test cross should be predictable. That is, in a cross of **Pp Rr** × **pp rr**, the offspring should be 1/4 **Pp Rr** and be *walnut*, 1/4 should be **Pp rr** and be *pea*, 1/4 should be **pp Rr** and be *rose*, and 1/4 should be **pp rr** and be *single*. When this test cross was made, the offspring were 139 *walnut*, 142 *rose*,

P phenotype   *rose*   **X**   *single*

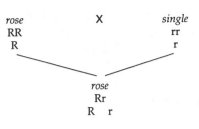

P	phenotype	*rose*	X	*single*
	genotype	RR		rr
	gametes	R		r

F$_1$ phenotype   *rose*
      genotype   Rr
      gametes   R r

F$_2$

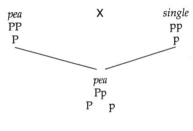

	R	r
R	RR *rose*	Rr *rose*
r	Rr *rose*	rr *single*

P	phenotype	*pea*	X	*single*
	genotype	PP		pp
	gametes	P		p

F$_1$ phenotype   *pea*
      genotype   Pp
      gametes   P p

F$_2$

	P	p
P	PP *pea*	Pp *pea*
p	Pp *pea*	pp *single*

54 *Genetic diagram for crosses of chickens with different types of combs.*

112 *pea*, and 141 *single*. Thus the prediction was confirmed (figure 55).

These crosses involving comb shape showed that some structures are controlled by at least two different genes. These alleles abide by the Mendelian rules of dominance, segregation, and independent assortment. They are puzzling only because both pairs are affecting the same character—comb shape. Further research was to show that this is the general rule for complex characters. There is no single gene for hair, eyes, ears, stomach, height, weight, arms, or brains, for example.

The following crosses represent still another variation on Mendelian ratios, and it proved even more difficult to understand. Bateson and

P phenotype	*pea*	X		*rose*
genotype	PP rr			pp RR
gametes	Pr			pR

F_1 phenotype *walnut*
genotype Pp Rr
gametes PR Pr pR pr

F_2

	PR	Pr	pR	pr
PR	PP RR *walnut*	PP Rr *walnut*	Pp RR *walnut*	Pp Rr *walnut*
Pr	PP Rr *walnut*	PP rr *pea*	Pp Rr *walnut*	Pp rr *pea*
pR	Pp RR *walnut*	Pp Rr *walnut*	pp RR *rose*	pp Rr *rose*
pr	Pp Rr *walnut*	Pp rr *pea*	pp Rr *rose*	pp rr *single*

F_2 ratio: 9 *walnut*; 3 *rose*; 3 *pea*; 1 *single*.

55 *Diagram of a cross of* pea comb × rose comb *chickens showing the production of* walnut.

others had done numerous crosses with confusing results with both sweet peas *(Lathyrus)* and stocks *(Matthiola)*.

(1) When two different varieties with *white* flowers were crossed, all of the F_1 plants were *colored*.
(2) When the F_1s were crossed it appeared, at first, that there were equal numbers of *white* and *colored*. This was a strange ratio for an F_2. The experiments were continued and when large numbers of individuals had been scored, the ratio was 9 *colored* to 7 *white*.

Bateson and his coworkers concluded that the white flower color of the two original *white* varieties was not controlled by the same genotype. One was shown as **CC rr** and the other as **cc RR**. For the flowers to be *colored* at least one **C** and one **R** is necessary. The F_1 would all be **Cc Rr,** and hence *colored*. In the F_2 9/16 were predicted to be of this genotype. And 7/16 would not have both **B** and **C** and, hence, would be *white*.

This survey of the genetics of the first few years of the twentieth century makes us acutely aware of the complexity and confusion that existed. Those who wished the world to be as outlined for Mendel's peas were to find that it was not. This does not mean that the original Mendelian story was "wrong." It means only that it was incomplete and was being replaced with a deeper understanding of the nature of inheritance.

Not one of the original Mendelian rules was found to be valid for all cases. It can be argued that the remarkable progress of genetics was based on an attitude that might seem "unscientific." That is, from the time that Mendel's work became known, it was clear that the original Mendelian hypothesis did not apply to all organisms. Nevertheless, the "true believers" ignored the exceptions and slowly found what could be explained in the original Mendelian terms. As they came to know more and more about breeding experiments in different species, it became possible to expand theory to accommodate the new data.

It was eventually found that some of the most intractable problems had a chromosomal basis. One such problem had to do with those puzzling accessory or "X" chromosomes noted before, so we should now check on what the cytologists were doing in the first few years of the twentieth century.

The Discovery of Sex Chromosomes

As Montgomery had suggested in his 1901 paper, it is important to study a variety of organisms since some may show variations in the behavior of their chromosomes and this will provide data not otherwise available and conclusions not otherwise possible. The accessory chromosomes turned out to be a case in point. In fact, it was by studying their behavior that the critical evidence that genes are parts of chromosomes was eventually obtained.

Recall the reason that Boveri experimented with polyspermy in sea urchins. His experiments provided a mechanism for allocating abnormal groups of chromosomes to the cells of the early embryo. As a consequence, the embryos died and the hypothesis that a set of normal chromosomes is necessary for normal development was supported. Nevertheless, this was not a fruitful type of experimentation. There was no way of recognizing individual chromosomes, of relating specific chromosomes to specific phenotypes, or of controlling which chromosomes entered which embryonic cell.

As it so often happened, nature was doing the required experiment all along. And as it so often happens, it took a considerable length of time for cytologists to realize that was so. In 1891 H. Henking published some observations on the behavior of chromosomes in spermatogenesis in the bug *Pyrrhocoris* (figure 56). This species has a diploid number of 23 chromosomes—11 pairs plus an extra one, which he called the "X." At synapsis the 11 homologous pairs formed 11 tetrads. But the behavior of the X was different. Having no homologue, it had nothing to synapse with but did replicate to form a dyadlike structure. At the beginning of meiosis, therefore, the cell would have 11 tetrads plus the X dyad. At the first meiotic division the 11 tetrads separated, a dyad from each going to each daughter cell. The X-dyad, however, went entirely to one pole of the spindle and hence was included in only one of the daughter cells.

In the second meiotic division of the cell with only the 11 dyads, separation of the dyads was observed and one chromosome of each went to each daughter cell. The cell with the 11 dyads plus the X-dyad divided and one chromosome of each of the 11 dyads went to opposite poles of the spindle. The X-dyad divided also and each of the daughter cells received one X chromosome.

Thus, of the 4 cells produced in equal numbers by meiosis, 2 would have 11 chromosomes and 2 would have 11 chromosomes plus an X.

Two types of sperm would be formed, again in equal numbers, one type with an **X** and the other without.

Henking reported what he had seen and left it at that. Thereafter other observers noted similar peculiar chromosomes in many different species. They were noticed either because they stained differently from the other chromosomes, or they moved to the poles of the spindle earlier or later than the other chromosomes, or they lacked a mate for synapsis, or they were distributed to only half of the sperm. The vast majority of the observations were made on males since, for technical reasons, spermatogenesis is easier to study than oogenesis.

In 1901 the American cytologist C. E. McClung suggested that the **X** chromosome was in some way connected with the determination of sex.

> Upon the assumption that there is a qualitative difference between the various chromosomes of the nucleus, it would necessarily follow that there

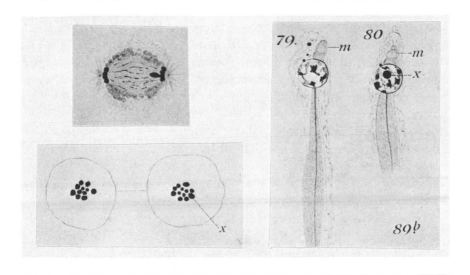

56   *H. Henking's (1891) illustration of meiosis in the "bug" Pyrrhocoris. The upper left figure is of a spermatocyte in telophase of the second meiotic division. The **X** chromosome, lagging behind the rest, is going to the pole at the right. The resulting daughter cells are shown at the lower left figure with the **X** in the right cell only. Two sorts of sperm will be formed, as shown in the rightmost figure— one with an **X** and one without.*

are formed two kinds of spermatozoa which, by fertilization of the egg, would produce individuals qualitatively different. Since the number of each of these varieties of spermatozoa is the same, it would happen that there would be an approximately equal number of these two kinds of offspring. We know that the only quality which separates the members of a species into these two groups is that of sex.

This hypothesis provided an explanation for those odd chromosomes that were being found in more and more species. Montgomery had observed several cases. Sutton had described the same condition for *Brachystola*. At first it was believed that the accessory chromosomes were extra and restricted to males. Sutton had reported that the chromosomes of ovarian cells resembled those in the testis except for the lack of the accessory. But he made a mistake—subsequently it was discovered that the female of *Brachystola*, far from lacking the accessory chromosome, has two. Sutton had made a very important claim and, as is usual in science, important claims are checked by other investigators.

During the first decade of the twentieth century the study of sex chromosomes exhibited a pattern not uncommon in science. An important hypothesis, presumably of wide applicability, is proposed—although on inadequate evidence. This was McClung's suggestion that the accessory chromosome might determine maleness, which initiated a period of active research. There emerged conflicting observations, and it was clear that the original hypothesis that males have an extra chromosome did not hold for all species. There were also conflicting conclusions. Some investigators failed to find accessory chromosomes. Those who did suggested a variety of hypotheses to account for them. Some believed them to be degenerating chromosomes, others believed them to be a special type of nucleolus, and still others thought that McClung was probably correct.

The final stage in this scenario is when one or a few individuals, careful of their supporting data and cautious in their conclusions, bring conceptual order to the subject being investigated. And, again, as so often happens, two or more individuals, working independently, reach essentially the same conclusion at the same time. E. B. Wilson was the person mainly responsible for solving the riddle of the accessory chromosomes, but the announcement of his discovery coincided with a report reaching similar conclusions by Nellie M. Stevens.

Wilson's most important contributions are contained in eight long papers, *Studies on Chromosomes I–VIII*, published between 1905 and 1912.

His own observations, together with those of others, revealed a complexity not imagined by McClung or Sutton. In most groups of animals the female has a pair of homologous X chromosomes and she is designated as **XX** (figure 57). All of her eggs will have one **X**. The males of various species, on the other hand, vary considerably. Some have only a single **X** and are designated as **XO**—the "O" indicating the absence of a chromosome. Half of their sperm will have one **X** and the other half will lack a sex chromosome. In other species the males may have two chromosomes, one like the **X** of the female and the other, usually differing in size or shape, called the **Y**. These males are designated **XY**. With respect to the sex chromosomes, the males in this case produce two sorts of sperm in equal numbers, **X**-carrying sperm and **Y**-carrying sperm, and thus are heterogametic. The females produce a single type of ovum and hence are homogametic. (It was discovered later that both human beings and *Drosophila* are of the **XX** female and **XY** male type.)

These two patterns of sex chromosomes, while the ones most commonly encountered, do not exhaust the range of possibilities. Some species may have multiple sex chromosomes. In birds and butterflies the females are heterogametic and the males homogametic for the sex chromosomes.

These are some of the conclusions that can be drawn from the numerous studies of Wilson, Stevens, and others.

(1) The sex of an offspring is determined at the time of fertilization.
(2) The sex of an individual will be irreversible if it is based solely on the sex chromosomes—unless we can alter the chromosomes or their expression.
(3) If meiosis is normal and fertilization is random, the two sexes should be produced in approximately equal numbers.
(4) The relation between sex and chromosomes, firmly established by 1910, is additional evidence supporting Sutton's hypothesis that chromosomes are the physical basis of inheritance.

At the end of the first decade of the twentieth century an increasing proportion of the cytologists and geneticists who were studying inheritance were becoming convinced that chromosomes must have some important role in inheritance. Biologists in other fields were not so sure. Wilson, always cautious, wrote (1911):

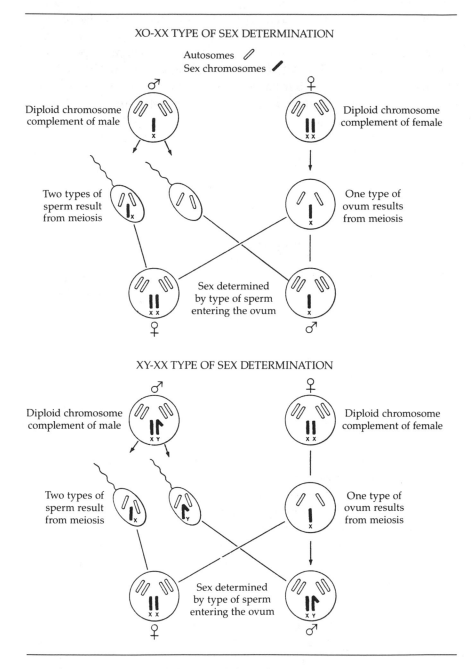

57 *The two main types of chromosomal sex determination. **XO** males have only a single **X** chromosome, whereas **XY** males have an **X** chromosome and a **Y** chromosome, which differs in size and shape from the **X**.*

From any point of view it is indeed remarkable that so complex a series of phenomena as is displayed, for example, in sex-limited inheritance can be shown to run parallel to the distribution of definite structural elements, whose combinations and recombinations can in some measure actually be followed with the microscope. Until a better explanation of this parallelism is forthcoming we may be allowed to hold fast to the hypothesis, directly supported by so many other data, that it is due to a direct causal relation between these structural elements and the process of development.

The hypothesis derived from "this parallelism" allows deductions to be made and so to test further the correctness of the hypothesis. Here is one: *If some genes are part of the X chromosome, we predict that inheritance of these genes will follow the inheritance of sex chromosomes.*

Postulate, for example, a gene of the X chromosome of a species with XX females and XY males. A male offspring can receive his X only from his mother (if he received an X from his father, he would be a daughter). Daughters receive one of their X chromosomes from the mother and the other from the father. Thus, if a gene is part of an X chromosome, males can inherit that gene only from their mothers. That is a testable deduction, as we shall see in the next section.

chapter 16

The Genetics of the Fruit Fly

The most famous individual fly in the history of science is a male fruit fly with the name *Drosophila melanogaster.* It became famous because it had white eyes instead of the normal red ones, but most importantly because it happened to appear in Room 613 of Schermerhorn Hall at Columbia University in the spring of 1910. This was the "Fly Room," the laboratory of Thomas Hunt Morgan and a remarkable group of young students. Down the hall was the laboratory of E. B. Wilson, who was finishing up his series *Studies on Chromosomes.*

That fly had chosen the proper time and place to spin out its short life and achieve immortality. Morgan (1910) tells the story:

> In a pedigree culture of *Drosophila* which had been running for nearly a year through a considerable number of generations a male appeared with white eyes. The normal flies have brilliant red eyes.
>
> The white-eyed male, bred to his red-eyed sisters, produced 1,237 red-eyed offspring (F_1), and 3 white-eyed males. The occurrence of these three white-eyed males (F_1) (due evidently to further sporting [mutation]) will, in the present communication, be ignored.
>
> The F_1 hybrids, inbred, produced: 2,459 red-eyed females, 1,011 red-eyed males, 798 white-eyed males.
>
> *No white-eyed females appeared.* The new character showed itself therefore to be sex limited in the sense that it was transmitted only to the grandsons. But that the character is not incompatible with femaleness is shown by the following experiment.
>
> The white-eyed male (mutant) was later crossed with some of his daughters (F_1), and produced: 129 red-eyed females, 132 red-eyed males, 88

white-eyed females, 86 white-eyed males. The results show that the new character, white-eyes, can be carried over to the females by a suitable cross, and is in consequence in this sense not limited to one sex. It will be noted that the four classes of individuals occur *roughly* in equal numbers (25 per cent.).

What was one to conclude? The original cross of the *white-eyed* male with the *red-eyed* females gave an F_2 ratio of 4.3 to 1. This might be accepted as a 3 to 1 ratio since it seemed clear that the *white-eyed* flies were less viable than their *red-eyed* siblings (as shown in the later cross of the F_1 daughters with the *white-eyed* male). But one could not interpret the data as showing them to be a typical F_2 ratio of 3:1. White eye color was not evenly distributed among females and males as it should be in a normal Mendelian cross. In the F_2 of the original cross there were no *white-eyed* females. This association of inheritance with sex hinted that a critical test of Sutton's hypothesis might be in the making. Back to Morgan:

Morgan's First Hypothesis

An Hypothesis to Account for the Results.—The results just described can be accounted for by the following hypothesis. Assume that all of the spermatozoa of the white-eyed male carry the "factor" for white eyes "W"; that half of the spermatozoa carry a sex factor "X" the other half lack it, *i.e.,* the male is heterozygous for sex. Thus the symbol for the male is "WWX," and for his two kinds of spermatozoa WX—W.

Assume that all of the eggs of the red-eyed female carry the red-eyed "factor" R; and that all of the eggs (after reduction) carry one X, each, the symbol for the red-eyed female will be therefore RRXX and that for her eggs will be RX—RX.

It is of the greatest interest to note how Morgan indicated the genotype of both adults and gametes. He treated both genetic "factors" *and* chromosomes as though they were independent phenomena. His symbolism of "R" for the allele for red eyes and "W" for the allele for white was eventually replaced by the Mendelian scheme for using upper- and lower-case symbols for dominant and recessive alleles, so, for clarity, I will alter Morgan's original notation and use **w** for the allele for *white eyes* and **W** for the allele for *red eyes*. Another point to be noted is that the male was assumed to have only one **X** chromosome, that is, to be of an **XO** type of male. Subsequently it was realized that the *Drosophila* male has a **Y** chromosome as well.

Figure 58 uses Morgan's hypothesis to explain the results of the first cross of the *white-eyed* male with a *red-eyed* female. The scheme fits the data, that is, the F_1 is predicted to consist only of *red-eyed* daughters and *red-eyed* sons. Continuing to the F_2, the hypothesis predicts that all of the females will have *red eyes* and that half of the sons will have *red eyes* and half will have *white eyes*. So far the hypothesis predicts what was found. This is not in the least surprising since it was formulated to do so.

But the hypothesis explained the data only with one important qualification. Note the F_1 individuals in figure 58. When gametes are formed by the female, half are shown with the **W** going with an **X** and half with the **w** going with an **X**. However, the hypothesis demanded a very different situation for the F_1 male. The male is shown as **WwX**. One should expect, therefore, four classes of gametes: **WX, wX, W** (or **WO**), and **w** (or **wO**). Morgan recognized only two classes of sperm: **WX** and **w**. He explains:

> It is necessary to assume . . . that when the two classes of spermatozoa are formed in the F_1 red male (WwX), W and X go together—otherwise the results will not follow (with the symbolism here used). This all-important point can not be fully discussed in this communication.

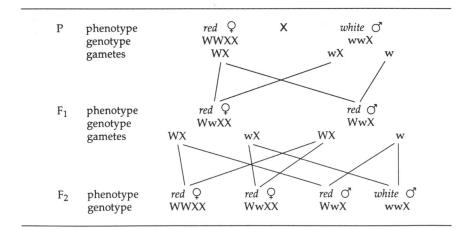

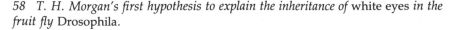

58 *T. H. Morgan's first hypothesis to explain the inheritance of* white eyes *in the fruit fly* Drosophila.

The value of a hypothesis is not only to explain the data at hand but also to predict what will happen in new situations. Morgan undertook four tests of his hypothesis.

(1) If the genotype of the *white* males is **wwX** and of the *white* females **wwXX**, their offspring should consist of *white* males and *white* females only. The cross was made and the results were according to predictions (figure 59, **1**).

(2) The F_2 *red-eyed* females in the first cross were predicted to be of two genotypes, **WWXX** and **WwXX**, even though all were identical in appearance. If several of these females were crossed individually with *white-eyed* males, one would expect two results: Approximately half of the crosses should result in all of the offspring having *red* eyes and the other half should produce four phenotypes among the offspring. These crosses were made and the predicted results were observed (figure 59, **2** and **3**).

(3) The genotype of the F_1 female of the original cross was predicted to be **WwXX**. If so, the cross of such a female with a *white-eyed* male should give the same results as shown in figure 59, **3**. Again the cross was made and the predicted outcome was observed.

(4) The hypothesis requires that the original F_1 males (figure 58) be **WwX**. If such a male is crossed with a *white-eyed* female, the prediction would be for *red-eyed* females and *white-eyed* males, as shown in figure 59, **4**. The crosses were made and the prediction verified. Once again, *however, the hypothesis required an unusual type of meiosis in the **WwX** males: the **W** factor would always be with the **X**, to form **WX** sperm; there could be no **wX** sperm.*

True beyond all reasonable doubt? Well, maybe. Few new hypotheses in the early days of genetics, apart from Mendel's, were tested so thoroughly as this one. Nearly all of Morgan's first hypothesis was based on well-substantiated genetic principles: dominance and recessiveness, segregation, and the behavior of sex chromosomes. His four deductions were explicit and critical. In every case the experiments to test the deductions provided data that verified the predictions and, hence, supported the hypothesis.

To be sure there was that qualification about spermatogenesis in **WwX**

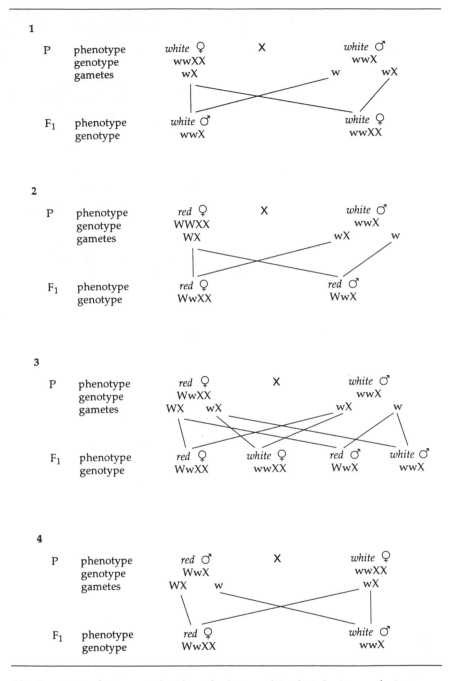

59 *Four tests of Morgan's first hypothesis to explain the inheritance of* white eyes *(see text).*

males, but by 1910 his colleague Wilson and other cytologists were reporting all sorts of strange behavior of chromosomes in meiosis. There was no *a priori* reason to exclude the hypothesis of the association of **W,** but never **w,** with the **X** in males.

Morgan reported another discovery that was a real puzzle:

> A most surprising fact appeared when a white-eyed female was paired to a wild, red-eyed male, *i.e.,* to an individual of an unrelated stock. The anticipation was that wild males and females alike carry the factor for red eyes, but the experiments showed that all wild males are heterozygous for red eyes, and that all the wild females are homozygous. Thus when the white-eyed female is crossed with a wild red-eyed male, all of the female offspring are red-eyed, and all of the male offspring white-eyed.

These data presented a difficulty. If all males in natural populations are heterozygous for these eye-color alleles, one would expect numerous *white-eyed* flies to be present in wild populations and in cultures. Yet Morgan had been raising *Drosophila* for many months and had observed no such thing. "As yet I have found no evidence what white-eyed sports occur in such numbers. Selective fertilization may be involved in the answer to this question."

There are many interesting points about this famous paper which started the line of experimentation that revolutionized genetics. The most puzzling is why Morgan failed to realize that the data could be explained more simply by assuming that the alleles for eye color were *parts* of the **X** chromosome. In 1910 he was still most suspicious of the Suttonian hypothesis, but wouldn't he have discussed the data with his colleague Wilson? To be sure the paper had been written in haste. Allen (1978, p. 153) estimates that the *white-eyed* male was discovered in January of 1910. Then the experiments were done. The paper was finished July 7, 1910, after Morgan had gone to the Marine Biological Laboratory in Woods Hole, and was published in the July 22, 1910, issue of *Science*.

Of considerable pedagogical interest is the fact that the paper is written in a form that corresponds to the popular view of "the scientific method." First there are the observations of some natural phenomenon, in this case the crosses involving the strange new fly with the white eyes. Then a hypothesis is formulated. Finally, deductions are made from the hypothesis and these are tested. In Morgan's case the tests seemed to have supported the hypothesis. These steps are rarely mentioned in published reports, even though something like "the scientific method" is happening in the mind of the investigator. Morgan's paper is unusual in that these steps are explicitly stated in the published report.

Morgan's Second Hypothesis

It took Morgan only a few months to realize that his first hypothesis to explain sex-limited inheritance of eye color was fundamentally flawed. Several additional mutations were found, and these were inherited in the same manner as that *white-eyed* allele. The results were first announced in a public lecture given in the Marine Biological Laboratory at Woods Hole, Massachusetts, on July 7, 1911. The new hypothesis was simplicity itself: instead of thinking of the sex-limited alleles as being associated with, but not part of, the **X** chromosome—the first hypothesis—assume that they are part of the **X** chromosome (figure 60).

> The experiments on Drosophila have led me to two principal conclusions:
>
> First, *that sex-limited inheritance is explicable on the assumption that one of the material factors of a sex-limited character is carried by the same chromosomes that carry the material factor for femaleness.*
>
> Second, *that the 'association' of certain characters in inheritance is due to the proximity in the chromosomes of the chemical substances (factors) that are essential for the production of those characters.* (Morgan 1911a, p. 365)

Therefore, if one assumes that the allele for *white eyes* and the dominant allele for *red eyes* are parts of the **X** chromosome, the results of all the crosses correspond to what would be expected from the distribution of the **X** chromosome in meiosis and fertilization. It would then be unnecessary to invoke subsidiary assumptions, such as the **w** allele not being able to associate with the **X** in meiosis of the **WxX** males or that all wild males must be heterozygous.

Morgan's second hypothesis has withstood every conceivable test, and it can be accepted as true beyond all reasonable doubt. Figure 60 shows how it explains the inheritance of *white eyes*. This figure also shows a **Y** chromosome because it was soon realized that the male *Drosophila* is **XY**, not **XO**. The data indicated that the **Y** chromosome did not have an allele at the locus for *white eyes* and, as we now know, it has only a very few active gene loci of any sort.

Once again we find an example of the "obvious" not being obvious at all. More often than not, things become obvious after the fact. Morgan was working in the Zoology Department where a short seven years earlier Sutton had maintained that genes must be parts of chromosomes. His colleague E. B. Wilson had continued to work within the Suttonian paradigm. However, Morgan had not accepted the chromosomes as the physical basis of inheritance and was not to do so until his own experiments convinced him. In fact, he had a poor opinion of the explanations being used by geneticists to account for the data of inheritance. In

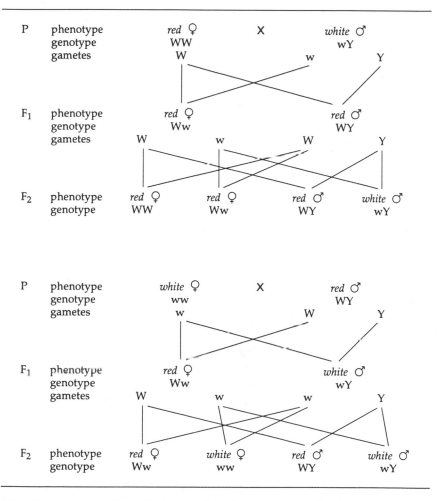

60   *Morgan's second hypothesis to explain the inheritance of* white *eyes.*

January 1909, the year before his classic first paper on the *white-eyed* fly, he had this to say in a lecture to the American Breeders' Association:

> In the modern interpretation of Mendelism, facts are being transformed into factors at a rapid rate. If one factor will not explain the facts, then two are invoked; if two prove insufficient, three will sometimes work out. The superior jugglery sometimes necessary to account for the results may blind us, if taken too naively, to the common-place that the results are often so excellently "explained" because the explanation was invented to explain them. We work backwards from the facts to the factors, and then, presto! explain the facts by the very factors that we invented to account for them . . . I cannot but fear that we are rapidly developing a sort of Mendelian ritual by which to explain the extraordinary facts of alternative inheritance. (p. 365)

Such was the opinion of the one who, in a few short years, was to be recognized as the giant of genetics of our century.

The Fly Room

In the decade after 1910 a medium-sized room in the Zoology Department of Columbia University, occupied by Morgan and his students, became the center of genetics. In 1911 Morgan (1866–1945) was 45 years old. He had come to Columbia in 1900 as a world-class embryologist. Throughout this formative decade he had three close associates, who began as his students and remained as coworkers: Sturtevant, Bridges, and Muller. Alfred Henry Sturtevant (1891–1970) was to receive his Ph.D. in 1914 for using linkage data to construct the first genetic map. Calvin B. Bridges (1889–1938) received his Ph.D. in 1916 for a classic paper on nondisjunction, widely regarded as the final and conclusive proof that genes are parts of chromosomes. Herman J. Muller (1890–1967) received his degree in 1915 for a definitive study of crossing-over. Biologists came from all over the world to visit or to do research in the Fly Room.

The basis for all these discoveries was the fruit fly, *Drosophila melanogaster*. It appears to be an immigrant from the Old World and is frequently found in homes, stores, and garbage dumps—wherever there is fruit. It has also spread into more natural habitats and in some areas is the most abundant species of its genus. Morgan began to use *Drosophila* because he was unable to obtain funds for experiments on mammals. *Drosophila* could be raised in large numbers on inexpensive food, at first bananas, in small milk bottles, which Morgan apparently appropriated from those brought to his home by the local milkman. A few other laboratories were using *Drosophila* at the same time, and there has been much speculation about where Morgan obtained the stocks of those famous flies. There is no reason to believe that he had a single source. When I was a student at Columbia in the 1930s, the source was remembered as a pineapple on the window sill outside Morgan's laboratory.

The discovery of the white-eyed male was credited to Calvin Bridges. He was at the time a Columbia College undergraduate, hired to wash the dirty fly bottles. Just before washing one he noticed a fly with white eyes. Bridges is remembered as the person in the Fly Room with the sharpest eyes for detecting new mutants, and in just a few years there were 85. All in the Fly Room were active and successful in discovering mutant alleles. Sturtevant discovered many new mutant alleles even though he was color blind.

If one searches for mutants of *Drosophila*, or any other organism, merely by keeping a culture going and examining the individuals of each new generation, new mutants are exceedingly rare. *Drosophila* with *white eyes* or *vestigial wings* are encountered only when thousands of individuals are examined. One suspects that the dedication, focus, and discipline of those working in the Fly Room was the major reason that so much was discovered in such a short time.

One might ask, "Why study so many?" Once it had been established that the Mendelian scheme worked for alleles on the autosomes (all chromosomes except the sex chromosomes) and with modification for alleles on the **X**, why pile confirmation upon confirmation? The answer was simple: the mutant alleles could be used to gain more information about the physical basis of inheritance—the relation of genes to chromosomes, the location of genes, genetic maps of the chromosomes, and various alterations of the structure of the chromosomes themselves.

Linkage and Crossing-Over

When Sutton started it all in 1902, he argued that there must be more pairs of alleles than there were pairs of homologous chromosomes. "We must, therefore assume that some chromosomes at least are related to a number of different [alleles]. If then, the chromosomes permanently retain their individuality, it follows that all the [alleles] represented by any one chromosome must be inherited together."

That statement is a necessary deduction from the hypothesis that genes are parts of chromosomes. That means that the number of these groups of alleles inherited together cannot exceed the number of pairs of homologous chromosomes. The deduction can be tested by determining the number of chromosomes by microscopic examination and the number of linkage groups by breeding experiments.

It was found that *Drosophila melanogaster* has four pairs of chromosomes—three pairs of autosomes and a pair of sex chromosomes. In mitotic metaphase there are two pairs of long bent autosomes and one pair of tiny dot-shaped autosomes. In females the two **X** chromosomes are rods of medium length and in males there is one **X** and a hook-shaped **Y**.

In the early months of experimentation, the Morgan group quickly found that a number of different genes were linked and that their pattern of inheritance suggested strongly that they were part of the **X** chromosome. That is, their pattern of inheritance followed that of the *white-eyed* allele shown in figure 60. It was reasonable to conclude, therefore, that these sex-linked genes were part of the **X** chromosome seen in cytological preparations.

The test for linkage in the autosomes was to see if in a dihybrid cross one observed a 1:1:1:1 ratio in the cross of an F_1 with the pure recessive. For example, in a cross of **AA BB** × **aa bb**, the F_1 would be **Aa Bb**. Such an individual crossed with **aa bb** should produce 1/4 offspring with both the A and B characteristics, 1/4 with the A and b characteristics, 1/4 a and B, and 1/4 a and b. If **A** and **B** were parts of the same chromosome, however, one would expect the half of the offspring to show the A and B phenotype and half the a and bb phenotype. When the rapidly increasing number of new mutations were tested in this manner, the results were equivocal. Although the AB and ab predominated, there was a small percentage of the recombinant phenotypes, Ab and aB. That is, linkage was strong but not complete. For the moment we will ignore these exceptions to complete linkage.

Two other linkage groups, in addition to the sex-linked genes, were found, and it was assumed that these were associated with the two pairs of long autosomes—it seemed more likely that the large chromosomes would have more genes than the tiny dot autosomes, which might consist of only a few undiscovered genes or even none at all. The latter seemed to be the case for the **Y** chromosome.

Eventually one mutant fly was discovered and, when crossed with flies with mutant alleles of the three known linkage groups, there was independent assortment. It was highly probable, therefore, that this new mutant gene was part of the dot-shaped autosomes. Eventually other genes were discovered to belong to this fourth linkage group, but to this day there have been very few.

By 1915 the Morgan group had worked out the inheritance of 85 genes. These fell into four linkage groups, as shown in table 2, which also shows the metaphase chromosomes. The parallelism between the number of chromosomes as determined by cytological examination and the number of linkage groups as determined by genetic experiments was strong evidence not only that genes are parts of chromosomes but also that those on the same chromosome will be inherited together.

The data of table 2 are instructive in another way. Notice that many different genes affect the same character: 13 influence eye color; 33 modify the wings in some manner; 10 affect the color of the body. What, then, determines the normal *red eye* color? The answer is that the wild-type alleles of all these 13 eye-color genes, together with many discovered later and others yet to be discovered, act together to produce the normal wild-type *red eyes*. If an individual is homozygous for the mutant allele of any one of these genes, the eye color will not be *red*, but *white*, or *peach*, or *sepia*, or one of many other colors. We should think of the normal *red eye* color as the end product of a series of gene actions. If any of these actions is altered, the eye color will be different.

It is important to realize also that there is more to the compound eye of an insect than its color. There are many other genes that influence the morphology of the eyes—some drastically as in the case of *eyeless* in the fourth linkage group or the *bar eye* mutant of the X chromosome.

The mutant alleles were named for their most visible effects—the *white eye* allele produces white eyes. When the *white-eye* flies were carefully examined, however, it was found that not only was the pigmentation of the eye affected but the coloration of some of the internal organs as well. This is not an unusual case— many genes are pleiotropic, that is, they affect more than one structure or process. Some geneticists in the early days even went so far as to suspect that every gene affects, at least in some small way, all aspects of structure and function of the body.

But now back to those exceptions to complete linkage. They simply could not occur if the two different genes are part of the same chromosome. Thus as in our model, if **A** and **B** are on the same chromosome and **a** and **b** on the other homologue in an **Aa Bb** individual, there would be complete linkage. The few individuals with the *Ab* and *aB*

Table 2. Linkage groups in the genes of Drosophila.

GROUP I		GROUP II	
Name	*Region affected*	*Name*	*Region affected*
Abnormal	Abdomen	Antlered	Wing
Bar	Eye	Apterous	Wing
Bifid	Venation	Arc	Wing
Bow	Wing	Balloon	Venation
Cherry	Eye color	Black	Body color
Chrome	Body color	Blistered	Wing
Cleft	Venation	Comma	Thorax mark
Club	Wing	Confluent	Venation
Depressed	Wing	Cream II	Eye color
Dotted	Thorax	Curved	Wing
Eosin	Eye color	Dachs	Legs
Facet	Ommatidia	Extra vein	Venation
Forked	Spines	Fringed	Wing
Furrowed	Eye	Jaunty	Wing
Fused	Venation	Limited	Abdominal band
Green	Body color	Little crossover	II chromosome
Jaunty	Wing	Morula	Ommatidia
Lemon	Body color	Olive	Body color
Lethals, 13	Die	Plexus	Venation
Miniature	Wing	Purple	Eye color
Notch	Venation	Speck	Thorax mark
Reduplicated	Eye color	Strap	Wing
Ruby	Legs	Streak	Pattern
Rudimentary	Wings	Trefoil	Pattern
Sable	Body color	Truncate	Wing
Shifted	Venation	Vestigial	Wing
Short	Wing		
Skee	Wing	GROUP III	
Spoon	Wing	*Name*	*Region affected*
Spot	Body color	Band	Pattern
Tan	Antenna	Beaded	Wing
Truncate	Wing	Cream III	Eye color
Vermilion	Eye color	Deformed	Eye
White	Eye color	Dwarf	Size of body
Yellow	Body color	Ebony	Body color
		Giant	Size of body
GROUP IV		Kidney	Eye
Name	*Region affected*	Low crossing over	III chromosome
Bent	Wing	Maroon	Eye color
Eyeless	Eye	Peach	Eye color
		Pink	Eye color
		Rough	Eye
		Safranin	Eye color
		Sepia	Eye color
		Sooty	Body color
		Spineless	Spines
		Spread	Wing
		Trident	Pattern
		Truncate intensf.	Wing
		Whitehead	Pattern
		White ocelli	Simple eye

phenotype are enough to disprove the hypothesis because they would seem to indicate that genes can migrate from one chromosome to another. Thus, if the hypothesis is to be maintained, a mechanism must be discovered that permits genes to be exchanged between chromosomes. The genetic data demanded a cytological explanation and, in fact, a cytological phenomenon had been described recently that might suffice.

In 1909—the year of the birth of that white-eyed fly—the cytologist F. A. Janssens (1863–1924) had described a chromosomal phenomenon, occurring during meiosis, that Morgan and his associates required for his hypothesis of the exchange of genes between homologous chromosomes (figure 61). It was called crossing-over. During synapsis the homologous chromosomes come close together with their long axes parallel. Both chromosomes replicate and a tetrad of four chromatids is formed. This much can be observed. Next, according to Janssens, there is considerable coiling of the chromatids around one another and, in some cases, two of the chromatids break at the corresponding place on each. The broken chromatids rejoin (recombine) in such a way that a section of one chromatid is now joined with a section of the other. As a result "new" chromatids are produced that are mosaics of segments of the original ones. The breaking and rejoining could not be seen, so this suggestion was but a hypothesis.

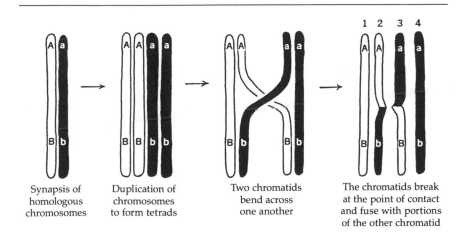

| Synapsis of homologous chromosomes | Duplication of chromosomes to form tetrads | Two chromatids bend across one another | The chromatids break at the point of contact and fuse with portions of the other chromatid |

61   F. A. Janssens's (1909) hypothesis of crossing-over.

The evidential basis for Janssens's hypothesis left much to be desired; nevertheless it was the only acceptable way Morgan saw to explain the data. In a one-page paper Morgan (1911b) proposed a new hypothesis to account for the observation that linkage may not be complete.

> I venture to suggest a comparatively simple explanation based on results of inheritance of eye color, body color, wing mutations and the sex factor for femaleness in *Drosophila*. If the materials that represent these factors are contained in the chromosomes, and if those factors that [are linked] be near together in a linear series, then when the parental pairs (in the heterozygote) conjugate [*i.e.* synapse] like regions will stand opposed. There is good evidence to support the view that [when the tetrad begins to separate] homologous chromosomes twist around each other, but when the chromosomes separate (split) the split is in a single plane, as maintained by Janssens. In consequence, the original material will, for short distances, be more likely to fall on the same side of the split, while remoter regions will be as likely to fall on the same side as the last, as on the opposite side. In consequence, we find [complete linkage] in certain characters, and little or no evidence at all of [linkage] in other characters; the difference depending on the linear distance apart of the chromosomal materials that represent the factors. Such an explanation will account for all of the many phenomena that I have observed and will explain equally, I think, the other cases so far described. The results are a simple mechanical result of the location of the materials in the chromosomes, and the method of union of homologous chromosomes, and the proportions that result are not so much the expression of a numerical system as of the relative location of the factors in the chromosomes. *Instead of random segregation in Mendel's sense we find "associations of factors" that are located near together in the chromosomes. Cytology furnishes the mechanism that the experimental evidence demands.*

The term *linkage* is used for cases where different genes that are parts of the same chromosome are inherited together. *Crossing-over*, which occurs in meiosis, consists of the homologous chromosomes coming together at synapsis, replicating, then breaking, and finally the chromatids rejoining in new ways that result in altered associations of genes.

Thus can we give credit to Th. H. Morgan for having established that those puzzling exceptions to simple Mendelian inheritance are a consequence of the fact that genes are parts of the same chromosome and at times they are reshuffled by crossing-over during meiosis?

In truth, we can do nothing of the kind. All that could be concluded was that linkage *could* be explained if two genes were parts of the same chromosome, and that crossing-over *could* be an explanation of the dissociation of previously linked genes. We credit Morgan with these important insights because later research showed that his hypothesis was correct.

The Cytological Proof of Crossing-Over

Critical proof for crossing-over would be cytological evidence for the actual breaking and recombination of homologous chromatids during meiosis. This may seem impossible since homologous chromosomes are identical in appearance, so even if crossing-over had occurred during synapsis, the chromatids would give no evidence that they had broken and recombined. What was required, therefore, was some experimental means to make the homologues different. In the 1910s there was no obvious way of doing this, and the *Drosophila* group accepted the hypothesis of crossing-over because it continued to explain their data. They had to wait almost 20 years before Curt Stern (1931) was able to provide cytological proof of crossing-over (figure 62).

By the time Stern began his work, *Drosophila* geneticists had a large number of stocks of mutant flies, including numerous ones with chromosomal abnormalities. Some of these had appeared spontaneously, but others were the offspring of flies that had been exposed to radium or X-rays.

Stern used this extensive library of mutants to construct stocks that provided the test material that he needed—female flies with structurally and genetically different homologous X chromosomes. One of the X chromosomes was in two portions: one portion was an independent chromosome and the other was attached to a tiny fourth chromosome. The other X had a piece of a Y attached to it. These structural differences were so great that it was possible to identify the chromosomes in fixed and stained cells.

The two X chromosomes carried genetic markers as well. One part of the divided X carried the recessive allele *carnation* (**c**), which when homozygous produces eyes of a dark ruby color, and the dominant allele *bar* eyes (**B**), which reduces the normal nearly round eyes to a narrow band. The X with the piece of the Y carried the *wild-type* alleles C and **b**, which produce *red* eyes and *normal-shaped* eyes.

During meiosis in the female there was crossing-over between the two loci in some instances but not in others. As a consequence, four types of gametes were produced. Each of these was unique both genetically and structurally. When such a female was crossed with the double recessive—a *carnation, normal-eyed* male—the alleles of each of the gametes of the female were expressed.

The critical evidence was provided by the F₁ females, which were in four phenotypic classes. Furthermore, each phenotypic class had predictably different chromosomes. Stern had set up the crosses in such a manner that the crossover individuals with *carnation, normal-shaped* eyes had two long **X** chromosomes. The other crossover class had eyes that were both *red* and *bar*. Their chromosomes showed one long **X** and one **X** with the piece of the **Y**. The other two phenotypic classes, the noncrossovers, also had unique chromosomal configurations.

Stern checked the chromosomes of nearly 400 of the females of all four classes and found that the phenotypes corresponded to the predicted cytological configurations. This was an elegant demonstration that Morgan's hypothesis of chromosomal crossing-over as the mechanism for genetic recombination was indeed correct.

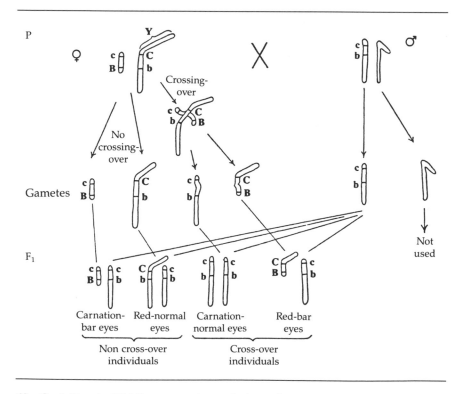

62   *Curt Stern's (1931) cytogenetic proof of crossing-over.*

And history was repeating itself. In a paper published a few weeks earlier Harriet Creighton and Barbara McClintock (1931) had demonstrated that crossing-over has a cytological basis in maize *(Zea mays)*. Their method was similar to that used by Stern for *Drosophila*. They had developed strains of maize with genetically and cytologically different 9th chromosomes. Their evidence was the presence of predicted cytological configurations in the plants with the different phenotypes.

Mapping the Chromosomes

Much of *Drosophila* genetics was anticipated in Morgan's first full-length paper (1911*a*). There he stressed two principal hypotheses: genes are localized in definite places in the chromosome, and they are in a linear order. If one assumes that crossing-over can occur at any place along a chromosome, the chance of one occurring in any one segment will depend on the length of that segment—the longer the segment, the greater the probability that crossing-over will occur somewhere within it. This suggested to Morgan the possibility of mapping the chromosome. As his student Sturtevant noted, "The proportion of 'cross-overs' could be used as an index of the distance between any two factors. Then by determining the distances (in the above sense) between A and B and between B and C, one should be able to predict AC. For, if proportion of cross-overs really represents distance, AC must be approximately, either AB plus BC, or AB minus BC, and not any intermediate value" (1913, p. 45).

Sturtevant then began experiments in which he crossed flies with the mutant alleles on the X with wild-type flies. The F_1 females were then crossed with males carrying the recessive alleles. F_1 males were not used, since Morgan had discovered earlier that crossing-over does not occur in them. The offspring were then scored to see if recombination of the alleles had occurred. Such recombination would indicate that a cross-over had occurred between the loci.

The percentage of crossovers between **y** (*yellow* body) and **v** (*vermilion* eyes) was found to be 32.2 and between **y** and **m** (*minature* wings) to be 35.5. Morgan's hypothesis would suggest that **v** would be slightly closer to **y** than **m** would be. But what could be concluded about the relative positions of **m** and **v**? Sturtevant's prediction was that the distance between **m** and **v** must be either 67.7 (35.5 + 32.2) or 3.3 (35.5 − 32.2). Figure 63 is a diagram of the deduction.

The hypothesis allows a critical deduction and test to be made—

measure the amount of crossing-over between **v** and **m** and see whether the value is approximately 3.2 or 67.7 percent. Sturtevant did the experiment and found the value to be 3 percent. This indicated that **v** and **m** were close together and on the same side of the chromosome as **y**. The close correspondence between the actual and the expected results was strong support for the correctness of the hypothesis.

Similar experiments with other sex-linked genes were the basis of the first genetic map, shown as C in figure 63. Is the genetic map, constructed in this manner, an accurate reflection of the positions of the genes on the chromosomes? Sturtevant has this to say (the genetic symbols have been updated):

> Of course there is no knowing whether or not these distances as drawn represent the actual relative spacial distances apart of the factors. Thus the distance **wv** may in reality be shorter than the distance **yw** but what we do know is that a break is far more likely to come between **w** and **v** than between **y** and **w**. Hence, either **wv** is a long space, or else it is for some reason a weak one. The point I wish to make here is that we have no

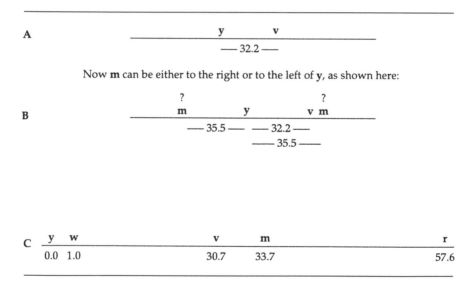

63 *Alfred H. Sturtevant's (1913) method for determining the linear order of gene loci. A shows **y** and **v** separated by a distance equal to the percentage of crossing-over between them. B shows the percentage of crossing-over between **y** and **m** and the impossibility of knowing whether **m** is on the side with **v** or not. C shows the relative position of five loci on the X chromosome.*

means of knowing that the chromosomes are of uniform strength, and if there are strong or weak places, then that will prevent our diagram from representing actual relative distances—but, I think, will not detract from its value as a diagram. (p. 49)

Sturtevant reaches the following conclusion:

These results are explained on the basis of Morgan's application of Janssens' chiasmatype hypothesis to associative inheritance. They form a new argument in favor of the chromosome theory of inheritance, since they strongly indicate that the factors investigated are arranged in a linear series, at least mathematically.

When Sturtevant conducted the experiments that led to the construction of the first genetic map of the X chromosome, he identified the locus of the mutant *eosin,* which when homozygous produces eyes colored like the cytological stain of that name. He also found that, so far as he could see, *eosin* and *white* gave the same crossover percentages relative to adjacent loci. This would seem to indicate that *eosin* and *white* are at the same place on the chromosome.

This is how the Morgan group explained the data:

This example indicates that the conception of allelomorphs should not be limited to two different factors that occupy identical loci in homologous chromosomes, but that there may be three, as above, or even more different factors that stand in such a relation to each other. Since they lie in identical loci they are mutually exclusive, and therefore no more than two can occur in the same animal at the same time. On *a priori* grounds also it is reasonable to suppose that a factor could change in more than one way, and thus give rise to multiple allelomorphs . . .

On the chromosome hypothesis the explanation of this relation is apparent. A mutant factor is located at a definite point in a particular chromosome; its normal allelomorph is supposed to occupy a corresponding position (locus) in the homologous chromosome. If another mutation occurs at the same place, the new factor must act as an allelomorph to the first mutant, as well as to the "parent" normal allelomorph. (Morgan, Sturtevant, Muller, and Bridges 1915, pp. 155–157)

As the years went by, many more mutants were discovered that mapped at the same *white* locus. This is no longer an isolated case. Multiple alleles of the same gene are a common genetic phenomenon.

It should be emphasized, once again, how new insights into the mechanisms of heredity were obtained as the body of information about this one species increased. It was far more profitable for that very active group in the Fly Room to have concentrated their efforts on one species than for them to have studied the genetics of a dozen. With the extensive library of mutant alleles available as early as 1915, all sorts of questions could be asked with a good chance of obtaining acceptable answers. Years later the fact that the bacterium *E. coli* received such concentrated attention meant that its biology was to become the best known of any species.

The Final Proof

During the last two decades of the nineteenth century, the hypothesis that the factors responsible for inheritance were associated with chromosomes was held by only a few prominent cytologists. That hypothesis was given new life by Sutton in 1903, and after 1910 the investigations on *Drosophila* by Morgan, Sturtevant, Bridges, and Muller made it increasingly probable that the hereditary factors—genes—are parts of chromosomes. Yet Bateson and many other experts remained totally unconvinced.

Calvin Bridges is credited with providing the "final proof" of Sutton's hypothesis. Once again, an exception to a known rule was to provide increased understanding. Figure 64 shows the normal inheritance of sex chromosomes in *Drosophila*. The chromosomes of the female are shown in large letters and those of the males in small letters. The x chromosome of the male is transmitted only to his daughters and his y only to his sons. The X chromosomes of the female are transmitted to both sons and daughters. Thus a daughter receives one X from her mother and one x from her father. The son receives his X from his mother and his y from his father. Thus, in a cross of a *white-eyed* female

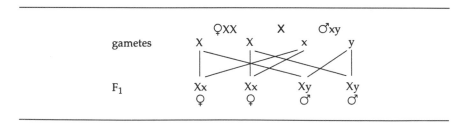

64 The normal inheritance of sex chromosomes in Drosophila melanogaster. *Female chromosomes are in large letters and male chromosomes in small letters.*

and a *red-eyed* male one would expect only daughters with *red* eyes and sons with *white* eyes, nothing more.

Among the many new mutations that appeared, there was a *white-eyed* female that behaved in a most unusual manner. When Bridges (1916) crossed her with a *red-eyed* male, there were some *white-eyed* daughters and some *red-eyed* sons (figure 65). There was no way this could have occurred, given normal inheritance of the sex chromosomes. Those daughters with the *white* eyes could not have received the X of the father; since it carried the dominant allele for *red* eyes, its influence would have prevailed. Therefore these exceptional daughters must have inherited their sex-linked genes from the mother alone. The exceptional *red-eyed* sons demanded a similar explanation. Since their X chromosome would normally come only from the mother, and both of her Xs

P	Nondisjunctional white eye ♀ XXY		X	Normal red eye ♂ XY	
Gametes	XY (46%): X (46%) XX (4%): Y (4%)			X (50%) Y (50%)	
F₁	XY (46%)	X (46%)	XX (4%)	Y (4%)	

	XY (46%)	X (46%)	XX (4%)	Y (4%)
X (50%)	**1** XXY 23% Red eye ♀ Would show nondisjunctional behavior if crossed.	**2** XX 23% Red eye ♀ Normal chromosome behavior.	**3** XXX 2% Triploid X. ♀ Usually dies.	**4** XY 2% Red eye ♂ The X has come from the father and the Y from the mother. This is the reverse of the normal situation.
Y (50%)	**5** XYY 23% White eye ♂ With extra Y chromosome.	**6** XY 23% White eye ♂ Normal chromosome behavior.	**7** XXY 2% White eye ♀ Would show nondisjunctional behavior if crossed.	**8** YY 2% Always dies.

65 *Calvin Bridges's (1916) experiment with nondisjunction females.*

carried the allele for *white* eyes, the **X** of the *red-eyed* male must have come from his *red-eyed* father.

The hypothesis Bridges proposed was that the female parent with the exceptional offspring had not only two **X** chromosomes but also one **Y**. During meiosis in this hypothetical **XXY** female, four classes of gametes were imagined to be produced: **X, XX, XY,** and **Y.** (A normal female produces only one class of gametes with a single **X.**) There was no way of predicting the percentages of these gametes, but experiments suggested that 46 percent were **X,** 46 percent were **XY,** 4 percent were **XX,** and 4 percent were **Y.**

These **XXY** flies were called nondisjunction females. The term refers to the fact that in some of the ova there is no segregation, or disjunction, of the two **X** chromosomes and both remain in the ovum, with the **Y** going to the polar body. In normal meiosis, one **X** would pass to the second polar body and the other **X** remain in the ovum.

It must have taken considerable courage to postulate this preposterous hypothesis; yet, if one were to continue to maintain that genes are parts of chromosomes, some such hypothesis was required. The fact of key importance in Bridges's hypothesis is that it could be tested and hence its degree of preposterousness estimated. These are the main deductions:

(1) If the hypothesis is true, we would expect 50 percent of the F_1 daughters to be nondisjunctional females (classes 1 and 7 of figure 65; the percentages shown in the figure are for all the flies, so when females alone are being considered, the values should be doubled). All of the *white-eyed* females (class 7) should be nondisjunctional. The vast majority of the females were predicted to have *red* eyes (classes 1 and 2). These could not be distinguished by their phenotypes but, if used in genetic experiments, half would be normal (class 2) and half nondisjunctional (class 1). Bridges made the crosses and found that this deduction was true.

(2) If the hypothesis is true, we would expect the exceptional males (class 4), that is, those males that inherited their **X** chromosome from their fathers, not to transmit the power of producing exceptional offspring in later generations. They were predicted to behave like normal males. They were tested, and this was found to be true.

(3) If the hypothesis is true, we would expect 46 percent of the males to be **XYY.** These would be expected to produce four

types of sperm: **X, YY, XY,** and **Y.** Such a male crossed to a normal female should produce no exceptional offspring, that is, males inheriting their sex-linked characteristics only from the father and females inheriting theirs only from their mothers. However, every **XY** sperm entering a normal egg with its single **X** would produce an **XXY** daughter. She would be nondisjunctional. These deductions were tested and found to be true. (That last short sentence gives no notion of the huge amount of work involved in this test as well as the others.)

(4) If the hypothesis is true, we would expect that 50 percent of the daughters (classes 1 and 7) would be **XXY**. This deduction was tested by making slides of the chromosomes of many of the females. Figure 66 shows what he found. Approximately half of the females had the normal chromosome set with two **X**s. The other half had the normal autosomes but two **X**s and one **Y**.

These were demanding deductions and elegant tests. Young Bridges concluded "there can be no doubt that the complete parallelism between the unique behavior of the chromosomes and the behavior of the sex-linked genes and sex in this case means that the sex-linked genes are located in and borne by the X-chromosomes." That is a brave, though properly restricted, statement. The only thing the experiments had shown was that, at the time they were conducted, it was true beyond all reasonable doubt that the *white* and *red* alleles were parts of the **X**

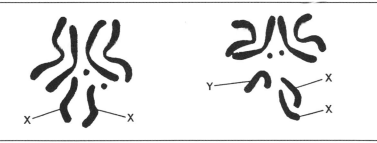

66 Bridges's drawings of the chromosomes of female offspring in the cross shown in figure 65. Approximately half of the females checked had the normal chromosomal complement shown on the left. They would have been class 2. The remaining females (classes 1 and 7) were **XXY**.

chromosome in the strain of *Drosophila melanogaster* used in the experiments.

What, then, is the basis for claiming that these experiments were the final proof that genes are parts of chromosomes—implying that this is true for all genes in all species at all time? Had this been the first genetic experiment done with any organism, Bridges's conclusion just quoted would have been as much as could be said. But it was not the first. In the sixteen years since 1900, an enormous amount of genetic information had accumulated. Many species of animals and plants showed a pattern of inheritance that appeared to be based on simple rules. In fact, there was an underlying uniformity of genetic systems among the species, in contrast to the vast differences in their structure and physiology.

Thus it was reasonable to extend the conclusions based on nondisjunction experiments to the other genes of *Drosophila melanogaster* and to genes of other species and hypothesize that genes are parts of chromosomes in all species.

In 1921 Bateson visited the Fly Room. One of the main events during his visit was a demonstration by Bridges of the chromosomal preparations from the nondisjunction experiments. Bateson, who knew next to nothing about cytology, is said to have gone from microscope to microscope dropping ashes from his pipe over everything. Eventually he announced that he was convinced that genes were parts of chromosomes.

In his lecture at the 1922 meeting of the American Association for the Advancement of Science (AAAS) in Toronto Bateson had this to say:

> We have turned still another bend in the track and behind the gametes we see the chromosomes. For the doubts—which I trust may be pardoned in one who has never seen the marvels of cytology, save as through a glass darkly—can not as regards the main thesis of the Drosophila workers, be any longer maintained. The arguments of Morgan and his colleagues, and especially the demonstration of Bridges, must allay all skepticism as to the direct association of particular chromosomes with particular features of the zygote. The transferable characters borne by the gametes have been successfully referred to the visible details of nuclear configuration.
>
> The traces of order in variation and heredity which so lately seemed paradoxical curiosities have led step by step to this beautiful discovery. I come at this Christmas Season to lay my respectful homage before the stars that have arisen in the west.

The Determinants of Sex

These experiments on *Drosophila*, plus those on many other species, showed that the sex of an individual is determined by the sex chromosomes it receives when the ovum and sperm combine at fertilization. (We now know this general rule is not true for all species.) Thus by 1915 it appeared that the full explanation of sex determination was at hand. An **XX** zygote would become female and **XY** would become a male. But is a female a female because she has two **X** chromosomes or because she has no **Y**? Is a male a male because he has a **Y** or because he has only one **X**? Or is sex determination the consequence of more complex phenomena?

The answer to these questions would seem to require the ability to juggle the chromosomes in ways that would provide tests of deductions. But, once again, a survey of many organisms showed that some had already done the experiment. Recall that in some species males are **XO,** that is, they have only a single sex chromosome. Obviously they do not require a **Y** to be a male. Furthermore, no genes had then been discovered for the **Y** in *Drosophila* so it was thought to be genetically inert. Therefore, the hypothesis that males are males because they have only one **X** and females are females because they have two has some support.

This hypothesis was strengthened by some remarkable flies that appeared in the Fly Room—they were female on one side of the body and male on the other. The same abnormality had been reported in other species, and they were called gynandromorphs. Detailed analysis had not been done, however, and the underlying cause remained unknown.

Males and females in *Drosophila* differ externally in several ways. The males have groups of bristles, the sex combs, on the forelegs, and the posterior portion of the abdomen is solid black, whereas it is barred in the females, which lack sex combs. The genitalia of the two sexes differ considerably. In addition, males are smaller than females.

Cytological study indicated that these gynandromorphs began as normal **XX** females but, through some cytological accident at the very beginning of development, one of the **X** chromosomes was lost from a cell that would form half of the body. The descendants of this cell would have only a single **X** and, hence, have the genotype of a male. The other cell, which would give rise to the other half of the body, would have both **X** chromosomes and be genotypically female. As a consequence, such individuals were males on one side of the body and females on the other. The male side had the sex combs and the solid black posterior abdomen. The difference in body size resulted in this gynandromorph having a bent body—the larger female side bent the body considerably, making the male side concave. The genitalia were typically male on one side and abnormal on the female side. A most interesting group of gynandromorphs was heterozygous for the **X** chromosome alleles *red eyes* and *white eyes*. When these were bilateral gynandromorphs, the flies had a red eye on one side and a white eye on the other.

The hypothesis that an individual *Drosophila* was male or female depending on the number of **X** chromosomes in its cells was made highly probable by these observations. Through some accident of development there had been a juggling of the chromosomes and an important test of deductions had become possible.

The work of Bridges on nondisjunction (1921, 1939) showed that accidental events were producing even more striking chromosomal variations. As a consequence it became feasible to test in new ways the relation of the number of **X** chromosomes to the sex of the individual.

As we have seen, Bridges's **XXY** fly was a structurally normal and fertile female. She was among the first of many individuals discovered in the Fly Room that had abnormal chromosomes. After careful study Bridges gradually came to believe that sex was not determined solely by the number of **X** chromosomes but by some relation between the **X**s and the autosomes. His data suggested little role for the **Y**. The following is a simplified version of his hypothesis:

Recall that a *Drosophila melanogaster* female has three pairs of autosomes and two **X**s. We will use the term "autosomal set" and the letter **A** to apply to the monoploid group of autosomes—three chromosomes, one of each homologous pair. The normal female, therefore, will have two sets of autosomes and a pair of **X**s. The ratio of **X**s to sets of autosomes will be 2 **X**/2 **A** = 1.0. The male will have a single **X** and two sets of autosomes. His ratio will be 1 **X**/2 **A** = 0.5.

A female was discovered that proved to be triploid—three of each kind of chromosome. What would that extra **X** do? A superfemale? Not at all. She was normal and, on the scheme just described, she would be 3 **X**/3 **A** = 1. It seemed, therefore, that the ratios 1.0 = female and 0.5 = male was the rule. Were other combinations possible?

Once a fertile triploid female was available, the possibility of creating chromosomal havoc was at hand. Such a female crossed with a diploid male would produce various sorts of abnormal chromosomal combinations. If any of these new combinations were fertile, they could be used in crosses to further perturbate the chromosomal system.

Some of the various combinations are shown in figure 67. So long as the number of **X** chromosomes equals the number of autosomal sets, the ratio is 1 and the fly is a female. If a fly has two **X** chromosomes but 4 sets of autosomes then the ratio is 0.5 and its sex is male. Thus the hypothesis **XX** = female is true only if there are also two sets of autosomes.

But what would happen if the ratio were between 0.5 and 1.0? The extraordinary thing is that such a question could be asked and answered—the answer being that such flies are intermediate in their sex characteristics. They are called intersexes.

It was possible to increase the ratio to values higher than 1.0 by having more **X** chromosomes than sets of autosomes. These flies, called superfemales, tended to have the female characteristics exaggerated.

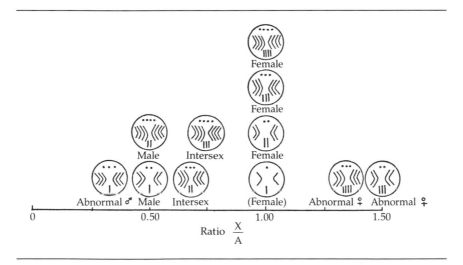

67 *The various combinations of X chromosomes and autosomes obtained by Bridges and others. The lowest circle at ratio 1.00 is a monoploid female. Bridges did not observe such an individual, but some diploid flies were discovered to have monoploid areas in their bodies. If such areas happened to include sex structures, they were of the female type.*

The combinations shown in figure 67, plus others, form a consistent pattern. Sex in *Drosophila*, far from being determined by genes associated with the "sex" chromosome alone, is the result of interactions between genes in the X chromosomes and in the autosomes. The autosomal genes have a net male-forming tendency and the X chromosomes a net female-forming tendency. In a normal male the genes of the two autosomal sets overbalance the genes of the single X. In a normal female the double dose of genes provided by the two Xs overbalance the genes on the autosomes.

And a very general phenomenon received further support: Complex characteristics are based on complex interactions among many genes.

The Conceptual Foundations of Classical Genetics

After thousands of crosses had been made and millions of offspring classified, geneticists of the late 1930s had the satisfying feeling that the big questions that had been asked for centuries had acceptable answers. When genetically unknown species were studied for the first time, the rules of Mendel, Sutton, and Morgan, or acceptable extensions of these rules, accounted for the data. Genetics was the first branch of biology to reach this level of conceptual adequacy. It was inevitable that this would be so. Even though it deals with the most basic problems, genetics is the least complex of the biological sciences. The genotype must be simpler than the phenotype, since what is basic is less conceptually complex than what is derived. The genetic code is essentially universal in the realm of life, while the structure and functions of organisms take their myriad forms.

If one asks what had been accomplished in genetics and cytology the answer might seem unimpressive. It was little more than discovering the rules governing the transmission of genes from parent to offspring. These rules, seemingly universal, held for plants, animals, and microorganisms. They are:

(1) The basic morphology, physiology, and molecular biology of an individual are determined by its inheritance, acting in a defined environment.
(2) Although an individual's material inheritance, the genes, is small in quantity, those genes contain all the genetic informa-

tion necessary for the development of an organism like its parents.

(3) Genes are parts of chromosomes. (Later research was to show that some genetic information is contained in mitochondria, plastids, viruses, and some viruslike bodies.)

(4) Each gene usually occupies a definite site, its locus, in the chromosomes. Understandable exceptions to this concept—inversions and translocations—were known, and examples of the movement of genetic material from chromosome to chromosome have increased with time.

(5) Each chromosome has many genes, except for a few cases like the Y of *Drosophila*, and the genes are arranged in a linear order.

(6) The somatic cells (nonreproductive cells) contain two of each kind of chromosome; that is, they are in homologous pairs. This means that every gene locus is represented twice. There are some well-known exceptions. In some species—bees, for example—queens and workers are diploid females and the drones are monoploid males. Sex chromosomes are another exception where **XO** and **XY** males have only single copies of sex-linked genes. Also some of the cells in some tissues of some animals may be polyploid, as in our livers.

(7) During each mitotic cycle the genes are duplicated from the chemical substances in the cell. Cellular duplication involves a prior genic duplication.

(8) Although genes are characterized by great stability through time, inheritable changes—mutations—do occur. These are inheritable as alleles of the original gene. Mutations are very rare and normally occur for any one gene about once in a million cell generations.

(9) Genes can be transferred from one homologous chromosome to another by crossing-over. This is a normal part of meiosis, but there are a few exceptions, such as the male of *Drosophila melanogaster*, where crossing-over does not occur in genetically active regions.

(10) The meiotic process ensures that each gamete receives one chromosome of each homologous pair. *Which* of the two homologues it receives is a matter of chance. Thus the gametes will receive one or the other of each gene pair (segregation). Each

homologue, with the genes it contains, will be distributed to half of the gametes. **XO** males are an obvious exception.

(11) In the formation of gametes, the segregation of the chromosomes of one homologous pair, with the genes it contains, has no effect on the segregation of the other pairs of homologous chromosomes with their genes. Thus there is independent assortment of chromosomes and the genes they contain.

(12) Fertilization consists of the random union of ova and sperm, each with one chromosome of every homologous pair. Therefore, the zygote receives one chromosome of each homologous pair from the mother and one from the father. Again, sex chromosomes introduce an understandable exception.

(13) When two different alleles of the same locus are present, the individual is heterozygous for that gene. The allele with the greater phenotypic effect is known as the dominant and the other as the recessive. In most cases the heterozygote appears to be identical with individuals homozygous for the dominant allele. Less frequently the heterozygotes are intermediate.

(14) Finally, genes must produce their effects through the production of chemical substances, which in turn control the biochemical reactions of the cell. In the 1930s this was little more than a tentative hypothesis but no other alternative seemed possible. Some geneticists suggested that the genes' major function is to produce specific enzymes, which in turn control the life of the cell.

These 14 propositions account for most of the phenomena of classical, or transmission, genetics. They formed a satisfying conceptual whole. But this was not enough. The inquisitive human mind is more stimulated by what is unknown than by what is known. One knew with great precision how the genes for eye color were inherited but essentially nothing of the structure of those genes or their mode of action. One could sense what would be the concern of the next major paradigm of genetics—carrying the analysis to the level of cells and molecules.

We end this chapter with a homily. Recall Medawar's statement, "The reasons that have led professionals without exception to accept the hypothesis of evolution are in the main too subtle to be grasped by laymen." It is unfortunate that this is the case not only for evolutionary biology but for most of science. If you have labored through the experiments and arguments, especially Bridges's "Final Proof that Genes are Parts of Chromosomes," you will have a good idea whether or not the proofs would be acceptable to a person with no background in biology who maintained, "I do not believe that genes are parts of chromosomes."

Laypersons will usually accept a scientific statement, even if they cannot understand the evidence for it themselves, believing scientists know what they are talking about. The difference in acceptability between statements in evolutionary biology and genetics reflects the fact that no one claims that Genesis is a textbook of heredity.

But acceptance of new and radically different evidence is not easy even for the professionals. Recall Darlington's estimate that, in the mid 1920s and a decade after Bridges's experiments, few biologists in England had accepted the chromosome theory of heredity.

Something to think about.

chapter 17

The Structure and Function of Genes

B y the end of the 1930s no big questions remained in transmission genetics, so the emphasis switched to the difficult questions of "What do genes do?" and "What is the chemical nature of genes?" Of course there had been interest in these questions from the turn of the century but the techniques available offered little possibility of obtaining detailed answers. None of the routine technology of today, such as electron microscopes, radioactive isotopes, computers, chromatography, and unbelievably sophisticated analytical instruments, was available. There was no National Science Foundation, no overhead, and little external source of support for research. Laboratory assistants and post-docs were scarce. Teaching and research were accepted as of equal importance in the operations of the great universities, so less time was available for research. Nevertheless, E. B. Wilson and T. H. Morgan were able to carry out vigorous programs of research and publication while carrying a teaching load that would astonish most cutting-edge biologists today.

Despite these constraints, the crude probes available before 1953 made possible important discoveries in gene function. Among the probes were those developed for studying enzymes. During the first half of the twentieth century one of the most vigorous fields of cell biology and biochemistry was the study of enzymes. Enzymes were viewed as major factors in making life possible. The sorts of reactions that were known or suspected to occur in cells simply could not take place without these organic catalysts, which vastly speed up the rate of reactions.

In one of those strange episodes in the history of ideas, genes and enzymes were first linked at a time when very little was known about either. An infant patient of the English physician Archibald E. Garrod (1857–1936) had a rare disease—alkaptonuria. It was so named because the urine of people with this disease has alkapton bodies, which consist largely of homogentisic acid. That substance becomes dark red or black when oxidized. A clue to the patient's problem was stained diapers or, since the baby was British, nappies.

Garrod knew that the baby's parents were first cousins, and he wondered if alkaptonuria might be an inherited disease. Garrod spoke of alkaptonuria and similar ailments as "inborn errors of metabolism." In 1902 he consulted Bateson, who suggested that the disease might be due to recessive alleles. Bateson proposed an explanatory hypothesis:

> Alkaptonuria must be regarded as due to the absence of a certain ferment which has the power of decomposing the substance alkapton. In a normal body that substance is not present in the urine, because it has been broken up by the responsible ferment; but when the organism is deficient in the power to produce that ferment, then the alkapton is excreted undecomposed and the urine is coloured by it. (1913a, p. 233)

Neither Garrod nor alkaptonuria is mentioned in any of the books written by the Morgan school in their years of active discovery. Even if Morgan knew of Garrod's hypothesis, he may have ignored it. Morgan was so pro experimental science and anti all else—including nonexperimental science—that he would have viewed Garrod's hypothesis as useless, for he had written:

> It is the prerogative of science, in comparison with the speculative procedures of philosophy and metaphysics, to cherish those theories that can be given an experimental verification and to disregard the rest, not because they are wrong, but because they are useless.

But part of the answer may lie elsewhere. When research programs are developing rapidly and productively, as they were for the *Drosophila* workers, there is little stimulus to look for new things to do. It was not until the 1930s, with transmission genetics satisfactorily explained, that geneticists began an intensive study of the sorts of problems that interested Garrod.

One Gene, One Enzyme

George W. Beadle (1903–1989), Edward L. Tatum (1909–1975), and Boris Ephrussi (1901–1979) were leaders in the quest for information on how genes act. By the late 1930s there was considerable information about cell metabolism. That fundamental reaction of all life,

$$C_6H_{12}O_6 + 6O_2 \rightarrow 6H_2O + 6CO_2,$$

had been resolved into several dozen separate reactions, each controlled by a specific enzyme.

The elucidation of this one metabolic pathway had required the efforts of many scientists for many years. One of the major problems is the speed of cellular reactions, often requiring but a fraction of a second. How was one to study a reaction that would be over before the investigator even knew it had started? The standard way was to use chemical substances ("enzyme poisons") that would block the action of a specific enzyme. The result would be that the substrate for that enzyme would then accumulate in the cell and could possibly be detected and identified.

Assume, for example, that one metabolic pathway in cells involves molecule A being changed into molecule B and then B into molecule C and then down the alphabet to molecule Z. We will assume that the change from A to B is controlled by enzyme A-ase (*ase* is the ending which identifies a chemical molecule as an enzyme) and from B to C by B-ase and from Y to Z by Y-ase. All we know at first is that the cell changes molecule A to molecule Z. That is, the conversion might be accomplished by a single enzyme in a single reaction.

If the well-known enzyme poison, cyanide, is used, no Z is formed, and instead a previously undetected molecule, M, is found. What can we conclude? We say that the cell converts A to Z in a minimum of two steps: A is converted to M and then M to Z—of course, there may be other steps between A and M and M and Z. Other poisons could be tried, and with time more and more could be learned about normal metabolism by throwing these chemical wrenches into the biochemical gears of the cell.

Some early studies of Beadle and Ephrussi on the way that eye-color genes of *Drosophila* produce their effects had indicated that the hypothesis "one gene, one enzyme" might be a fruitful approach. The biochemistry of *Drosophila* proved to be too complex to test that hypothesis, and for the first time that noble animal let a geneticist down. So a long-standing experimental technique was invoked: if the experiment cannot be done with one organism, search for another one that is suitable. By this time Beadle was at the California Institute of Technology with

Morgan. Before Morgan left Columbia, Bernard Dodge of the New York Botanical Garden gave him a culture of the red bread mold *Neurospora crassa,* in the belief that it might be useful in genetic experiments. But Morgan stuck to *Drosophila* and never used *Neurospora.* It was still being cultured in his laboratory when Beadle and Tatum sought an organism for their research.

Beadle and Tatum reasoned that lethal mutations change alleles so that they are incapable of producing enzymes essential for the life of the organism. Thus they intended to induce lethal mutations with radiations and to study their biochemical effects. This might appear to be a considerable problem since, if the lethal mutation kills the individual, there would not appear to be much to investigate. But Beadle and Tatum solved that problem in what was surely one of the most innovative and productive lines of experimentation in the late 1930s and 1940s. Others must have thought so too, because Beadle and Tatum shared a Nobel Prize, with Joshua Lederberg, in 1958 for this work.

For reasons that will shortly become apparent, they first had to determine exactly the minimum variety of molecules required for the normal growth of *Neurospora*—the minimal medium. The menu was surprisingly simple: air, water, inorganic salts, sucrose, and the vitamin biotin. *Neurospora* is, of course, composed of innumerable organic compounds, all interacting as the life of that organism. Yet from those few raw materials it is able to synthesize the amino acids, proteins, fats, carbohydrates, nucleic acids, vitamins, and other substances of its body.

As an example of the many experiments done by Beadle and Tatum, we will discuss those concerned with the synthesis of the amino acid arginine. The working hypothesis was that specific genes control the production of specific enzymes, which catalyze the reactions that end with arginine. Presumably these genes could mutate to allelic forms that would either be unable to make the enzyme or not be able to make it in needed amounts. Since arginine is essential for the life of *Neurospora*, such mutations would be lethal.

Beadle and Tatum then devised a method for the production of these lethal mutations, for identifying them as related to the synthesis of arginine, and for maintaining them in culture in order to work out the metabolic pathway of arginine synthesis. This may sound impossible, especially when we realize that for most of its life cycle *Neurospora* is monoploid and hence lethal mutations could not be carried as heterozygotes.

This was their game plan. First, X-rays were used to induce mutations. They assumed that all sorts of mutations would be produced, but by chance some might be involved with the production of arginine. When we remember how rare any specific mutation would be, the chance of obtaining the desired mutations would be exceedingly small.

Huge numbers of spores from irradiated *Neurospora* were placed on the minimal growth medium. Most of them grew, showing that whatever mutations may have occurred, none was so serious as to prevent the *Neurospora* from synthesizing all of its substance from the few chemicals in the minimal medium. Other spores did not germinate, and among these might be some biochemical mutants that could not produce the enzymes necessary for normal growth and development. And somewhere among them might be genes involved in the synthesis of arginine. How could one find them? The spores were not germinating, so they were for practical purposes "dead."

The solution of this difficulty was elegant in its simplicity and effectiveness. If the spores could not grow because they could not synthesize their own arginine, it would be simple to provide them with that amino acid. This was done and the spores that had not germinated on minimal medium were checked to see if they would now. Again most of the spores did not grow but a precious few did. Among these precious few might be mutants of genes involved in arginine synthesis.

The next, and critical, step in the analysis was to make sure that whatever was wrong with the spores was inherited. It could not be concluded that, just because the otherwise "lethal" spores could grow on arginine, a mutational event was the cause.

The life cycle of *Neurospora* makes it ideal for some sorts of genetic analysis. The colonies are monoploid for nearly their entire life. There are two mating types, A and a, which cannot be distinguished except by their mating behavior. If colonies of A and a are grown together, parts of each will fuse and A nuclei will unite ("fertilize") with a nuclei to form diploid zygotes. Meiosis occurs immediately and 4 monoploid spores are formed. These divide, by mitosis, to produce 8 monoploid spores, which are enclosed in an elongate spore sac (ascus). The spore sacs can be opened under a microscope and the individual spores removed and placed in culture media. Thus one can obtain all of the products of meiosis of a single zygote.

The presumed mutant strains were crossed to normal strains. Meiosis occurred immediately afterwards and monoploid spores were formed. These were then isolated. Half were found to grow on the minimal medium and half only if arginine was added. These results were consistent with the hypothesis that the wild-type *Neurospora* had a gene **A,** which was necessary for the synthesis of arginine. The radiation treatment had caused a mutation of **A** to **a,** and **a** was unable to play some essential role in arginine synthesis.

The experimental procedure appeared to be working and numerous genetic strains were isolated that required arginine for growth. Were all the genetic strains alike or had different genes mutated to alleles that could not synthesize arginine?

There were two possible answers: First, all of the mutant strains could be due to changes at a single gene locus. Second, many different loci could have mutated. In this case one would suspect that many genes are involved in arginine synthesis: A_1, A_2, A_3, A_x, and so on. Any one of these could have mutated to a_1, a_2, and so on. In all these mutants the same phenotype would be observed—inability to grow on minimal medium without arginine, yet their inability to grow would be for different reasons. Crosses could test the alternatives, as follows:

If a single locus was involved, a cross of two strains would produce spores unable to grow without arginine. Alternatively, if different loci are involved, some of the spores will grow as wild-type colonies for the following reason. Assume that different genes are involved and we are crossing a_1 × a_2. If a mutation had occurred at a different locus in each strain, which is overwhelmingly probable, the mutated strain would have a normal allele at the other locus. Thus, mutant strain a_1 would be expected to have A_2. Strain a_2 would be expected to have A_1. Thus a cross of a_1A_2 × A_1a_2 would produce diploid zygotes with a genotype A_1a_1 A_2a_2. Meiosis then occurs and monoploid spores are produced. If the two loci are on different chromosomes, the isolated spores should give these results:

1/4 should be A_1A_2 and grow on minimal medium
1/4 should be A_1a_2 and will require arginine since a_2 cannot function.
1/4 should be a_1A_2 and require arginine since a_1 is not functioning.
1/4 should be a_1a_2 and require arginine since neither allele can function.

From among the many thousands of spores tested, Beadle and Tatum discovered seven genetically different mutants, each requiring supplemental arginine if it was to grow normally. Various interpretations of the data were possible, but Beadle and Tatum preferred the hypothesis that the synthesis of arginine required that at least seven normal genes be present—each producing an essential enzyme. When any one of these genes mutated in such a way that its specific enzyme could not be produced, the synthesis of arginine was blocked. There was no reason to believe, of course, that there are only seven steps in the synthesis of arginine in *Neurospora*. We can conclude only that seven was the minimum number.

It was possible to extend the analysis by taking advantage of what was already known about the synthesis of arginine. In 1932 the biochemist Hans A. Krebs had discovered that in some vertebrate cells arginine is formed from citrulline, citrulline from ornithine, and ornithine from an unknown precursor. A specific enzyme is required for each transformation.

$$? \rightarrow \text{ornithine} \rightarrow \text{citrulline} \rightarrow \text{arginine}$$

If *Neurospora* has a similar metabolic pathway, one should be able to determine how the seven mutant strains are involved. This could be done by seeing which, if any, of the seven would grow if either citrulline or ornithine could replace arginine.

Many experiments were done. Four of the mutant strains would grow if either ornithine, citrulline, or arginine was added. This suggested that these four mutants were involved in reactions before the ornithine stage. If ornithine or the other substances were added, the remaining enzymatic down-stream steps, being normal, could carry the reactions to arginine.

Two of the strains would not grow if only ornithine was added, but they would grow if either citrulline or arginine was added. In these cases the block was between ornithine and citrulline. Since two genetically different strains were both blocked between ornithine and citrulline, it is reasonable to conclude that there are at least two steps between these molecules.

Finally, one strain was found that would grow only if arginine was added. This suggests that some enzyme between citrulline and arginine was deficient or defective.

Thus, Beadle and Tatum were able to conclude that, for *Neurospora* to synthesize arginine, a minimum of seven enzyme-controlled reactions are required and a minimum of seven kinds of molecules are involved. Two of these are known: ornithine and citrulline.

The hypothesis that one function of genes is to control the production of specific enzymes was supported. One could not conclude that this is the only thing genes do. Beadle and Tatum had designed their experiments solely to detect enzymes involved in metabolic pathways.

Much as Sutton had linked cytology and genetics in the early 1900s,

Beadle and Tatum effectively linked genetics and biochemistry in the early 1940s. Their type of experimentation was used immediately by numerous other investigators on other molds, yeasts, and bacteria. This approach led directly to the molecular biology of today.

While all this was going on still another attempt to study genetics at the molecular level was under way. This was a line of investigation that began in the 1920s and ultimately led to the positive identification of the gene as DNA. That will be the next topic, bringing us to the current paradigm of genetics formulated by Watson and Crick in 1953.

The Substance of Inheritance

The dynamics of scientific discovery elude us to this day. There is no way of knowing the who, the what, and the where. Important discoveries are nearly always made by scientists active in the field, but the breakthrough may be made by an outstanding scientist or by a newcomer. Among the latter, Mendel, Sutton, Morgan, Watson, and Crick were not leaders in the field of inheritance when they began to make their notable contributions. Many of the advances in genetics that followed from Watson and Crick's famous discovery of the structure of DNA were due in part to scientists from other fields, mainly physics, deciding that the problems in biology were more exciting than those in physics. Many prominent molecular geneticists of midcentury remember being made aware of new possibilities for genetic research by a slender book *What Is Life?* written by Schrödinger (1945), himself a physicist. It could be that it is easier for those not steeped in the data and traditions of a field to see problems and solutions.

But some important discoveries are the outcome of deliberate attempts to find answers to specific questions. The elegant experiments of Beadle and Tatum are an example of experiments planned to test a specific hypothesis. In other cases discovery is more of an accident. Progress along the road to DNA started with some chance observations on disease-causing bacteria.

In 1928 Frederick Griffith, a Medical Officer with the British Ministry of Health, was a bacteriologist studying diseases of human beings. His publications give no evidence of an interest in genetics, yet his experiments were an important step in identifying DNA as the genetic material.

Pneumonia in human beings and many other mammals is caused by the pneumococcus bacterium, properly known as *Diplococcus pneumon-*

iae. As in many disease-causing microorganisms, there are numerous genetic strains called Type I, Type II, and so on. They are distinguished by chemical composition of the polysaccharide capsule that surrounds the cell.

If capsulated cells are grown on culture plates, they form colonies that are *smooth* and shiny. In any large number of colonies, some have a different appearance—they are *rough*. There was considerable medical interest in this phenomenon because the *smooth* cells cause pneumonia but the *rough* cells do not. The *smooth* cells have the polysaccharide capsules but the *rough* cells do not.

Griffith knew that if he injected mice with capsulated Type II *smooth* cells, they would die but that noncapsulated Type II *rough* cells would not cause their death. Was the capsule the cause? In a further experiment, the pathogenic *smooth* cells were killed with heat and injected into mice. The mice did not die. Since heat did not destroy the coat, it was concluded that the polysaccharide coat was not the cause of death.

The next experiment gave a most unexpected result. Griffith gave four mice a double injection of Type II cells: living nonpathogenic *rough* cells plus dead pathogenic *smooth* cells. Survival was expected, since the *rough* cells are not pathogenic and the pathogenic *smooth* cells had been killed. Nevertheless, all four mice died after five days. Type II *smooth* cells were found in their blood, whereas the only live cells that had been injected were *rough*. Thirty control mice injected only with living *rough* cells remained healthy.

This was an unbelievable result—but the experiment was repeated and confirmed. It appeared that the ability to synthesize a capsule had been transferred from the dead capsulated cells to the living noncapsulated cells. Any geneticist of 1928 who might have known of these experiments would have shuddered and rededicated himself to *Drosophila melanogaster*.

But during those years geneticists ignored microorganisms almost entirely and microbiologists ignored genetics. It was not suspected by either group that microorganisms possessed a genetic system remotely similar to that of higher organisms.

This line of research was taken up by many bacteriologists, including Oswald T. Avery and M. H. Dawson of the Rockefeller Institute (now University) in New York. They became convinced that transformation must be due to some chemical substance, and it was reasonable to suspect the polysaccharide of the capsule. Nevertheless, that proved not to be so. It was found that the transforming principle could be extracted from capsulated cells and that transformation could occur *in vitro*—no need that mice be used. After a decade Avery, MacLeod, and McCarty reported that they had purified a substance that would transform *rough* cells into *smooth* cells. It was almost certainly deoxyribonucleic acid, or DNA. They came to this conclusion after many tests. For example, the overall chemical composition of the transforming principle agreed closely with that of DNA. The molecular weight was judged to be about 500,000. The substance was highly active—one part in 600 million was effective in causing transformation. Treatment with trypsin and chymotrypsin left activity intact, indicating that the substance was not protein. The enzyme ribonuclease, which denatures RNA, was also without effect. However, a crude preparation of deoxyribonuclease, which acts on DNA, destroyed the activity of the transforming substance.

This is how Avery, MacLeod, and McCarty (1944) interpreted their experiments:

> Various hypotheses have been advanced in explanation of the nature of the changes induced. In his original description of the phenomenon Griffith suggested that the dead bacteria in the inoculum might furnish some specific protein that serves as a 'pabulum' and enables the [noncapsulated] form to manufacture a capsular carbohydrate.
>
> More recently the phenomenon has been interpreted from a genetic point of view. The inducing substance has been likened to a gene, and the capsular antigen which is produced in response to it has been regarded as a gene product. In discussing the phenomenon of transformation Dobzhansky has stated that "If this transformation is described as a genetic mutation—and it is difficult to avoid so describing it—we are dealing with authentic cases of induction of specific mutations by specific treatments . . ."
>
> It is, of course, possible that the biological activity of the substance described is not an inherent property of the nucleic acid but is due to minute amounts of some other substance adsorbed to it or so intimately associated with it as to escape detection. If, however, the biologically active substance isolated in highly purified form as the sodium salt of deoxyribonucleic acid actually proves to be the transforming principle, as the available evidence strongly suggests, then nucleic acids of this type must be regarded not merely as structurally important [at the time biochemists could not discover any function for the nucleic acids] but as functionally active in determining the biochemical activities and specific characteristics of [the bacterial] cells. Assuming that the sodium deoxyribonucleate and the active principle are one and the same substance, then the transfor-

mation described represents a change that is chemically induced and specifically directed by a known chemical compound. If the results of the present study on the chemical nature of the transforming principle are confirmed, then nucleic acids must be regarded as possessing biological specificity the chemical basis of which is as yet undetermined.

Was DNA only an inducing agent or was it something else? Most geneticists would probably have agreed with Dobzhansky that DNA could not be the genetic material but that it could induce genetic change. The evidence was fairly convincing. Enough was known about DNA to realize that it was a rather simple molecule—composed of a few bases, a simple sugar, and phosphate. Presumably an extremely complex substance would be required to be the genetic material and control the life of cells. Proteins were far more likely candidates than DNA to be genes. They could be huge and were composed of a number of amino acids about equal to the number of letters in our alphabet. Just as the combinations of a few letters can give us the uncounted numbers of words in the languages of the world, that same number of amino acids combined to form huge protein molecules should be adequate to supply all the genetic variation required.

The solution came in less than a decade: DNA *is* the gene, not a mutagenic agent. One of the more important experiments was done in 1952 by A. D. Hershey and Martha Chase when much more sophisticated experimentation was possible. As the result of work on the atom bomb in World War II, many sorts of radioactive substances had been produced that could be used to study intracellular reactions. Methods were developed for culturing different sorts of microorganisms, and they were becoming the favorite experimental organisms for geneticists. There was also very much more research being done. The extraordinary contributions of scientists to the war effort were recognized in Washington, and scientific research began to be supported on a lavish scale. It was estimated that in the 1950s the number of active scientists was equal to all the scientists who had ever lived. Big Science was national policy and a national activity.

An extremely minute form of life, the bacteriophage, or phage, is incapable of an independent existence. Phages are parasites of bacteria, upon which they depend for their own reproduction. Hershey and Chase took advantage of the peculiar life cycle of bacteriophage to ascertain whether or not DNA contains the genetic information of that organism.

If the bacterium *Escherichia coli* is infected with a phage called T$_2$, the bacterium is killed in about 20 minutes. Before entrance of the phage, the bacterial cell was synthesizing its own specific molecules: bacterial proteins, bacterial nucleic acids, and so on. The phage changes all this. It assumes control of the bacterial synthetic machinery and diverts it to producing phage molecules instead of *E. coli* molecules. About 100 phages are made in about 20 minutes, at which time the bacterium bursts and liberates them. Phages then enter other bacterial cells and repeat the process.

Structurally, phages are simple, being composed of a protein coat and a DNA core. The protein coat contains sulfur but little or no phosphorous. The reverse is true for the DNA core. The coat and core can be labeled differently, the coat with radioactive sulfur S^{35} and the core with radioactive phosphorous P^{32}.

The Hershey–Chase experiment was as follows: One group of bacteria was grown in a medium with P^{32}, which became incorporated in the bacterial molecules. Later phages were introduced. When the bacterium then began to synthesize new phages, the latter's DNA became tagged with the P^{32}. The phage protein coat would have little or no label. In a parallel experiment bacteria were grown in a medium containing S^{35}. This became incorporated in some of the bacterial proteins. Later phages were introduced and in this case the protein coats of the phages became labeled with S^{35}.

These two sorts of phages, one labeled for the protein coat and the other for the DNA, were then used in separate experiments. They were introduced into cultures of bacteria, and Hershey and Chase found that the labeled phage DNA entered the bacterial cells. The labeled protein remained on the outside.

These observations, together with others, suggested that the phage attaches itself to the cell wall of the bacterium and injects its DNA core, the coat remaining on the outside. The phages in both experiments reproduced and destroyed the bacterial cells. The experiments had shown that, since only the phage DNA enters the bacteria, the entire genetic information on "how to make phage" is contained in the phage DNA.

The Watson–Crick Model of DNA

The experiments of Avery and his colleagues and those of Hershey and Chase were not immediately accepted as proving that the gene was DNA. Genes have to do awesome things, and it was far from clear that

DNA could do much of anything. Convincing proof that DNA is the gene would have to await proof that DNA could do the things that genes are known to do.

Classical genetics had shown that genes are transmitted from generation to generation, are parts of chromosomes, can replicate with great precision, can mutate to new alleles, and control the life of the cell. But classical genetics had not provided evidence for their chemical nature. When it was clear that genes are parts of chromosomes, the next step was to ascertain the chemical nature of chromosomes. Of course it would have been more direct to study the chemistry of genes, but there was no way to do that. The best that could be done was to ascertain the chemical nature of chromosomes, hoping that genes were their main component.

But that was not a simple problem. Chromosomes form a minute fraction of the total mass of a cell, and before the middle of the twentieth century there was no way to obtain a large sample of chromosomes that was not contaminated by other cellular substances. But, as is so often the case, a broad search will reveal that nature can provide the required material. Sperm was one such material. Cytological work had shown that the nucleus is the major component of sperm and that chromosomes are the major component of the nucleus. Thus sperm were analyzed and it was found that their nuclei, and presumably their chromosomes, are composed mainly of proteins and DNA.

In the 1950s optical techniques were developed that made it possible to measure the amount of DNA in cells. It was found that tetraploid somatic cells have twice the amount of DNA as diploid cells. Biochemical measurements showed that sperm have half as much DNA as diploid somatic cells. This was important but not critical information.

From the hypothesis that DNA is the hereditary material four deductions follow:

(1) Cells must have the ability to replicate DNA, that is, make exact copies of it. That requirement sounds simple enough, but how can molecules be reproduced? The most prominent suggestion was that the cell must have a mold, or template, that could stamp out copies of DNA.

(2) It must be shown that DNA carries the hereditary information. But how can a molecule carry information? Among the better-known molecules, hemoglobin carries oxygen, and enzymes

speed the reactions in cells, but one had the impression that they had been "told" to do those things—it was not their idea.

(3) DNA must be able to translate this information in ways that control what cells are and do. Again, this requirement called for totally new and unknown chemical processes.

(4) DNA must be able to mutate. That is, it must be able to change in such a manner that it ceases to carry one type of genetic information and begins to carry another.

It was difficult to see how a chemical molecule could do such complex things, and the first step in finding out would necessitate knowing the chemical structure of DNA. James D. Watson, an American biologist born in 1928, and his English associate, Francis Crick, born in 1916, found out, and their discovery initiated the incredible progress in the biological sciences that has characterized the last half of the twentieth century. They worked together at Cambridge University in England, where there was also a group using X-ray diffraction techniques to obtain information about the shape of DNA molecules.

Watson and Crick neither performed a single experiment nor made a single observation on DNA. In this stage in their careers they were pure theoretical biologists. Their extraordinary contribution in formulating a hypothesis for the structure of DNA was the brilliant analysis of the few available facts. And they were indeed few.

(1) Biochemists had found that DNA consists of just six kinds of molecules: four bases—adenine, guanine, thymine, and cytosine—plus the sugar deoxyribose, and phosphoric acid. These molecules combine to form four kinds of nucleotides as follows: adenine, guanine, thymine, and cytosine each combine with deoxyribose and phosphoric acid to form adenine nucleotide, guanine nucleotide, and so on. These four nucleotides are so commonly mentioned, and their names so long, that they are usually abbreviated A, G, T, C.

(2) X-ray diffraction data suggested that DNA consists of two long strands twisted around one another to form a double helix. These data suggested also that the double helix had a uniform thickness of about 20 Ångstroms.

(3) Another key bit of information, discovered by Erwin Chargaff, was the relative amounts of A, G, T, and C. Although the amounts varied from species to species, in the cells of any one

species the amounts of A and T were equal, as were the quantities of G and C. T and C were known to be small molecules and A and G were about twice as large. Thus, the combinations that were equal in amount, that is, T + A or G + C, would consist of one small and one large nucleotide. The two combinations would be the same size.

The problem then was to put all these molecules together and end up with a structure that could do what genes do—be replicated, mutate, carry genetic information, and translate that genetic information so it could be used by cells. Watson and Crick assembled the available facts about the chemical composition of DNA and the information that X-ray diffraction gave about size and shape. They made paper cutouts in the shapes and relative sizes of the molecules to see how they could be fitted together within the size limits known from X-ray diffraction data.

Their model, or hypothesis, was as follows: The main axis of each strand of the double helix consists of alternating phosphate and deoxyribose units. One nitrogenous base, either A, G, C, or T, is attached to each sugar (deoxyribose; figure 68). The two strands of the double helix are held together by hydrogen bonding—the chemical attraction between the atoms on the bases of one strand with atoms of the bases of the other strand through shared hydrogen atoms. Since the X-ray diffraction data suggested that the double helix of DNA had a uniform diameter, that meant that one small base, C or T, and one large base, G or A, must be opposite one another (figure 69). That hypothesis provided an explanation for the observation that the relative amounts of bases T = A and G = C. Thus T bonds with A and G with C.

Watson and Crick argued (1953a) that the two strands were complementary in the sense that what is present on one automatically specifies what is on the other:

> In other words, if an adenine forms one member of a pair, on either chain, then on these assumptions the other member must be thymine; similarly for guanine and cytosine. The sequence of bases on a single chain does not appear to be restricted in any way. However, if only specific pairs of bases can be formed, it follows that if the sequence of bases on one chain is given, then the sequence on the other chain is automatically determined.

Thus, if the sequence of bases on one strand is A-A-C-T-G-T, that on the other must be T-T-G-A-C-A (figure 70). Then follows a modest suggestion:

It has not escaped our notice that the specific pairing we have postulated immediately suggests a copying mechanism for the genetic material.

And here it is (1953*b*):

Previous discussions of self-duplication have usually involved the concept of a template, or mould. Either the template was supposed to copy itself directly or it was to produce a "negative," which in its turn was to act as a template and produce the original "positive" once again. In no case has it been explained in detail how it would do this in terms of atoms and molecules. Now our model for deoxyribonucleic acid is, in effect, a *pair* of templates, each of which is complementary to the other. We imagine that prior to duplication the hydrogen bonds [between the A and T and G and C on the two strands] are broken, and the two chains unwind and separate. Each chain then acts as a template for the formation on to itself of a new companion chain, so that eventually we shall have *two* pairs of chains, where we had only one before. Moreover, the sequence of the pairs of

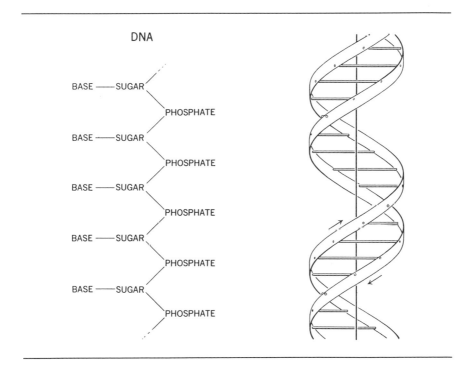

68 *The structure of DNA, redrawn from the original paper of Watson and Crick:* Nature *(1953) 171:965.*

bases will have been duplicated exactly . . . Despite [some] uncertainties we feel that our proposed structure for deoxyribonucleic acid may help to solve one of the fundamental biological problems—the molecular basis of the template needed for genetic replication. The hypothesis we are suggesting is that the template is the pattern of bases formed by one chain of the deoxyribonucleic acid and that the gene contains a complementary pair of such templates.

Thus the Watson–Crick model accounts for the important genetic fact that genes can make exact copies of themselves.

Their model for DNA also suggested a mechanism to carry genetic information. Since the sugar-phosphate axis is always the same and

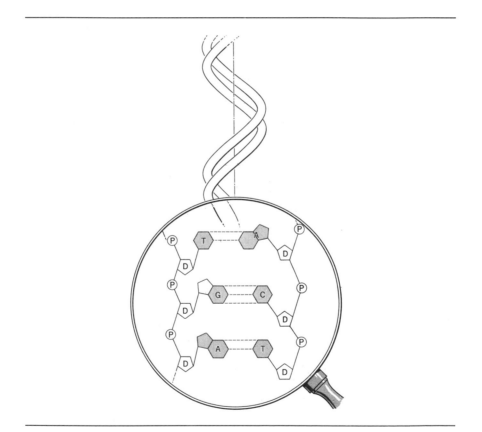

69   *A highly schematic representation of the DNA double helix. A = adenine, T = thymine, G = guanine, C = cytosine, D = deoxyribose, and P = phosphoric acid.*

only the sequence of the bases can vary, "it therefore seems likely that the precise sequence of bases is the code that carries the genetical information." During the next decade additional chemical and genetic data showed that the hypothesis was true beyond a reasonable doubt.

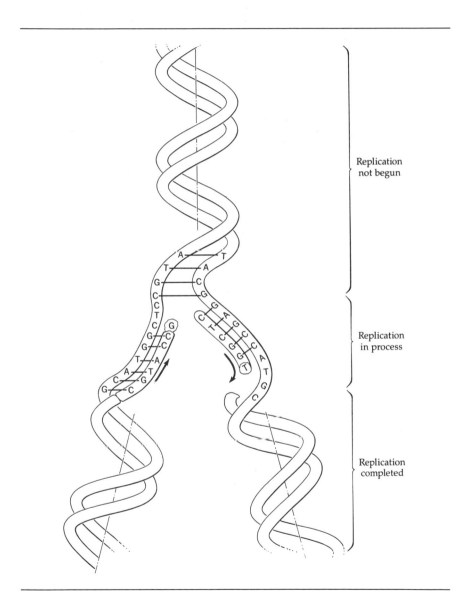

Replication not begun

Replication in process

Replication completed

70   *The replication of DNA. The two strands of the double helix separate, and each strand serves as a template for the formation of a complementary strand.*

In 1962 Watson and Crick shared a Nobel Prize with Maurice Wilkins, who had supplied much of the X-ray data indicating that DNA is a double helix of uniform diameter.

Three important, and closely interrelated, attributes of genes remained to be explained: how genetic information, the code, could be stored in a sequence of nucleotides, how this information of the code could be translated in ways that direct molecular events in cells, and how the code could be changed by mutation.

Genes and the Synthesis of Proteins

When Watson and Crick proposed their model for the structure of DNA, the hypothesis that the chief function of genes is to control protein synthesis was becoming increasingly probable. A large body of work, much of it on *Neurospora*, implicated genes in the production of enzymes, which are proteins. Another well-documented case was related to the genetic disease sickle cell anemia. This research involved a technique known as electrophoresis by which different proteins can be separated. The technique causes molecules to migrate at different speeds depending on their electrical charge. Genetic studies had previously shown that sickle cell anemia is caused by an autosomal recessive allele of a gene involved in the synthesis of hemoglobin. Linus Pauling and his colleagues in the late 1940s found that hemoglobin of healthy humans and the hemoglobin of those with sickle cell anemia differed slightly from each other. Heterozygous individuals produced both kinds of hemoglobin.

Further studies by Vernon Ingram in the late 1950s discovered the molecular basis. Hemoglobin consists of about 600 amino acids linked in two alpha chains and two beta chains. Ingram found that the normal and sickle cell hemoglobins differ by a single amino acid in each of their beta chains: at a specific site where normal hemoglobin has a glutamic acid, sickle cell hemoglobin has a valine.

The next step in the analysis would be to understand how a sequence of nucleotides of DNA can control the synthesis of proteins. Could DNA not only serve as a template for enzymes to make copies of itself but also for enzymes to link amino acids together to make proteins? That hypothesis did not seem likely for one important reason. The chromosomes with their DNA are in the nucleus, whereas most of the proteins are in the cytoplasm—and separated from the nucleus by the nuclear membrane, which is impermeable to large molecules. If proteins were made in the nucleus they could never escape to the cytoplasm, except during mitosis, and somatic cells almost never divide.

There was also a suggestion that another kind of nucleic acid, ribonucleic acid or RNA, might be involved. Fairly specific staining methods had shown that RNA is restricted mainly to the cytoplasm and DNA mainly to the nucleus. The two nucleic acids are much alike except that RNA has uracil instead of thymine as one of its four bases and its sugar is ribose instead of deoxyribose.

In the early 1950s there was really no way to obtain accurate information on how genes make proteins. Whole cells were just too complex. That is, each of the many hundreds of reactions would be a minute fraction of the entire metabolism of the cells. That meant that the substances interacting and their end products would be in such minute amounts that detection would be essentially impossible. Furthermore, the reactions occur with such speed as to make them essentially undetectable. Very special methods and materials were necessary to dissect single reactions—such as how genes make proteins—in the turbulent life of a living cell. And when the problem was solved it was found that, far from there being one reaction, genes → proteins, there were many different steps.

During the mid 1950s new and valuable methods for studying cellular reactions were made possible by the invention of the ultracentrifuge. The tremendous forces generated by an ultracentrifuge can fractionate substances in aqueous media even though their densities are nearly equal. If a large mass of cells is ground up—homogenized—and then centrifuged, the heavier particles, such as nuclei and unbroken cells, are thrown to the bottom of the centrifuge tube. Smaller and less dense particles form fractions above the nuclear layer. The successively higher fractions are mitochondria, endoplasmic reticulum plus ribosomes, and finally a supernatant, which is a liquid free of all but the smallest particles.

It was discovered that some of the fractions, or combinations of them, can carry out some of the reactions that occur in normal cells. For example, the fraction containing the mitochondria is capable of carrying out some oxidative reactions. Of greater interest for the argument being developed here was the discovery that protein synthesis could occur if the fraction containing the ribosomes was added to the supernatant.

In parallel experiments by a very large number of molecular biologists, it was discovered that DNA serves not only as a template for making more DNA during replication but for making three kinds of RNA:

messenger RNA (mRNA), ribosomal RNA (rRNA), and transfer RNA (tRNA). Messenger RNA carries the message "how to make a protein" from DNA to the cytoplasm. There it comes in contact with the ribosomes with their rRNA. Transfer RNA combines with amino acids and brings them to the mRNA bound to the ribosomes. It is there that the amino acids separate from the tRNA and are linked with one another to form the huge proteins. Specific enzymes are required for all these steps.

That briefly shows how the hereditary information of DNA is translated in ways that direct events in the cell. With the aid of three sorts of RNA and enzymes, DNA guides the production of enzymes and other proteins that control the reactions in cells. The DNA message becomes encoded in mRNA in a process similar to DNA making a copy of itself. However, important differences are that uracil, U, replaces thymine, T, and ribose replaces deoxyribose. Thus if the sequence of bases of the DNA is G-T-A-C-A-A, the sequence on the corresponding part of mRNA will be C-A-U-G-U-U.

The Genetic Code

The biochemical stage was now set to answer the question of all questions, "What is the message?" This was made possible by the discovery of an enzyme that permitted RNA to be made *in vitro*. That suggested the possibility that the four RNA nucleotides A, G, C, and U could be joined to make an RNA strand, which might act like real mRNA. W. M. Nirenberg and J. H. Matthaei (1961) reasoned that the message would consist of unique sequences of nucleotides, as Watson and Crick had hypothesized, and each sequence would be related to one amino acid. There was no way of knowing how many nucleotides were required for a message, so they constructed an artificial mRNA in which the number would be irrelevant—it was a long strand consisting only of U, hence called poly U. To the centrifuged cell-free fractions that could make protein they added poly U, plus an abundance of each of the 20 amino acids to see which one(s) would be used to form protein. Protein did form, and it consisted solely of long chains of the amino acid phenylalanine. No other amino acid was incorporated in the protein when the message was U U U . . . That message carried all the information needed to tell the cell machinery to make a protein composed solely of phenylalanine.

These experiments did not indicate how many uracils were needed for each phenylalanine. However, it would have to be more than one because there are 20 amino acids and only 4 RNA nucleotides. Nor would 2 RNA nucleotides be sufficient because there are only 16 (4^2) combinations of 2 nucleotides (the sequence of the nucleotides could be significant also; that is, A U and U A could be different). The minimum number of nucleotides is 3 because this will provide a total of 64 combinations (4^3). Three proved to be correct—three nucleotides, a triplet, code for an amino acid. The triplet, or codon, for phenylalanine is UUU.

Table 3 shows which amino acid each of the codons will insert into a protein. In most cases two, or even up to five, different codons can specify the same amino acid. Only methionine and tryptophan are specified by a single codon. Leucine, arginine, and serine can be coded by any one of five codons. Three codons—U A A, U A G, and U G A —do not code for any amino acid. These are "stop" codons, and when one of them is reached as a protein chain is being formed on a ribosome, synthesis stops.

Once the genetic code had been worked out, a hypothesis for mutation became possible—it is just a mistake in the DNA nucleotide sequence. And a mistake can be serious. When the single codon GAG, among the hundreds present in the nucleotide sequence of DNA that controls the production of hemoglobin, mutates to GTC, the phenotypic consequence is sickle cell anemia.

By the 1960s it had been proven beyond reasonable doubt that DNA had all the basic properties of genes: the ability to be replicated, to carry hereditary information, to translate that information, and to mutate. Far more information about the properties of DNA was obtained subsequently, and is still being obtained, but with those accomplishments we can end the story.

So all the basic questions about inheritance seem to have a general answer. But caution must be exercised with any such broad and seemingly final statement. Without a knowledge of what happens, the conclusion might be drawn that both genes and DNA can alone replicate themselves out of the molecular equivalent of whole cloth. But such is not the case. DNA is synthesized from four sorts of nucleotides, which are themselves the end products of numerous reactions within cells. As Arthur Kornberg's experiments in the late 1950s showed, these reactions require specific enzymes whose production resulted from numerous other complex reactions. Formation of these enzymes involved mRNA, tRNA, and rRNA, and they in turn were stamped out on the DNA templates. The replication of DNA involves other enzymes, including those that repair (nearly all) of the mistakes that are sure to happen. Energy is required for these innumerable reactions, and the cell has a most complicated series of enzymatically controlled reactions and intermediate products that transfer energy in a step-wise manner.

Table 3.   The 64 codons of messenger RNA and the amino acids they specify.

Codon	Amino acid coded	Codon	Amino acid coded
UUU	phenylalanine	CUU	leucine
UUC	phenylalanine	CUC	leucine
UUA	leucine	CUA	leucine
UUG	leucine	CUG	leucine
UCU	serine	CCU	proline
UCC	serine	CCC	proline
UCA	serine	CCA	proline
UCG	serine	CCG	proline
UAU	tyrosine	CAU	histidine
UAC	tyrosine	CAC	histidine
UAA	(stop)	CAA	glutamine
UAG	(stop)	CAG	glutamine
UGU	cysteine	CGU	arginine
UGC	cysteine	CGC	arginine
UGA	(stop)	CGA	arginine
UGG	tryptophan	CGG	arginine
AUU	isoleucine	GUU	valine
AUC	isoleucine	GUC	valine
AUA	isoleucine	GUA	valine
AUG	methionine	GUG	valine
ACU	threonine	GCU	alanine
ACC	threonine	GCC	alanine
ACA	threonine	GCA	alanine
ACG	threonine	GCG	alanine
AAU	asparagine	GAU	aspartic acid
AAC	asparagine	GAC	aspartic acid
AAA	lysine	GAA	glutamic acid
AAG	lysine	GAG	glutamic acid
AGU	serine	GGU	glycine
AGC	serine	GGC	glycine
AGA	arginine	GGA	glycine
AGG	arginine	GGG	glycine

In fact, there appears to be nothing in a living cell that is not in some way required for DNA to be replicated. The unique thing about DNA is that it is replicated on its own template. In the last analysis *the cell* is the smallest unit that can self-replicate. Even viruses cannot self-replicate—they must get a cell to do it for them.

In the last few years RNA has been moving to center stage. Not only is it essential for syntheses that occur in cells, but we now know that some microorganisms have RNA, instead of DNA, as their hereditary material. Some biologists reasoned that this might be evidence of degeneration, but there is a much more exciting possibility. It now seems as though RNA might have been the controlling molecule at the dawn of life on earth, and under some circumstances had the ability to control its own replication.

The nucleic acids will continue to provide us with a deeper understanding of life, and already we recognize a unity of life never anticipated. The general principles of inheritance have been found to apply to all organisms—from microbes to mammals. Possibly the most startling discovery of all is that the genetic code is universal. From the simplest bacterium to you and me, UUU in mRNA puts a phenylalanine into a protein chain. And that can lead to an equally startling deduction: for all creatures living today there was a single origin and, therefore, all are related to one another.

part four

THE ENIGMA OF
DEVELOPMENT

chapter 18

First Principles

The basic problem of developmental biology is to explain how a single cell, the fertilized ovum or zygote, can give rise to the many different types of cells of adults, all molded into a functioning individual. Typically the ovum is a very generalized spherical cell with a monoploid set of chromosomes from each parent. By comparison, the cells of the adult exist in groups—the tissues—that may differ markedly from one another. The explanation of how this differentiation can occur has been long sought by philosophers and scientists. For millennia they could only speculate because the required information, techniques, and tools for experimentation were not available.

Developmental biology was not unique in this regard. Many important questions about nature remained unanswered for millennia, and many remain so today, mainly because there were or are no methods for initiating the inquiry. Some of the early questions about disease, for example, could not be answered until microscopes had been invented and previously invisible pathogens could be observed. In fact, satisfying answers to many basic biological questions were unobtainable until the invisible world of life could be entered with the techniques of microscopy and biochemistry.

The Peripatetic Stagirite

Aristotle, that Greek of universal intelligence, not only established the discipline of embryology but posed the major questions that have lasted to this day. His success was due not only to his own genius but also to the fact that he adopted the naturalistic approach favored by the Ionians.

Aristotle's *Historia Animalium,* a general biology of animals, discusses the structure, breeding habits, reproduction, behavior, ecology, distribution, and relationships of animals. It contains a large amount of factual material that is used for the more theoretical considerations of *De Generatione Animalium,* his treatise on developmental biology.

Aristotle knew that many animals produced visible eggs and that from these the young slowly developed. But since other species did not seem to have eggs, there must be "something" more basic. Aristotle was familiar with the early embryos of numerous animals, and his most complete description is of the developing chick. When this is read by a biologist today, it sounds so familiar that one tends to forget the tremendous intellectual steps that Aristotle took when he wrote: "Development from the egg proceeds in an identical manner in all birds." This implies that Aristotle was familiar with development in at least a few other species and, assuming a basic uniformity of natural phenomena, felt secure in extending the conclusions based on a few species to all species of birds.

This belief that nature is not capricious is a necessary premise for all scientists. That is, similar phenomena have similar causes. Today we feel confident that, for all intents and purposes, the genetic code is universal, yet that confident feeling is based on acceptable data for no more than a trivial fraction of one percent of all species—and today we know that of the many species of birds studied, development is fundamentally the same in all.

Of great importance was Aristotle's use of data from as many different species as he could obtain. This is so basic to biology today that we accept it as the obvious thing to do. When the comparative method is used, one sees variations in the phenomenon being studied—with some species giving a glimpse into one part of the process and another species giving a glimpse into a different part. Each species is an experiment in living and, when all of the observations have been made, there is a better chance of understanding the fundamentals of the phenomenon. In the early days of cytology and genetics this procedure was basic for establishing the concepts of those fields. Observations on the kidney of the goosefish were basic for the discoveries of how the mammalian kidney functions. Time and time again it has been found that if you cannot obtain an answer from one species, try another and you may succeed.

Aristotle reported that the first visible indications of an embryo in an incubated hen's egg came after three days, but that it is earlier in small species of birds and later in large species. At this time the heart appears as a tiny red spot, it pulsates, and what we now call the vitelline veins are seen to be carrying blood. A little later the body differentiates and the head, with very large eyes, becomes visible.

All these observations were made without a microscope, so he was working at the limits of the unaided eye. His description of the much larger embryo at 10 days is more complete. He reported that the head and eyes are relatively large and that the main internal organs are visible. He provides a fairly accurate description of the embryonic membranes. He even dissected the eye of the 10-day chick. In Chapter 7 of *Historia Animalium* he described a human embryo of 40 days, when it was as big as one of the larger ants. His observations appear to have been on an aborted embryo.

The analytical mind of Aristotle reasoned that the embryo must be formed out of something, by something, into something. The "out of something" is the life-giving material in the semen of the male plus the material substance supplied by the female. The "by something" is assumed to be carried in the semens of both parents. When it turns "into something," that is, develops, Aristotle considered two possibilities. Some philosophers held that all parts of the embryo's body form at the same time. Aristotle refuted this hypothesis by observation—the heart in the chick embryo appears before the lungs. One cannot deny the validity of this observation, said Aristotle, by suggesting that the lungs are too small to see because, in fact, they are larger than the heart and hence should be visible first. New things do appear in the course of development.

Thus, in this longest of any debate in embryology—preformation versus epigenesis—Aristotle came down on the side of epigenesis. Preformation assumes that all parts of the adult body are already present in the ovum, whereas epigenesis assumes that the different parts appear in sequence.

He made clear his belief that the semen transmits only the potential for the embryo's structures, not the actual structures themselves. The potential is in the female's contribution, to which the male provides the mechanism for potential to become actual.

But Aristotle obviously was not always correct in his biology. He was convinced, for example, that spontaneous generation was the rule for some creatures, implying that embryonic development is not the rule for all species. This belief was based on many observations of the apparent generation of some insects from decaying matter and the appearance of marine invertebrates on pots and other objects placed in the sea.

If one accepts these observations, which Aristotle probably obtained from others, one must conclude that inheritance and development are extremely labile and easily influenced by external conditions. Like need not beget like all of the time, and genetic continuity cannot be true for all species. It took a large amount of careful observation and experimentation by many naturalists, from Redi (1626–1698?) to Pasteur (1822–1895), to bell that cat—spontaneous generation.

Aristotle sought to understand by observing as broadly as possible; realizing that general concepts might emerge from the study of the same phenomenon in a variety of species; recognizing the fundamental similarity of development in fish, bird, and mammal; arguing for a physical basis of inheritance; providing argument and observation to support epigenesis; and suspecting that problems of development and regeneration are similar. And there are a host of minor observations of great interest. For example, one reads with incredulity his realization that nails, hair, and horns all form from skin. But as Dante was to say in the *Divine Comedy* (*Inferno*, Canto IV), Aristotle was the "Master of them that know."

Joseph Needham, the famous embryologist and even more famous historian of Chinese science and technology, credits Aristotle with extraordinary accomplishments:

> [Aristotle] stood at the very entrance into an entirely unworked field of knowledge; he had only to examine, as it were, every animal that he could find, and set down the results of his work, for nobody had ever done it before . . . The extraordinary thing is that building on nothing but the scraps of speculation that had been made by the Ionian philosophers, and on the exiguous data of the Hippocratic school, Aristotle should have produced, apparently without effort, a text-book of embryology of essentially the same type as Graham Kerr's or Balfour's [1919 and 1880] . . . The depth of Aristotle's insight into the generation of animals has not been surpassed by any subsequent embryologist, and, considering the width of his other interests, cannot have been equalled. (1959, p. 42)

One might have anticipated that the combination of those "right-thinking" Ionians and that omnivorous observer and speculator about nature, Aristotle, would have stimulated a vigorous investigation of embryonic development, as well as other biological problems. Not at all. A peak was reached with Aristotle, followed by a decline of interest and accomplishment for roughly two millennia.

The next major figure who wrote on embryological matters, five centuries after Aristotle, was the Greek physician Galen. He was interested mainly in human anatomy and physiology, but he did have a few things to say about development. Though he added little to Aristotle's work, his views were important since Galen was *the* authority in Western Europe during the early Medieval period when Aristotle's biology was not known. The following quotation gives the flavor:

> Genesis [=development], however, is not a simple activity of Nature, but is composed of alteration [=tissue formation] and of shaping [=organ formation]. That is to say, in order that bone, nerve, veins, and all other tissues may come into existence, the *underlying substance* from which the animal springs must be *altered*. In order that the substance so altered may acquire its appropriate shape and position, its cavities, outgrowths, attachments, and so forth, it has to undergo a *shaping* or formative process. One would be justified in calling this substance which undergoes alteration the *material* of the animal, just as wood is the material of a ship, and wax of an image. (1916, p. 19)

Looking back, with the knowledge of what was to come, we can say that Galen was defining the fundamental problem of development—differentiation. The formative material must be "altered" since the early embryo, which is to become that adult, lacks the tissues and organs characteristic of the adult. The conversion itself would involve a variety of morphogenetic movements. Thus, novelty would appear in the course of development; therefore, Galen was an epigeneticist.

The Death and Rebirth of Scientific Thought

Galen's death in AD 200 marks the end of embryology for over thirteen centuries. The American Revolution seems remote to most of us, yet those thirteen centuries were more than six times the interval between our national birth and today.

The political and social stability sustained by the Roman Empire had been swept aside by degeneration from within and invasion from without. The rise of Christianity and the establishment of the Christian Church as the only effective institution in the West changed the topics for serious thought. The problems of the natural world were replaced by those of the supernatural world. The ability to read and write became a rare skill—to be sure there was very little to read apart from theology. What education there was consisted mainly of instructions for those seeking a career in the Mother Church. Those with interests in science were infrequent, as they always had been, and insufficient to form the critical mass that is essential for sustained scientific progress. There were no universities where science was taught, no scientific academies, and few libraries. Essentially no Greek science, except for Galen, was available in the West.

But even if these constraints had not existed, *what* was one to do in order to extend Aristotle's and Galen's analysis of development? The answer was far from obvious. The major questions they had raised were not really approachable until the nineteenth century, when it first became possible to work at the cellular level.

It is most unlikely, however, that fascination with the mystery of development, especially human development, ever ceased. Cleopatra (69–30 BC), for example, sponsored some gruesome research in developmental biology using some of her pregnant slaves as material: "It happened that Cleopatra, the Queen of Alexandria, presented to the physicians some of her maids who had been condemned to death and they were dissected. It was found that the male embryo is complete after forty-one days and the female embryo after eighty-one days" (quoted in Kottek 1981). The dissections were done at "known intervals of time from conception, following the precepts of Hippocrates with regards to the hen's eggs" (Needham 1959, p. 65). Other versions of the account mention no differences in the times males and females are "complete."

There were isolated observations on embryos during the Middle Ages and early Renaissance, the developing hen's eggs being the usual object of study. By the time of Albertus Magnus (1193?–1280), the works of Aristotle were becoming available in the West from Arab sources, and Albertus was a close student of the Master's works. In a proper Scholastic manner, he described the development of the chick, but seemingly his information came only from Aristotle, not from an opened egg. That was standard procedure for the Middle Ages.

Leonardo da Vinci (1452–1519) observed human embryos and left us some beautiful drawings of them, and there are many other fragments of embryological observations and speculations. It is more realistic, however, to renew the narrative of embryology with Hieronymous Fabricius of Aquapendente (1533?–1619), who for most of his active life was a crusty professor of medicine at the University of Padua, where William Harvey was one of his students.

The basic problem of trying to formulate questions that could be answered still proved elusive, so the study of development concentrated on what was possible—describing normal development. This is not an unworthy goal. Science seeks to associate and conceptualize the phenomena of nature. That activity, quite obviously, depends on knowing what the phenomena are.

Fabricius's *De Formatione Ovi et Pulli*, mainly about the chick, and *De Formato Foetu*, mainly about mammalian development, date from about 1600. Chicken eggs were opened daily after the beginning of incubation and the embryos were studied and drawn. No magnification was used (compound microscopes were in the process of being invented). No wonder he said that in the 4-day chick the body looks like a very tiny flea. By the fifth day, however, Fabricius could make out the head, eyes, heart, arteries, veins, liver, and lungs (Adelmann, 1942).

In this description of the developing chick, Fabricius combined the work of Aristotle and Galen with his own. When we remember that these three span nearly 2,000 years, the advances made in embryology appear most modest. There is a strong Scholastic streak that makes Fabricius most reluctant to disagree with his illustrious predecessors, especially Aristotle. Nevertheless, he helped to keep alive an interest in the subject, he corrected some of the errors of Aristotle and Galen, and he added a few of his own.

Harvey and Malpighi

The advent of William Harvey (1578–1657) began "the transition from the static to the dynamic conception of embryology, from the study of the embryo as a changing succession of shape, to the study of it as causally governed organization of an initial physical complexity" (Needham 1959, pp. 116–117). Harvey was a student of Fabricius at Padua from 1598 to 1602, and this is where he first studied embryology. His *Exercitationes de Generatione Animalium* was published in 1651, half a century after his teacher's book on the same subject. In it he gives a detailed account of the chick's development in which the observations and opinions of earlier students, mainly Aristotle and Fabricius, are confirmed, extended, and corrected. Thereafter he discusses more general matters where the cogency of the argument replaces direct observation.

For example, he rejects the old Aristotelian view that the female contributes only the substance (menstrual blood) and the male the effective generating stuff (male semen): "For the egg is to be viewed as a conception proceeding from the male and the female, equally endued with the virtue of either, and constituting a unity from which a single animal is engendered" (pp. 270–271). After watching the daily changes in the developing chick, Harvey accepts epigenesis as the mode.

> The structure of these animals commences from some one part as its nucleus and origin, by the instrumentality of which the rest of the limbs are joined on, and this we say takes place by the method of epigenesis, namely, by degrees, part after part; and this is, in preference to the other mode, generation properly so called. (p. 334)

That other mode seems restricted to insects where there is a conversion of a caterpillar into a butterfly "already of a proper size, which never attains to any larger growth after it is first born; this is called metamorphosis. But the more perfect animals with red blood are made by epigenesis, or the superaddition of parts" (pp. 334–335). This hypothesis of epigenesis is strengthened by another belief that all parts of the body are derived from the same basic materials:

> For out of the same material from which the first part of the chick or its smallest particle springs, from the very same is the whole chick born; whence the first little drop of blood, thence also proceeds its whole mass by means of generation in the egg; nor is there any difference between the elements which constitute and form the limbs or organs of the body, and those out of which all their similar parts, to wit, the skin, the flesh, veins, membranes, nerves, cartilages, and bones derive their origin. For the part which was at first soft and fleshy, afterwards, in the course of its growth, and without any change in the matter of nutrition, becomes a nerve, a ligament, a tendon; what was a simple membrane becomes an investing tunic; what had been cartilage is afterwards found to be a spinous process of bone [a remarkable conjecture but based on what Harvey could see], all variously diversified out of the same similar material. (p. 339)

A final important hypothesis of Harvey was that all life comes from eggs. The frontispiece of the 1651 edition of *De Generatione Animalium* shows Zeus opening an egg from which emerge a bird, human being, katydid, porpoise (?), deer, snake, spider, lizard, and various plants. An inscription appears on the egg: *ex ovo omnia*. This is usually expanded to *omne vivum ex ovo*, but that precise phraseology does not appear. "Exercise the Sixty-Second" carries the title "An egg is the common origin of all animals." Harvey is not using "egg" in the customary restricted sense since, quoting Aristotle, he accepts spontaneous generation as a mode of origin for some creatures.

Apart from the possible researches of Cleopatra, embryological studies before Harvey were the work of males. As such they were well aware of the male's contribution to conception, but that of the female was confusing. Aristotle had recognized several patterns of generation. Oviparous females like the hen laid eggs from which the young hatch. Ovoviviparous females such as sharks and some snakes have eggs, but these were retained in the body until hatching. Viviparous females of human beings and other mammals, however, puzzled Aristotle and many who followed him. These females bear living young and seemingly have nothing that corresponds to eggs in the oviparous species. Menstrual blood or some other secretion was assumed to be the female's contribution to conception. Harvey refuted this notion because he could find nothing in the uteri of deer immediately after mating that might be the beginning of the new individual. This is not surprising since most mammalian eggs are about the size of the period at the end of this sentence. Nevertheless it was assumed that the female must contribute something.

The answer seemed to come from observations of de Graaf (1672) on the mammalian ovary, which at the time was called the *testis muliebris* (female testis). Its function, if any, was unknown. Harvey thought that it had no role in copulation or generation. De Graaf found that some ovaries had spherical structures, now known as Graafian follicles, and he suspected they might be the long-sought mammalian eggs or be "egg nests." This made it seem more reasonable, to many at least, that Harvey's dictum *ex ovo omnia* might be correct. Subsequently it was established by von Baer that Graafian follicles are not eggs but structures in which the eggs are formed.

Marcello Malpighi (1628–1694), an Italian biologist and professor at the University of Bologna, followed Harvey by a generation. His scientific contacts, however, were mainly with the Royal Society of London with which he corresponded actively—describing his latest discoveries in great detail and receiving encouraging letters from the Secretary.

The Royal Society published his two main works on the development of the chick (1672, 1675; see Adelmann, 1966). They consist of minute descriptions of what he could see not only with the unaided eye but also with magnification—he was one of the first biologists to use the rapidly improving microscopes of the day, being able to obtain magnifications as high as 143X. Malpighi had no deep interest in causal factors, and in this sense he contributed little. His descriptive embryology, however, was masterful.

Malpighi found that he could remove the early embryo from a chick egg and place it on a glass slide for study. This simple technique, followed to this day, made it far easier to use a microscope in making observations.

A Two-Millennial Summing Up

The study of development can be divided into two main categories: descriptive and analytical. Until recently the first has been primarily a morphological discipline. The course of development, from conception to maturity, was described in detail—how the embryo changes and grows. Included also were guesses about whatever it is that parents contribute to their offspring at the time of conception. Analytical or experimental embryology is concerned with the mechanisms of embryological change, that is, how whatever it is that parents contribute to their offspring is converted into a new individual. Thus descriptive embryology addresses the question "What happens?" and analytical embryology asks "How does it happen?"

What had Aristotle, Galen, Fabricius, Harvey, and Malpighi accomplished in descriptive and analytical embryology? Not much—nor could they, without good microscopes and histological techniques. Aristotle was the one most interested in concepts and causes (analytical embryology), Malpighi the least. At the conceptual level one could have passed directly from Aristotle to the eighteenth century and lost almost nothing. But the embryologists were not uniquely unsuccessful. Progress was slow in all fields of biology and, for that matter, in all fields of science. Notable progress was made in only some aspects of physics and astronomy.

There are valid reasons for this conceptual stasis. Concepts must be based on data, and during those long millennia the necessary data were unavailable. The data for biology were to come from a then invisible level of analysis—the level of cells and their parts. First there had to be microscopes and then came knowledge of cells. Microscopes, though inadequate, became available in the late seventeenth century. On April 15, 1663, Robert Hooke reported to the Royal Society his observations on cells in cork; yet nearly two centuries were to pass before cells became important in embryological explanations.

A few general principles of development seemed to be true beyond reasonable doubt. All were known to Aristotle, and this is a measure of the lack of significant conceptual progress. Here is the balance sheet.

(1) Sexual reproduction, the interaction of males and females, is required for the production of new individuals in many species. It was assumed that there must be some material contribution, but it was not known what it might be.

(2) Both sexes influence the characteristics of the offspring, but the mechanism of this influence was not understood. This means that not only was the basis of genetic continuity a complete mystery but so also were the mechanisms of transforming that basis into a new individual of the same type as the parents. There was no clear distinction between transmission of material and transformation of that material into a new individual.

(3) The embryos of different species of the same major group, birds for example, resemble one another closely. There are even resemblances among various species of vertebrates—mammals, birds, and fishes.

(4) Development appeared to be epigenetic, although Aristotle and later workers could not be sure, since the naked eye and, later, the microscopes available were not adequate to show any minute beginnings.

Clearly the Scientific Revolution did not produce any vast improvement in the understanding of development. In fact, its effect on the life sciences as a whole was slight. Vesalius produced a better human anatomy than Galen's, but no conceptual breakthrough was involved. Physiology started grandly with Harvey's observations and experiments on circulation, but thereafter progress was exceedingly slow.

One could argue that a knowledge of embryology did not have high priority among scholars and hence progress would be slow and episodic. True enough—but this cannot be the entire explanation since the same argument does not apply to medicine with its many practitioners. There had always been many individuals with a deep concern for learning about human ailments and how to ameliorate them; yet progress was slow and seemingly the physicians were as perplexed as the embryologists at the end of the seventeenth century.

The following quotation, by an English physician, Dr. James Cooke (1762), illustrates how much ideas in biology would have to change before modern understandings were to be possible. Cooke was a pre-formationist and an animalculist—one who believed that an already-formed body was located in the sperm, or animalcule. He was concerned with the fate of all those sperm that were present in semen but were not to be involved in conception:

> All those other attending Animacula, except that one that is conceived, evaporate away, and return back into the Atmosphere again, whence it is very likely they immediately proceeded; into the open Air, I say, the common Receptacle of all such disengaged minute sublunary bodies; and do there circulate about with other *Semina*, where, perhaps they do not absolutely die, but live a latent life, in an insensible or dormant state, like Swallows in Winter, lying quite still like a stopped watch when let down, till (they) are received afresh into some other male Body of the proper kind . . . to be afresh set on Motion, and ejected again in Coition as before, to run a fresh chance for a lucky Conception: for it is very hard to conceive that Nature is so idly luxurious of Seeds thus only to destroy them, and to make Myriads of them subservient to but a single one. (quoted from Punnett 1928, p. 506)

Not everyone would have accepted Cooke's analysis, but this quotation suggests that the Middle Ages were alive and well in his thought patterns in the late eighteenth century. Much must be unlearned before progress was to become possible.

It is interesting to note that science can remain in a relatively sterile period such as those two millennia from Aristotle to Malpighi. Progress in science is usually presented as a series of consecutive discoveries that, if the time scale is omitted, suggest rapidity and inevitability. This is certainly not always so. Consider that most elegant feature of the Scientific Revolution, Newton's theory of gravitation. Once it had been formulated and applied to various phenomena, progress seemed to cease. Physicists today are still struggling to think further about gravity—what is "it" that seemingly pulls bodies toward one another in relation to their mass and distance apart. We know, most precisely what gravity can do—not what it is.

The seemingly inexorable advance of science is not a reflection of continuous progress in solving problems but of one advance now, another later. Progress should not be visualized as a host of parallel arrows but as a network with a very irregular advancing edge. One small area of that edge will be pushed out, and only gradually will some of the adjacent areas be "pulled along." Progress in cytology and Mendelian genetics slowed until it was discovered that the data of one provided deep understanding of the other. Attempts to determine the age of the geological strata reached a stalemate until an advance in an entirely different field, radioactivity, provided new techniques and insights. Direct attempts to determine the nature of genes reached a dead end until further advances were made in biochemistry. And embryology remained in an eddy until the equipment and techniques for studying cells became available.

Preformation versus Epigenesis

The resolution of the conflicting hypotheses of preformation or epigenesis was the dominant theoretical problem of embryologists from the last quarter of the seventeenth century to the end of the eighteenth. This was also the first time that a critical number of individuals was alive at the same time and so could engage in dialogue. One could now argue with the living instead of solely with the dead—a process of enormous importance in resolving issues, detecting errors, comparing techniques, and making scholarship seem worthwhile. Science is a social enterprise.

In its most restricted sense, preformation means that the parts of the adult exist as much, albeit much smaller, at the very beginning of development. Some preformationists, also known as "evolutionists," reported that they could see tiny organisms in eggs or in sperm.

In epigenesis, on the other hand, the adult parts are not present at the beginning of development but appear seriatim as development proceeds. Some embryologists, from Aristotle to Harvey, believed that epigenesis was the more probable hypothesis. Since neither hypothesis could be proven beyond all reasonable doubt, cogent argument became the main method for defending one's position. Those who debated preformation versus epigenesis were concerned with the fundamental problem of differentiation. How could structures appear in the course of development from structureless material? What could be the stimulus that would convert structureless semen into heart, brain, legs, eyes, and all the complex parts of the body? A fifth-century BC Greek philosopher-scientist, Anaxagoras of Clazomenae (in Ionia), and some other philosophers, held that truly new things cannot originate. There could be no "coming-into-being out of non-existence," as Cornford (1930, p. 30) expressed it.

The hypothesis of preformation circumvented the problem of differentiation—structure was present from the very beginning so there was no problem of deriving form from a formless beginning. That profound philosophical difficulty of explaining how there could be a coming-into-being out of nonexistence caused most embryologists of the late seventeenth and the entire eighteenth centuries to reject epigenesis and espouse preformation. However, some of the deductions from the hypothesis of preformation proved exceedingly troublesome. If the egg of a horse contained a preformed horse, how could one account for a mule? When different varieties of plants are crossed, how can the offspring be intermediate? If there is a strict preformation, how can there be any variation among offspring at all if they are raised under the same conditions? If we assume that both ova and sperm (terms that we will use from now on to avoid confusion) have preformed bodies, why do not twins result from each conception? Could a single offspring be the result of the fusion of two little heads, hearts, skeleton, and all the other complex parts of the body? One had to assume some sort of amalgamation, otherwise twins should be the usual occurrence in human births.

This difficulty was circumvented with the assumption that *either* the sperm *or* ovum contained the tiny body—in the case of human beings a homunculus, or "little man." Not surprisingly, this resulted in two schools of thought among the preformationists: the ovists, who believed the homunculus to be in eggs, and the spermists (or animalculists), who believed the homunculus to reside in sperm. These were not silly aberrations of human thought but necessary deductions from the hypothesis.

How could these deductions be tested? By looking, but there was a severe problem here for the ovists: the true mammalian egg was not to be discovered until 1827, by von Baer. The spermists did not suffer this restriction. Leeuwenhoek had reported that semen contains microscopic animalcules, later to be given the name "spermatozoa." These were examined with the crude microscopes of the day and, as predicted, found to contain tiny bodies. Hartsoeker (1694) published an illustration of a severely cramped homunculus with a huge head and the fontanelle clearly indicated. Hartsoeker made no claim that he had observed this homunculus—merely that if he could see it that is what it would look like. Others described sperm as being of two sorts—some with a male homunculus and others with a female homunculus. It was a necessary deduction, of course, that the tiny bodies in sperm would be species-specific. And they were. For example, Gautier d'Agoty claimed to see tiny chickens, horses, and donkeys in the semen (not sperm) of those species.

During the seventeenth and eighteenth centuries information about regeneration began to become available. Some animals were found to have astonishing abilities to replace lost parts. Strict preformationism, however, would preclude the possibility of regeneration.

Another deduction from preformation was so necessary and so improbable to many that it contributed to the rejection of the hypothesis. Let us adopt the ovists' position and assume that the human egg has a completely formed homunculus—of a female. That homunculus must contain ovaries and those ovaries must have eggs with homunculi. Those homunculi again must have the next generation of homunculi and so on—like a set of Russian dolls. This deduction is a logical necessity from the hypothesis of preformation, since the possibility of anything new appearing, epigenesis, is excluded.

One cannot imagine an infinite series of ever smaller and ever encased homunculi, so eventually the supply would be exhausted and the species would become extinct. It was suggested that the entire future of the human race was included in the successively encased homunculi in the ovaries of Eve. In more senses than the metaphorical, the ovists thought of her as the Mother of Humanity.

Preformationism was based initially on an inability to see how epigenesis might work. Epigenesists, on the other hand, based their hypothesis on observations, crude as they were, that seemed to show that new things do appear during development. Moreover, they were able to advance objections to preformation, as in the case of hybrids. But pure epigenesis also raised serious problems. One could argue that there must be some sort of preformation in the sense of there being a transfer of "information." Offspring do resemble their parents—rabbits do not hatch from hen's eggs. This transfer of information could be imagined to occur either at conception or even later in the viviparous species. In oviparous species, however, especially those that broadcast their semen into the ocean, there could be no subsequent transfer of information from parent to offspring. So if there was some general rule that applied to all species, the transfer of information must occur at conception. Thus there must be preformed information, whether or not there are preformed structures.

The hypothesis of preformation accounted for a very great deal, but it could not account for everything. Slowly, efforts to refute it gained ground. In 1759 Caspar Friedrich Wolff published his *Theoria Generationis* based mainly on the chick. Wolff interpreted his observations as indicating a true epigenetic development. He observed embryos at a much earlier stage than Malpighi and saw no recognizable organs. Preformationists countered once again that just because a structure could not be seen, one could not conclude that it was not there.

But Wolff did make one strong and eventually convincing point. He emphasized that when organs first become clearly observable they are not in their final form. For example, the intestine of the chick embryo could be shown to start as a flat sheet and then become a tube. Epigenesis, therefore, was proven for individual structures. Thus it was not unreasonable to extend the hypothesis to development as a whole.

Then there was the seemingly well-established fact of spontaneous generation. From the Greeks onward it was generally held that some organisms arose spontaneously—in decaying meat, from excrement, and in decaying food. If organisms as complex as insects can arise spontaneously, the hypothesis of preformation becomes difficult to maintain. One cannot imagine that all rotting meat contains preformed primordia of insects, which will begin to develop once the meat begins to decay.

So, if we accept epigenesis, the awesome problem of differentiation still remains. Thus the structure of embryological theory at the end of the eighteenth century remained roughly as it was formulated by Aristotle.

chapter 19

The Century of Discovery

W e now cross a most important time line in the development of science and enter the nineteenth century. This will be the century when humanity first began to truly understand and predict the phenomena of nature. In roughly the first half of the century chemistry was to have its John Dalton (1766–1844), geology its Charles Lyell (1797–1875), and biology its Charles Darwin (1809–1882)—all Englishmen. There were to be radical changes in these three sciences, whereas astronomy and physics, already with notable accomplishments, were to continue their rapid evidential and conceptual development.

The early nineteenth century was also to witness radical and irrevocable changes in the ways people live once the arts of technology and transportation were unleashed. James Watt's (1736–1819) improved steam engine of the late eighteenth century powered the industrial revolution and was the basis of George Stephenson's (1781–1848) locomotive engine. Again, both Watt and Stephenson were Englishmen. Life in Western civilization could never be the same again after the early nineteenth century. In field after field the impossible became possible —leading to the ultimate impossibilities of our own times.

And so it was with developmental biology, although it was not until later in the nineteenth century that rapid and sustained progress in the "how's" of development became possible. The early decades saw a few outstanding embryologists building on previous discoveries and using the slowly improving technology to make notable advances in descriptive embryology, the chick embryo continuing to be the material of choice. There were no startling breakthroughs and no radical new theories that directed research programs in new ways.

Von Baer's Discovery of the Mammalian Ovum

One of these embryologists was Karl Ernst von Baer (1792–1876), an Estonian biologist. He and others of his time, such as his colleague Heinrich Christian Pander (1794–1865), a Latvian, began to study chick embryos in better ways. Malpighi's method of removing an early embryo from the egg and placing it on a glass slide for study continued to be used, together with Robert Boyle's suggestion for preserving embryos with alcohol or other substances. A method perfected by botanists for making thin slices of tissues with a very sharp razor was also employed. These thin slices, mounted on slides and studied with a microscope, revealed structures that could not be observed in whole embryos.

Von Baer begins his monograph *De Ovi Mammalium et Hominis Genesi* with a discussion of earlier attempts to discover something in mammals that corresponded with the well-known eggs of birds, fishes, reptiles, amphibians, and many invertebrates. These eggs were large and readily visible. Where could the mammalian egg be, or did it really exist? Interest centered on the structure earlier described by de Graaf:

> When I examined the ovaries before incising them, I clearly distinguished in almost all the [Graafian] vesicles a whitish-yellow point which was in no way attached to the covering of the vesicle, but as pressure exerted with a probe on the vesicle indicated clearly, swam freely in its liquid. Led on more by inquisitiveness than by the hope of seeing the ovules in the ovary with the naked eye through all the coverings of the Graafian vesicles, I opened a vesicle, of which, as I said, I had raised the top with the edge of a scalpel—so clearly did I see it distinguished from the surrounding mucus—and placed it under the microscope. I was astonished when I saw an ovule, already recognized from the [Fallopian] tubes, so plainly that a blind man could scarcely deny it. It is truly remarkable and astonishing that a thing so persistently and constantly sought and in all compendia of physiology considered as inextricable, could be put before the eyes with such facility. (O'Malley 1956, p. 132)

Von Baer was not the first to observe mammalian eggs. William Cruikshank had seen them in 1797 in the oviduct of rabbits three days after mating. In addition, in 1824 Prévost and Dumas had published similar observations of an egg in the oviduct. Von Baer was aware of these anticipations, but it was he who worked out some of the details of the relation of the Graafian follicles and eggs.

The 1827 paper closes with four main conclusions. It is interesting to list them because they show what a dominant mind of the time regarded as important.

> Every animal which springs from the coition of male and female is developed from an ovum, and none from a simple, formative liquid.
>
> The male semen acts through the membrane of the ovum, which is pervious by no foramen, and in the ovum it acts first on certain innate parts of the ovum.
>
> All development proceeds from the center to the periphery. Therefore the central parts are formed before the peripheral.
>
> The same method of development occurs in all vertebrate animals, beginning at the spine.

Note in the first conclusion the statement "every animal." Neither von Baer nor all the biologists since him have studied all species of animals. The important methodological principle here is that scientists assume that there are general rules that apply to natural phenomena— all is not chaos— and that one needs but a small sample to find rules with broad applicability. This first principle was not original with von Baer. Harvey had made such a statement two centuries before. The second conclusion is vague to the point of being meaningless; yet, in the decades to come, fundamental advances in cytology, genetics, and embryology would emerge from studies on the interactions of sperm and ova. The third conclusion is essentially correct, and it was to be explained satisfactorily with the experiments of the Spemann school in the twentieth century. In the final conclusion von Baer agrees with many students of development beginning with Aristotle. A few decades later, observations of this sort were linked to the theory of evolution and the reason for the similarity was understood.

A second major discovery often credited to von Baer is the recognition of embryonic layers. This conceptual advance was first made in 1817 by his colleague Pander but was greatly elaborated by von Baer, who extended it to include all major vertebrate embryos. It gradually took on its classical form, which is that embryos pass through a stage when they seem to be composed of three layers now known as ectoderm, mesoderm, and endoderm. The entire structure of the later embryo and adult is derived from these three layers. That, of course, is epigenesis —later structures being derived from earlier structures: "Each step forward in development is made possible only by the preceding state of the embryo, nevertheless the total development is governed and directed by the whole essence of the animal that-is-to-be. And thus conditions at any moment are not alone absolutely determining for the future" (p. 18).

Von Baer also believed that early embryos are generalized and only later do they become specialized:

The more special develops from a more general type. The development of the chick bears witness to this at every moment. In the beginning, when the back closes [that is, neural folds close], it is a vertebrate, and nothing more. When it constricts itself off from the yolk, and its gill clefts close and the allantois forms, it proves itself to be a vertebrate that cannot live in the water. Then later the two intestinal caeca form, a difference appears in the extremities, and the beak begins to appear; the lungs push upward, the rudiments of the airsacs are apparent, and we no longer can doubt that we are looking at a bird. While the character of the bird becomes still more evident through further development of the wings and airsacs, through fusion of the carpals, and so forth, the web between the toes vanishes and we recognize a land bird. The beak and feet proceed from a general shape to a particular one, the crop develops, the stomach has already divided into two chambers, the nasal shield appears. The bird attains the character of a gallinaceous bird, and finally, that of a domestic chicken. (quoted by Oppenheimer 1963, pp. 12–13)

That statement may seem like the theory of recapitulation—that the development of embryos of living species goes through, or recapitulates, the form of their ever more remote ancestors. Von Baer was a vigorous opponent of the then currently exaggerated concept of recapitulation. His extensive knowledge of embryos led him to emphasize these points, which over the years have proved satisfactory.

(1) The embryos of different species belonging to a major taxonomic group resemble each other more closely early in development than they do as older embryos.
(2) The embryos of higher species are like the embryos of lower species but not like the adults of lower species.
(3) Thus if one compares the course of development of embryos of different taxonomic groups, they are found to diverge progressively and not recapitulate different levels of adult organization.

Thus, living species of vertebrates, for example, recapitulate the embryonic structure, not the adult structure, of ancestors.

Darwin's Contribution to Embryology

One of the more striking aspects of early nineteenth-century biology was its unified approach to problems. We have spoken of Harvey, Malpighi, von Baer, and many others as "embryologists," but they were far more than that. They were general biologists, called "naturalists" in those days, who did not approach the phenomena of life as geneticists, evolutionary biologists, cell biologists, or developmental biologists. Those disciplines, so distinct today, were then part of a conceptual whole. This unity came not from the recognition of fundamental principles but from the general lack of such principles. Those who studied embryos were interested in the material contributions of parents to offspring (cytology), what might be the "information" transferred (genetics), how the course of development related to the *scala naturae* (evolutionary biology), as well as the details of development itself (developmental biology).

The Darwinian paradigm shift of 1859 changed not only what biologists did but also provided an explanation for what they observed. The new paradigm was able to offer a satisfying explanation for much that had already been learned. In fact, the data themselves seemed to be awaiting some organizing theory, and Darwin's basic idea provided it. But ideas are seldom simple until they have emerged.

Darwinism provided a new way of thinking about a cluster of major biological phenomena related to embryology:

(1) Living organisms seemed to form a continuum from the least complex to human beings—a "great scale of being" or *scala naturae*.
(2) Embryos of species in the same taxonomic group resemble one another.
(3) Embryos of higher forms (mammals, for example) go through stages that resemble the lower forms (fish, amphibians, and reptiles, in that order). This theory in its developed form would be known as recapitulation.
(4) Animals in major taxonomic groups seem to be built on the same general body plan. For example, the limbs of tetrapods, amphibians, reptiles, birds, and mammals seemed to have a

basic plan: the forelimbs have one proximal bone, the humerus, and two more distal; the "hand" has several bones forming a wrist, about five in the palm, and finally a few in each finger. Even the wings of birds and bats could be understood to be variations on this basic plan. The corresponding bones were said to be *homologous*. That is, the humerus is really the "same thing" in different species. It was recognized that homology is based on more than superficial resemblances. The wings of insects, for example, do not have bones and muscles and, hence, are not homologous to wings of birds and bats. Wings of birds (or bats) and insects were said to be *analogous*. Analogy, then, was restricted to structures that are functionally similar but morphologically different. Homology was restricted to morphologically similar structures that might be functionally similar (appendages of horses and frogs) or not (appendages of porpoise, bat, and monkeys).

Each of these phenomena was so striking and so pervasive that it was impossible not to think that there must be some underlying cause. Darwin put it all together in chapter 13 of the *Origin*, where he sought to

> explain these several facts in embryology, [1] namely the very general, but not universal differences in structure between the embryo and the adult; [2] of parts in the same individual embryo, which ultimately become very unlike and serve for diverse purposes, being at this early period of growth alike; [3] of embryos of different species within the same class, generally, but not universally, resembling each other; [4] of the structure of the embryo not being closely related to its conditions of existence, except when the embryo becomes at any period of life active and has to provide for itself; [5] of the embryo apparently having sometimes a higher organisation than the mature animal, into which it is developed. (pp. 442–443; I added the numbers in brackets)

The explanation of these five phenomena, which we hardly recognize as problems today, was not obvious in Darwin's time. The argument starts as follows: "There is no obvious reason why, for instance, the wing of a bat, or the fin of a porpoise, should not have been sketched out with all the parts in proper proportion, as soon as any structure became visible in the embryo" (p. 442). That is, fin and limb could have been preformed. Yet when they first start to develop in the embryo, the wing and fin are nearly the same. They diverge later.

Darwin thought that this similarity of fin and wing and the other problems he listed could be explained on the basis of three assumptions: first, that species evolve; second, that the modifications which occur in the course of evolution "may have supervened at a not very early period in life" (that is, late in embryonic life); and third, "that at whatever age any variation first appears in the parent, it tends to reappear at a corresponding age in the offspring" (p. 444).

Darwin's hypothesis of descent with modification—that is, natural selection acting on the hereditary differences among individuals of a species—did far more than make some otherwise confusing embryological phenomena understandable. It accounted for the grand phenomenon of organisms belonging to sets or taxonomic groups.

> As all the organic beings, extinct and recent, which have ever lived on this earth have to be classed together, and as all have been connected by the finest gradations, the best, or indeed, if our collections were nearly perfect, the only possible arrangement, would be genealogical. Descent being on my view the hidden bond of connections which naturalists have been seeking under the term of the natural system. On this view we can understand how it is that, in the eyes of most naturalists, the structure of the embryo is even more important for classification than that of the adult. For the embryo is the animal in its less modified state; and in so far it reveals the structure of its progenitor. In two groups of animals, however much they may at present differ from each other in structure and habits, if they pass through the same or similar embryonic stages, we may feel assured that they have both descended from the same or nearly similar parents, and are therefore in that degree closely related. Thus, community in embryonic structure reveals community of descent. It will reveal this community of descent, however much the structure of the adult may have been modified and obscured; we have seen, for instance, that cirripedes [barnacles] can at once be recognized by their larvae as belonging to the great class of crustaceans. As the embryonic state of each species and group of species partially shows us the structure of their less modified ancient progenitors, we can clearly see why ancient and extinct forms of life should resemble the embryos of their descendants,—our existing species. (pp. 448–449)

Here Darwin was applying his concept of evolution to the widely accepted but poorly understood hypothesis of recapitulation. He was cautious in accepting an extreme formulation of recapitulation; nevertheless, he saw in embryonic development hints of remote ancestors. Thus, Darwin gave embryologists a mission of first-rate theoretical importance—the search for lineages in the minutiae of development. To be sure, embryos could do no more than reflect these lineages, but when fossil evidence was so meager, there was no alternative.

Many triumphs with this approach were already known to Darwin. Barnacles were of special interest to him since he had produced the definitive monographs on these creatures. Barnacles are sessile animals enclosed in a shell. They resemble mollusks more than any other group of invertebrates, and Cuvier, the most respected naturalist of the early nineteenth century, had included the barnacles in the Mollusca. Darwin notes, "Even the illustrious Cuvier did not perceive that a barnacle was, as it certainly is, a crustacean; but a glance at the larva shows this to be the case in an unmistakable manner. So again the two main divisions of cirripedes, the pedunculated and sessile, which differ widely in external appearance, have larvae in all their several stages barely distinguishable" (p. 440). Among the vertebrates there were many examples known to Darwin of embryos apparently recapitulating stages found in putative ancestors. Birds and mammals, with only a single aorta in the adult, have in the embryo the 6 pairs characteristic of fishes (figure 28). Bird and mammalian embryos develop in succession a pronephros, mesonephros—the kidneys of the lower vertebrates—and finally their own adult metanephric kidney (figure 29). Possibly the most dramatic example is that of the development of the malleus, incus, and stapes of the mammalian ear from the jawbones of lower vertebrates (figure 27). This was predicted, on the basis of careful embryological work, as being highly probable long before paleontologists unearthed the mammallike reptiles that provided absolute proof of the evolution of ear bones from jawbones.

Thus by Darwin's time embryology had come to be more than the detailed study of successive stages in the development of organisms. The data of descriptive embryology could be used to suggest the course of evolution, a concept in Haeckel's hands that would become *ontogeny recapitulates phylogeny*.

Homology was also clarified by embryology. The meaning of the "same thing" could now be understood. A common structure in an ancestor would change in the course of the evolution of different daughter species. The structure would still be, in a very general way, the "same thing," although variously modified. In the case of hard structures, such as bones, it might be possible to trace the changes in fossils of different geological ages (jaw bones and ear ossicles, for example). Such would not be possible for soft parts (vertebrate kidneys and aortic arches), but often the embryos provided a clue. Homology, then, was defined as identity of embryonic origin.

Haeckel and Recapitulation

In the decades after Darwin, the fundamental theorem of evolutionary embryology was recapitulation. This concept was formulated well before Darwin, and it expressed the relationship of embryogenesis to classification and to the *scala naturae*. When the Darwinian paradigm reinterpreted the *scala naturae* as the consequence of descent with modification, the data of embryology were reinterpreted as well. As Darwin suggested, many facts about development are inexplicable without the concept of recapitulation.

It is fascinating to note how the concept of recapitulation, which itself suggests evolution, was used in the years immediately before the publication of *On the Origin of Species*. The case of Louis Agassiz (1807–1873) is especially interesting. He was a Swiss naturalist of great ability who spent much of his life in the United States. His famous *Essay on Classification* was first published in 1857 as part of his grandiose *Contributions to the Natural History of the United States*. The *Essay* was republished in 1859—the year of the *Origin*. There he noted:

> It may therefore be considered as a general fact, very likely to be more fully illustrated as investigations cover a wider ground, that the phases of development of all living animals correspond to the order of succession of their extinct representatives in past geological times. As far as this goes, the oldest representatives of every class may then be considered as embryonic types of their respective orders or families among the living. (p. 174)

And the cause?

> It exhibits everywhere the working of the same creative Mind, through all times, and upon the whole surface of the globe. (p. 175)

Thus even before the acceptance of Darwinism it was recognized that the positions of organisms in the *scala naturae* might parallel the times of their first appearance in the fossil record, and their patterns of development.

Recapitulation became a truly baroque edifice in the hands of Ernst Heinrich Phillip Haeckel (1834–1919), a dominant personality in German science of the last half of the nineteenth century. Haeckel's theory was proposed in his *Generelle Morphologie* of 1866 revised in *Natürliche Schöpfungeschichte* (1868) and in *Anthropogenie* (1874). The concept of recapitulation he developed was eventually regarded as either wrong or useless, yet it remains true to this day that some extraordinary phenomena "make sense" on the basis of a more restrained restatement of the concept. The rejection of the concept of recapitulation in the late nineteenth and early twentieth centuries is probably a case, as Gould notes (1977) of throwing the baby out with the bath water.

In the 1860s when Haeckel began to speculate about the phylogeny leading to the human species, the fossil record was mainly gaps (as it still is). Accurate knowledge of chromosomal structure, genetics, and biochemistry was essentially nil. Microscopists knew little about the biology of what we now call the prokaryotes, and even the mechanisms of evolutionary change were poorly understood. The most valuable of the then available information was to be found in comparative morphology and embryology. These two disciplines, therefore, provided the evidential basis for Haeckel's version of recapitulation.

Recapitulation was thought to reflect phylogeny, which in the case of the evolution of chordates to mammals, was assumed to be: prechordates → fishes → amphibians → reptiles → mammals. The birds were regarded as independently originating from reptiles. There are three diagnostic characteristics of all species that are classified as chordates: the possession of a notochord, gill slits or pouches in the pharynx, and a dorsal nerve tube. The human adult has only one—the dorsal nerve tube, which has become our brain and spinal cord. But as embryos we had all three. Our notochord lasts only through the early embryonic stages. The gill pouches also disappear in development, but the first one becomes the Eustachian tube of our ear.

We now know that all chordate embryos show this same basic pattern early in development. This discovery finds a ready interpretation if all chordates share a common ancestor. It is reasoned that the very early chordates possessed these three features and, as evolution proceeded, the embryos of higher forms continued to reflect the ancestral condition. The intense interest shown in *Amphioxus*, a living prechordate, is based on the fact that, in many aspects of its morphology, it has the features we predict for the earliest of chordates. The ammocoete larva of the lamprey is held in the same high regard—and for the same reason.

Jaw joints and ear bones provide one of the classic examples interpreted as recapitulation (figure 27). The ear has two main functions in the vertebrates. In fishes it is an organ of balance only. In tetrapods the ear retains its function as an organ of balance but adds the detection of vibrations in air—sound. The fishes lack both a tympanic membrane and ear ossicles. The living amphibians have one ear ossicle—the stapes—and, except for the urodeles, a tympanic membrane. Living reptiles and birds have basically the same structure as the amphibians. In mammals, however, there are two additional ear ossicles—the malleus and incus. What could be the origin of these ear bones?

A tentative answer was worked out almost entirely on the embryos of living vertebrates. Correlated with the formation of the "new" bones—that is, the stapes in amphibians and the two additional ones in mammals—there are important changes in the articulation of the jaws. In the fishes the hyomandibular bone supports the jaws. The amphibian embryo has a hyomandibular, but it becomes converted to the stapes of the adult. This is the condition in all adult tetrapods except the mammals. The mammalian embryo is similar, but as it develops the articular of the lower jaw and the quadrate of the upper jaw become reduced in size and move into the ear to form the malleus and incus of the adult.

The hypothesis was advanced that these events in development might be a reflection of what had occurred in evolution. One would then deduce that at some stage in the past, presumably when reptiles were evolving into mammals, there would be evidence of the loss of the quadrate and articular as the jaw articulation, and the modification of these bones into the incus and malleus. Concurrently, the jaw articulation must have switched to the mammalian pattern—the dentary of the lower jaw moving against the squamosal of the upper jaw. Now this suggests a very interesting and, seemingly, dangerous deduction for a paleontologist to make. Since there must always be some articulation point of the jaws if the animal intends to eat, how could there be a switch from the reptilian type to the mammalian type? It seemed unlikely that parents with the articular-quadrate joint would produce offspring with the dentary-squamosal type. But how else?

The answer was provided by the fossil synapsids, the mammallike reptiles. As their name suggests, they are intermediate in structure between reptiles and mammals. In recent years a very large number of fossils has been found—mainly in South Africa. The different kinds show all gradations from the characteristic reptilian jaw articulation to that of the mammals, and one genus from the Triassic is especially noteworthy. It was given the jawbreaking name *Diarthrognathus*, which means "two-jointed-jaws," in recognition of the fact that it had two functional jaw articulations—reptilian and mammalian. One could not ask for a better intermediate between reptiles and mammals.

The transformation of the aortic arches provides another example, also mentioned before. The embryos of all vertebrates have a circulatory system of almost diagrammatic simplicity—in contrast to the confusing patterns among the adults. In the anterior part of the body there is a ventral vessel that carries blood anteriorly to the region of the gill slits. There six pairs of aortic arches branch off and extend dorsally through the gill bars to the lateral dorsal aortae. The lateral dorsal aortae join posteriorly to form the dorsal aorta, which carries blood to all parts of the body behind the gill region. This simple pattern of the embryo is transformed in very different ways in different vertebrates to produce the circulatory systems of the adults. The frog embryo passes through a stage with the six aortic arches. With further development some are lost or modified until the adult pattern is formed. The human embryo also has six pairs of aortic arches (never all at one time) and these become transformed but in a manner quite different from that of the frog.

An even more famous case of ontogeny recapitulating phylogeny, mentioned earlier, is the vertebrate kidney. The early embryos of reptiles, birds, and mammals have a kidney on each side in the anterior portion of the coelom—the pronephros. Later in development the pronephros degenerates and a second kidney takes over—the mesonephros. It, too, disappears and is replaced by a third kidney—the metanephros, which remains as the kidney of the adult. The embryos of fishes and the agnathan lamprey start with a pronephros in the embryo and replace it with a mesonephros, which is the kidney of the adult. No metanephros develops.

Now back to Haeckel, who knew many examples of the sort just described. If the concept of recapitulation is accepted as a useful way of looking at these otherwise puzzling embryological phenomena rather than as a fundamental and relentless law of nature, it becomes a powerful heuristic device. But such a view demands that we modify Haeckel's striking formulation, "ontogeny recapitulates phylogeny," by adding "not quite" and "sometimes."

Haeckel did far more than formulate his aphorism; he attempted to provide a conceptual scheme for all of descriptive embryology and morphology—an overall theory which described the history of the development of individuals, paralleling Darwin's theory which described the history of the development of species. Furthermore, he suggested a close relationship between the two.

> These two branches of our science—on the one side ontogeny or embryology, and on the other phylogeny, or the science of race-evolution—are most vitally connected. The one cannot be understood without the other. It is only when the two branches fully co-operate and supplement each other that "Biogeny" (or the science of the genesis of life in the widest sense) attains the rank of a philosophic science. The connection between them is not external and superficial, but profound, intrinsic, and causal. This is a discovery made by recent research, and it is most clearly expressed in the comprehensive law which I have called "the fundamental law of organic evolution," or "the fundamental law of biogeny." This general law, to which we find ourselves constantly recurring, and on the recognition of which depends one's whole insight into the story of evolution, may be briefly expressed in the phrase: "The history of the foetus is a recapitulation of the history of the race"; or, in other words, "Ontogeny is a recapitulation of phylogeny." It may be more fully stated as follows: The series of forms through which the individual organism passes during its development from the ovum to the complete bodily structure is a brief, condensed repetition of the long series of forms which the animal ancestors of the said organism, or the ancestral forms of the species, have passed through from the earliest period of organic life down to the present day . . . (1905, pp. 2–3)

More and more data seemed to suggest that early embryos may retain some relics of the basic structure of the group to which they belong. If so, one might predict that a study of the embryos of species that were "problems" so far as their relationships were concerned would be productive. Three triumphs will be mentioned.

Sacculina was the name given to a baglike structure that could be found attached to various species of crabs. It resembles a tumor, espo cially since branching roots of the sac actually penetrate the host's abdomen—a cancer on *Cancer*. Closer study showed that the sac contains reproductive organs and some muscle and nerve tissue. Thus *Sacculina* could be considered a parasite and the branching roots that enter the host could be the mechanism for obtaining food.

One cannot deduce from the structure of *Sacculina* what its affinities might be. A study of the early embryos, however, provided the answer. The eggs were found to develop into a well-known larval type—the nauplius larva with three pairs of appendages—that is characteristic of many crustaceans. Later the nauplius larva transforms into the cypris larva, again a familiar crustacean larval type. After a period of independent life the cypris larva attaches to a crab, loses its appendages—and most of its anatomy for that matter—and becomes the adult *Sacculina.*

The story is similar for barnacles, mentioned before. Since they are covered by a shell, many early naturalists classified them as mollusks. A study of the embryos, however, showed that the barnacles are crustaceans, with a typical crustacean larval type.

There was a similar puzzle with the ascidians, or tunicates. These are marine organisms, and the most common ones look like an amorphous mass of "something" that is attached to wharf pilings, rocks, and so on. The adults consist mostly of a basket with perforated walls. Water enters an opening and passes through the walls of the basket and food particles are strained out. Again it would be difficult to establish the affinities of these creatures on the basis of the structure of the adult. The larvae provide the answer—they are chordates, with a nerve tube, pharyngeal gill slits, and a notochord. Clearly ontogeny was a powerful tool for discovering relationships among organisms.

Haeckel made a bold attempt to reduce a very large amount of data to a single concept. He speculated far beyond the information available to him but, in so doing, he provided hypotheses for others to test and suggested lines of research worth exploring (figure 71).

There are parallels between Darwin and Haeckel in the general acceptance of their main theses and the rejection of many of the details. Most biologists agreed that Darwin had shown beyond all reasonable doubt that evolution had occurred, but his suggested mechanisms—spontaneous variations acted upon by natural selection—were thought improbable or impossible until well into the twentieth century. Haeckel's synthesis of the data of descriptive embryology, evolution, and comparative anatomy is accepted as a general statement even though there has been a strong reaction against the details.

Many have claimed that Haeckel believed that embryos recapitulate the *adult* stages of ancestors. I read Haeckel differently, possibly being biased by admiration for his attempts at synthesis. He never suggested that we recapitulate a piscine stage by swimming around in the amnion, using pectoral and pelvic fins, possessing a fishy tail, and encased in a scaly skin. But von Baer came closer to an acceptable concept than did Haeckel. Taking the best-known examples, the chordates, we must admit that embryos do not in general recapitulate the adult stages of their ancestors. Chordate embryos do share a common plan of devel-

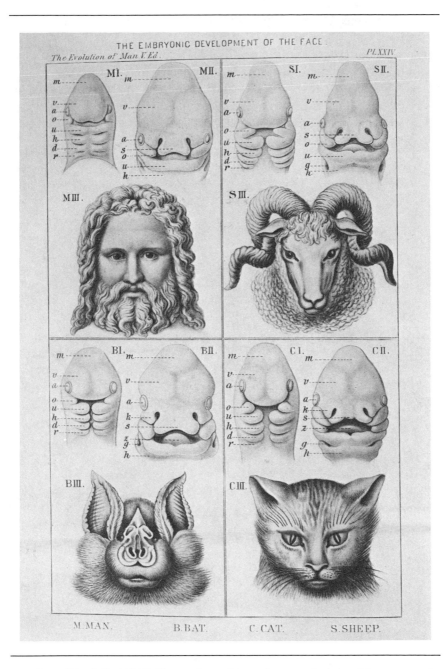

M. MAN.   B. BAT.   C. CAT.   S. SHEEP.

71 *Ernst Haeckel's (1905) illustrations of the embryos and adults of four mammals. He, as well as von Baer long before him, was struck by the fact that early embryos might be very similar even though the adults might differ greatly.*

opment that has an early stage with notochord, dorsal nerve tube, and pharyngeal gill slits or pouches separated by gill arches. The adults of the lower chordate classes change less from this fundamental plan than the adults of the higher classes. The higher forms retain these fundamental features in their early development and then differentiate in their special ways. Thus the organ systems of the amphibians, reptiles, birds, and mammals can be understood as variations on a pattern based on the morphology of agnathans and primitive fish.

chapter 20

Descriptive Embryology

I n the last half of the nineteenth century enormous progress was
made in descriptive embryology. There were more students of development and better tools and methods: better microscopes, better
histological techniques, and improved methods for obtaining and han-
dling embryos. As a result, data became available on the developmental
patterns of species belonging to all major groups of animals. There was
the hope that uniformities—that is, rules that would hold for many
different sorts of development—could be found. Had one known nothing of the varieties of ontogenies to be discovered, it might have been
predicted that patterns of development would be as varied as patterns
of adult morphology. Such proved not to be the case. The remarkable
similarities in development of vertebrates, known to observers from
Aristotle to Haeckel, were found to hold for invertebrates as well. There
were, indeed, rules, and the most diverse ends were achieved by variations on a fundamental pattern. One could usually recognize these
major steps in development:

(1) Development begins with the activation of an ovum by a sperm
    (sexual reproduction) or by nonsexual means (partheno-
    genesis).
(2) The activated ovum divides repeatedly, and comparatively rap-
    idly, about 8 to 12 times by mitosis. The result is a ball of cells,
    the *blastula*, with a cavity, the *blastocoel*.
(3) There then ensues a rearrangement of cells, with some moving
    inward, forming a cavity, the primitive gut or *archenteron*. The

archenteron has an opening to the outside, the *blastopore*. This cup-shaped structure with an inner and outer layer is the *gastrula*, which was regarded by Haeckel as the basic body plan from which all multicelled animals evolved. The rate of cell division slows by the gastrula stage and remains slow for the rest of development.

(4) A rearrangement of cells forms the embryonic layers. There are two of these in the coelenterates, an outer *ectoderm* and an inner *endoderm*. Other metazoans have an additional layer, the *mesoderm*, between the ectoderm and endoderm. When these layers first become delimited their cells are essentially the same.

(5) As development continues and cell numbers increase, the cells become visibly differentiated, and they rearrange to form organs and tissue.

(6) Throughout the metazoans there is considerable uniformity in the structures developing from each germ layer. Typically the skin, nervous system, and some types of excretory organs are derived from the ectoderm; the lining of the alimentary canal and the associated organs are derived from endoderm; the circulatory system, muscles, connective tissue, and some types of excretory organs are derived from the mesoderm.

This commonality of patterns of development found its formal explanation in the theory of evolution and its derivative, recapitulation. Nevertheless, it was discovered that there are two strikingly different patterns of development—direct and indirect—that result in similar end products. One of the main reasons appeared to be the quantity of yolk in the ovum or the availability of food directly from the mother. The ova of some species—human beings and sea urchins, for example—contain very small amounts of yolk. Such embryos must rely on external sources of food—either capturing it themselves or obtaining it from mother.

The sea-urchin mode of development, which is common among the invertebrates, is for the embryo to reach a free-living larval stage rapidly, in this case a pluteus. The pluteus is a microscopic larva that obtains its food from the ocean. It bears no resemblance to the adult. It swims, feeds, and grows. Eventually a complete restructuring of its anatomy, physiology, and life-style begins—it undergoes metamorphosis into the adult sea urchin. This is called *indirect development*, since the embryo does not develop directly into an adult but passes through a larval stage very different in structure, physiology, and behavior from the adult.

Human beings and other mammals rely on nourishment from the mother. They differentiate into the juvenile form without a free-living, food-capturing larval stage. These embryos have *direct development*. Direct development also occurs in birds, but there the source of the food supply is different—the ovum contains sufficient food to carry the embryo to the juvenile stage.

Germ Layers

The theory of germ layers has been one of the mainstays of descriptive embryologists. The concept grew slowly from the work of Wolff, Pander, and von Baer to its extensive development by Haeckel and Lankester. It has been almost as contentious as the concept of recapitulation. The arguments have been mainly about the applicability of the concept to embryos of different phyla and what is implied about the developmental potential of the layers themselves.

Let us consider those two aspects separately. If one is saying only that metazoan embryos consist of two (coelenterates) or three (the other major phyla) layers during an early embryonic stage, the concept has great heuristic value. It is conceptually satisfying to know that, in animals differing greatly from one another, the skin, the most anterior and most posterior parts of the alimentary canal, and the nervous system develop from the ectoderm; the muscles, connective tissues, skeletal and circulatory systems (if there are any) from the mesoderm; and, except for the two ends, the lining of the digestive system and its associated glands from the endoderm.

While "wide correspondence" exists between what the germ layers do, that does not signify that they are homologous. Nevertheless the origins of the three layers are so much alike throughout the vertebrates that it can be said that they are homologous in this phylum by virtue of identity of embryonic origin. The conclusive data—origin from the same part of an ancestral species—will most likely never be available.

One could go one step further and entertain the hypothesis that there is a basic homology of germ layers in all multicellular animals. As more information became available, it became clear that the germ layers arise in many different ways. Therefore one cannot use identity of embryonic origin as proof of homology, since the origins are not identical. One cannot maintain, for example, that the mesoderm is the "same thing" in all species. To what extent can the cells on the outside of an earthworm be considered homologous to the cells on the outside of a starfish? Any answer is as dubious as would be any clear notion of how one would find out. Once the germ layers have been formed, however, there is great uniformity in what they do. Therein lies their conceptual importance.

A second interesting problem has to do with the relation between what the germ layers form in the course of development and their innate abilities. Is there something "mesodermal" about the mesodermal cells, meaning that they produce only mesodermal organs, and that mesodermal organs are produced only by the mesodermal layer? Questions of this type can be formulated in hypotheses that can be tested. As we will learn later, the Spemann school found that there is no restriction on what mesodermal cells can form, and mesodermal structures can be formed from other than mesodermal cells.

Other evidence of the nonspecificity of the germ layers comes from studies of regeneration, where in some cases the structures of the regenerated individual are derived from different germ layers than those from which they were first formed in embryonic development.

The External Development of the Amphibian Embryo

When the main reason for studying embryos began to switch from seeking to learn about evolution to an analysis of the causal factors in development, the amphibians were found to provide excellent material. Prior to that switch, it was important to have embryological data from a broad sample of organisms. For analytical embryology, on the other hand, it was necessary to use species with embryos that could survive experimental manipulations. The mature eggs of many common European and North American frogs and salamanders are usually about 2 or 3 millimeters in diameter, hence large enough to be operated upon. Their embryos are hardy and heal well, and recovery can usually be expected following operations and other experimental procedures. Each fertilized egg has a supply of yolk granules sufficient to carry the embryo to a free-living stage. This is a great advantage since the difficult problem of supplying an external source of food is avoided.

A brief description of normal development is essential to understand the problems as well as the experiments performed to solve them. The following brief description is of embryonic development of *Rana pipiens* from Vermont, which have been widely used in experiments designed to unravel the causal factors in development. This species and other meadow frogs very similar to it are widely distributed in North and Central America. The external aspects of development will be described first (figures 72–77).

The rate of development depends on temperature—the embryos shown in the illustrations were kept at a constant temperature of 20° C. The numbers on each photograph give the time in hours after fertilization. Had the embryos been kept at 25°, development would have required about half the time, and if kept at 15° nearly twice as long. The lowest and highest temperatures for normal development are, respectively, about 5° and 28°. The embryos in the illustrations are magnified about 25 times.

Breeding

In the spring, spurred by warming days, moist nights, and hormonal changes, males and females congregate in ponds and swamps for a brief breeding period. The mature ova leave the ovary, pass into the coelom, and then enter the anterior openings of the oviducts. They move slowly along the oviduct where they are coated with thin layers of jelly. The ova accumulate in the posterior portion, or uterus, of each oviduct. When mating begins the male frog clasps the female in such a manner that his cloacal opening is directly over hers. The ova pass out of the female's body into the surrounding water and, concurrently, the male sheds sperm over them. The thin, almost invisible, jelly layers surrounding the fertilized ovum now imbibe water and begin to swell, eventually reaching a diameter about three times that of the egg. This jelly is at first sticky and adjacent eggs adhere to one another. As a consequence, all of the eggs, which may number more than a thousand, stick together and form a large globular mass in which the embryos develop, each in its own jelly envelope.

Meiosis and fertilization

Complex and important internal events have been occurring during this entire period. The ovarian egg has a very large nucleus, the *germinal vesicle*. (It is interesting to note the antiquity of some of the terms used in describing development. A large spherical object was seen in ovarian eggs before it was realized that it was the nucleus. Since it occurred in the "germ" it was named the "germinal vesicle.") When the ova start to break out of their follicles in the ovary, the nuclear membrane disappears and meiosis begins. The first meiotic division occurs by the time the ovum has reached the upper portion of the oviduct, and the first polar body is given off at that time. Metaphase of the second meiotic division occurs when the ova are in the uterus. Further nuclear changes are blocked at that stage.

A single sperm enters the ovum. Its head contains the paternal nucleus with the monoploid number of 13 chromosomes. A centriole is immediately behind the sperm head. It will become part of the first mitotic spindle. The entrance of the sperm removes the meiotic block in the ovum, and the second polar body is extruded in about a half hour after fertilization. The maternal pronucleus now has the monoploid number of chromosomes. The two pronuclei move toward the upper center of the egg and unite, restoring the diploid number of 26 chromosomes.

The uncleaved zygote

The just-fertilized ovum is a sphere approximately 1.7 mm in diameter. Somewhat more than half of the embryo, the *animal hemisphere,* is a dark chocolate-brown and the remainder, the *vegetal hemisphere,* is almost white (figure 72). The animal pole is on the surface in the center of the animal hemisphere. It is the site of polar body formation. The vegetal pole is 180° from the animal pole and in the center of the vegetal hemisphere.

Cleavage

Two and a half hours after fertilization the first spectacular event that is externally visible occurs. A short groove appears in the animal hemisphere, and it gradually lengthens to form the first cleavage furrow. The furrow slowly extends through the embryo until two cells are formed. Internally, mitosis began before the cleavage furrow appears. When the chromosomes are in telophase the furrow starts to form.

The second cell division starts at about 3.5 hours. The plane is again vertical and at right angles to that of first cleavage. Both cleavages pass through or very near to the animal and vegetal poles. The third cell division occurs at about 4.5 hours, and the plane is horizontal. It does not divide the embryo equally—the plane of cleavage is above the equator. Thus in the photograph one can see the 4 smaller animal

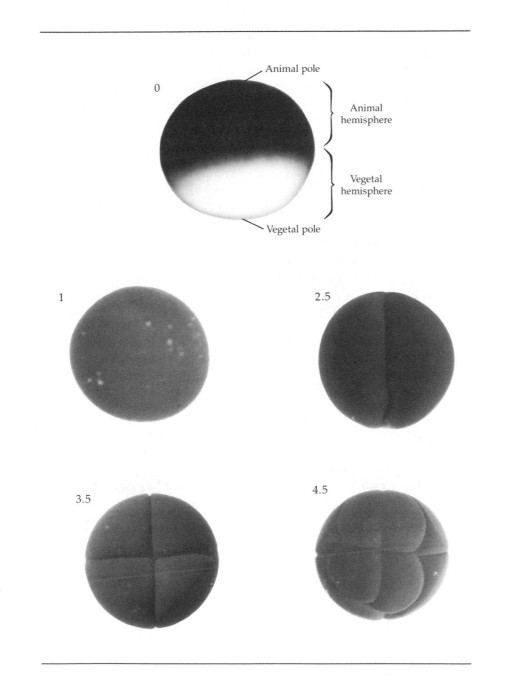

72 *Development of the frog's egg from fertilization to the 8-cell stage. The 0-hour embryo is in side view; the others are shown as though looking down on the animal hemisphere. The numbers to the upper left of each embryo here and in figures 73–77 are the hours after fertilization at 20° C.*

hemisphere cells and beneath them 4 larger cells. Each of these includes a lower part of the animal hemisphere and a quarter of the vegetal hemisphere.

In a group of embryos that were fertilized at the same time and kept together, the synchrony of development is awesome to observe. Each of the cleavages starts at almost exactly the same time. This synchrony is true of all early development—each stage being reached at almost the same time in all embryos. It is as though each cell has an internal clock that was started at the same time and ticks along together with the other clocks. Each cell must have some sort of biological clock, but as yet we know very little about what it is and how it works.

Mitosis continues, and soon the intervals between cell divisions increase. The embryo is divided into smaller and smaller cells—there is no obvious increase in total size. No food is entering the embryo, and its energy source is the yolk granules within each cell. As these are used in metabolism, the dry weight of the embryo decreases. Oxygen diffuses through the jelly layers and enters the embryo, and carbon dioxide and some waste materials take the opposite path.

The embryos from 9 to 22 hours are *blastulae*. The blastula stage is characterized by an internal cavity, the blastocoel, which, when fully formed, occupies most of the interior of the animal hemisphere. More will be said about it later when we consider the internal events of early development. All of the cells of the late blastula are essentially the same apart from differences in pigmentation and size. There is a gradient in size that runs from the smallest cells at the animal pole to the largest cells at the vegetal pole.

Thus the blastula has only a single axis—extending from animal to vegetal pole. There is no visible right or left that would enable an experimenter to ascertain a precise place on the surface of the embryo. If we compare the blastula with the earth, we could recognize a North Pole (animal pole) and a South Pole (vegetal pole). That would enable us to determine the latitude of any position on the blastula, but the lack of differences along the sides makes the determination of longitude impossible.

Gastrulation

The 22-hour blastula has a narrow groove of pigmented cells in the vegetal hemisphere just below the equator. By 25 hours this groove has become deeper and extended laterally. The groove itself is the blastopore, and its formation marks the beginning of gastrulation. The cells immediately above the blastopore are called the dorsal lip of the blastopore. They will play an extraordinary role in development (figure 73).

Gastrulation is a process that leads to a complete rearrangement of the cells of the embryo. Many of those on the outside of the blastula will move to the interior. This process of moving in at the lips of the blastopore is invagination. The blastopore of the 25-hour embryo leads into a tiny cavity, the archenteron. The pigmented surface area appears to be enlarging, caused by the dark cells of the animal hemisphere moving downward and the lighter cells of the vegetal hemisphere moving to the interior. By 27 hours the lateral lips of the blastopore have extended to the sides, and by 30 hours they have finally met to form a 360° blastopore. The area of light-colored cells has become much smaller, forming the yolk plug. Gastrulation continues until, by 36 hours, the animal hemisphere cells have almost overgrown the embryo, and of the original vegetal cells only those of the ever-smaller yolk plug are seen on the outside. Finally the yolk plug is overgrown and reduced to a tiny slit. This marks the end of gastrulation. Now the dark animal hemisphere cells cover the entire surface. The cells have become so small that at moderate magnification they are invisible.

With the formation of the *dorsal lip* of the blastopore at 25 hours, we are finally able to specify longitude. Therefore, once gastrulation has begun, we can describe any spot on the surface of the embryo in terms of its latitudinal distance from the animal or vegetal pole and its longitudinal distance from the dorsal lip. The center of the dorsal lip is the embryo's meridian of Greenwich—0° longitude.

At the end of gastrulation there is still essentially no obvious cellular differentiation. The diameter of the late gastrula is about the same as that of the uncleaved egg. The rate of metabolism has increased and, since there is still no external source of food, the dry weight is less than before. There are two cavities in the embryo—the vanishing blastocoel and the enlarging archenteron. These are filled with fluid.

Neurulation

The next prominent external change is the beginning of the formation of the nervous system, with the brain and spinal cord starting to develop on the outside of the body. In the 42-hour embryo the *neural folds* appear as low ridges on the dorsal side (figure 74). These are paired structures, extending from each side of the blastopore region anteriorly to where

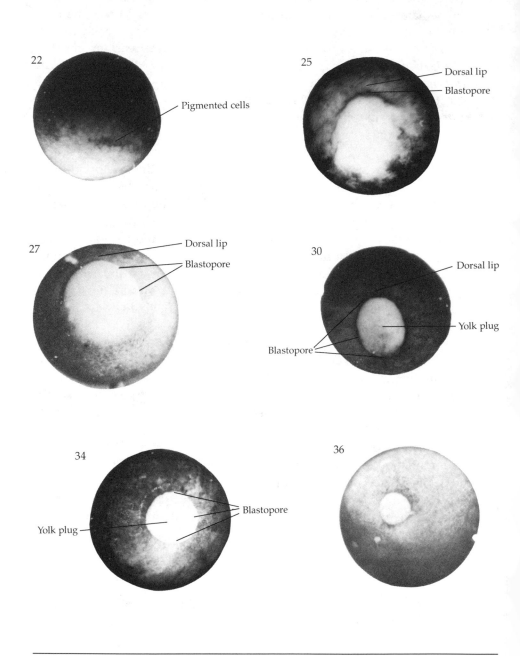

73   Gastrulation. The 22-hour embryo is in side view; the others are ventral views.

they connect in the region that will become the head. By 47 hours the folds are more elevated and begin to close, and by 50 hours the folds touch along their entire length. They close in such a manner as to form an internal tube, the *neural tube*. The walls of the broad anterior portion of the neural folds will become the brain, and the walls of the more posterior portion will become the spinal cord. The bore of the neural tube persists in the adult as the *neurocoel*.

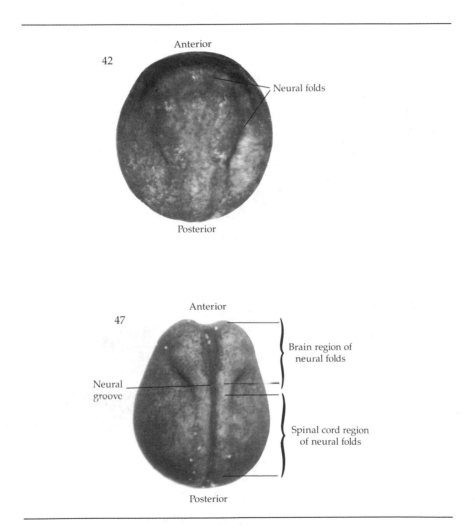

74 *Early and middle neurulation.*

The embryo at 42 hours is still nearly spherical, but it begins to elongate as neurulation proceeds. This growth in length is not the same in all areas, as can be illustrated by what happens in the central nervous system. At 47 hours the lengths of the brain and spinal cord regions of the neural folds are roughly the same. Later in development the spinal cord area will increase in length much more than the brain.

When the 50-hour embryo is turned over, one can see the beginnings of still another structure, the mucus glands. These secrete a sticky mucus that enables the larva to attach to various objects (figure 75).

Tailbud stage

After another day of development there are more external changes (figure 76). The elevated ridge along the back contains the brain and spinal cord, and paired bulges in the brain indicate the place where the eyes are forming. Large swellings behind the eye region are the beginnings of the gills. Still further back a small swelling marks the site where the *pronephros*, the embryo's first kidney, is beginning to form. On the ventral side the mucus glands are better developed.

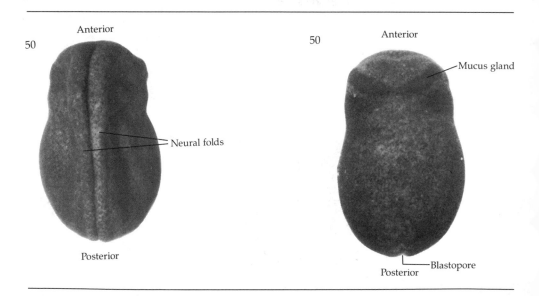

75 *Dorsal and ventral views of a late neurula with the neural folds closing.*

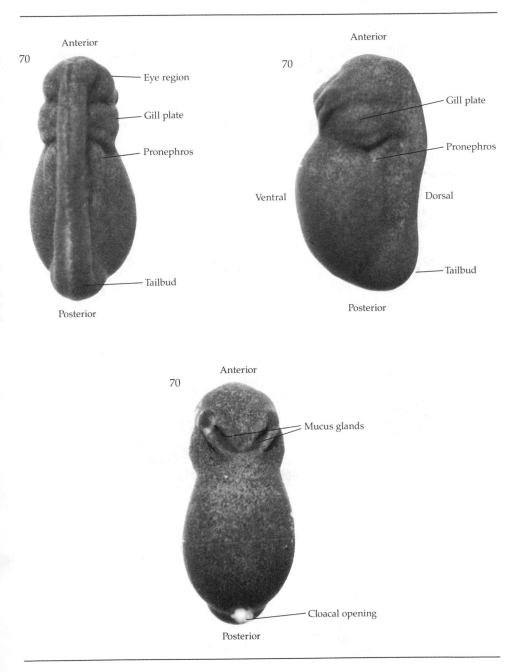

Anterior

70

Eye region

Gill plate

Pronephros

Tailbud

Posterior

Anterior

70

Gill plate

Pronephros

Ventral

Dorsal

Tailbud

Posterior

Anterior

70

Mucus glands

Cloacal opening

Posterior

76   *Tailbud embryo shown in dorsal, lateral, and ventral views.*

The 100-hour embryo

It is not too difficult to extrapolate from the 70-hour tailbud stage to the embryo of 100 hours (figure 77). By 100 hours all its organ systems are forming and some are beginning to function. For example, circulation has begun, and close examination shows blood cells moving through the gills. The embryo has begun to resemble a tadpole. The eyes are present as bumps on the side of the head, but they are not yet functional—the overlying skin is still deeply pigmented. The olfactory organs are paired pits at the anterior end, and between them is another pit, the *stomodaeum*. The stomodaeum will break through to the primitive alimentary canal, forming the mouth. There is a well-developed tail, that is, a portion of the body posterior to the cloacal opening.

The embryo hatches from its jelly envelopes at about this time. Most of its yolk will have been consumed; but before all the yolk has gone,

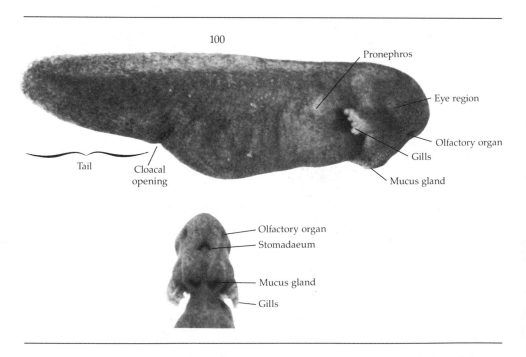

77 *A just-hatched larva shown in lateral view. Shown also is a ventral view of the head.*

the young creature will have begun to find and eat food in the pond where it is living.

Thus in a period of about four days, at 20° C, the frog-to-be will have started as a single cell and, with cell division, differentiation, and growth, produced a larva with all of its organ systems forming and some already functioning. The epigenetic changes will have occurred with a clocklike precision that continues to awe the observer today as much as it did when amphibian embryos were first studied. This is truly an astonishing phenomenon, all the more interesting to observe because our own development occurs in the almost inaccessible interior of the body—unavailable for easy study.

Although amazing external changes have occurred in the developing embryo, they are not as numerous nor as great as what is happening internally—our next topic.

The Internal Development of the Amphibian Embryo

The vital organ systems of all complex animals are internal—protected by a skin and often by scales, bone, an external skeleton of chitin, feathers, shells, or similar structures. Since these organ systems develop inside the embryo, their study presented a difficult problem at first. By the late nineteenth century, techniques had been developed for imbedding preserved embryos in paraffin wax and then making thin slices of them. These slices, or sections, could be mounted on glass slides, stained, and studied with a microscope. It even became possible to make serial sections of whole embryos—beginning at one end and making thin slices to the other end. The slices were then mounted in sequence on slides and the end result would be hundreds of slices of the entire embryo. One then had the cognitive task of deducing the whole internal structure from these thin sections.

An embryo in serial section is static and cannot provide a complete story of the movements of cells to their final sites where they form the various structures. One cannot determine, for example, how the archenteron develops from looking at slides. Does it involve an invagination of cells from the outside or is it a matter of new cells being formed at the advancing edge of the archenteron?

Consider the events in the early development of the frog. The changes from 22 to 36 hours can be explained as the downgrowth of the dark-colored animal hemisphere cells over the light-colored vegetal hemisphere cells. Alternatively, the events could be explained by assuming that the light-colored cells slowly become pigmented.

How could one decide between these two hypotheses? Some early experimental embryologists sought an answer by pushing a needle through the jelly membranes and killing some of the cells on the surface of the embryo. One could then trace the movements of the scar for as long as it persisted, which often was not very long because healing is usually rapid. Nevertheless, experiments of this sort made it seem true beyond all reasonable doubt that cells on the outside move down from the animal hemisphere.

By the 1920s the experimental analysis of the development of the amphibian embryo had reached the stage where it was necessary to know, with a high degree of accuracy, the direction of movement of the various parts of the embryo during gastrulation. The problem was to be able to describe all positions on the embryo and to be able to trace these positions throughout early development. The ability to determine latitudinal distance from the animal pole and longitudinal distance from the dorsal lip meant that any position on the surface of the early gastrula could be described accurately. But experimentalists needed to know not only where a given group of cells might be at the onset of gastrulation but where these same cells would be at various times thereafter. Would they be in the same place or would they have moved?

It took a German embryologist, Walther Vogt, many years of painstaking observation and experimentation to provide an acceptable answer. He prepared a *fate map* that showed where the cells on the surface of an early gastrula would be in the later embryos. That is, he determined the destiny, or fate, of the gastrula cells.

The technique for constructing a fate map is as follows (figure 78):

A layer of wax is put in a small dish and a pit, about the size of an early gastrula, is made in the surface. Tiny pieces of agar are stained with a variety of vital (nontoxic) dyes and placed in the sides of the pit. The outer jelly membranes are removed from an early gastrula, leaving only the vitelline membrane, and the embryo is pushed into the pit. It is held in place by a tiny piece of bent cover glass.

Some of the dye diffuses from the agar and stains the cells on the outside of the embryo. Differently colored vital dyes are used, thus allowing individual spots on the embryo to be recognized and traced. After exposure to the dyes, the embryo is removed from the pit and a drawing is immediately made of the exact position of the colored spots in relation to the animal pole and the dorsal lip. At frequent intervals thereafter the same embryo is studied and sketched.

Vogt placed 8 colored spots, 1 through 8, on the embryo along the meridian that passes through the animal pole and the dorsal lip of the blastopore. A short time later, he found that spot 7 had moved to the in-

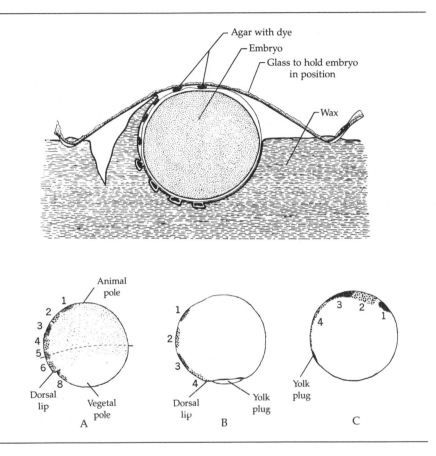

78 *Walther Vogt's (1925, 1929) technique for staining embryos (top). The three lower figures show one of the experiments. The vitally stained spots have been given numbers. A is an early gastrula with the dorsal lip below 6; spot 7 had already invaginated. B is a mid gastrula and spots 5, 6, 7, and 8 have invaginated. C is a late gastrula and only spots 1,2,3, and 4 remain on the outside. Compared to their positions in A, they have spread considerably. Compare the positions of all these spots with the fate map in figure 79.*

terior and that spot 6 was now the dorsal lip (diagram *a*). When the spots move to the interior, the embryo must be dissected to see where they go. In the middle gastrula, *b*, only spots 1–4 are still on the outside. They did not remain in that position, however, but became stretched to cover a much larger portion of the surface of the late gastrula, as shown in *c*. Hundreds of these experiments were necessary before Vogt could prepare a fate map of the early gastrula of the European toad *Bombinator*.

The presumptive ectoderm, that is, the cells that will form the ectoderm later in development, occupies nearly all of the animal hemisphere. Two main subdivisions are delimited in figure 79: the presumptive neural tube, an area consisting of those cells that will eventually form mainly the brain, spinal cord, and optic cup; and the presumptive epidermis, which occupies about a quarter of the surface of the early gastrula and will eventually spread to form the entire epidermis covering the embryo and later the adult.

The presumptive mesoderm forms a band of cells surrounding the embryo in the equatorial region. It, too, consists of two main areas: The cells immediately above the dorsal lip will form the notochord; the remainder of the presumptive mesoderm will form the muscular, skeletal, circulatory, reproductive, and excretory systems, as well as connective tissue and coelomic epithelia.

The presumptive endoderm occupies much of the vegetal hemisphere. Its cells will form the lining of the alimentary canal and structures derived from it such as the liver, pancreas, and bladder.

The presumptive mesodermal and presumptive endodermal cells that are on the outside of the early gastrula are all invaginated to the interior during gastrulation. The division between what goes in and what stays out is shown in figure 79 by the line that separates the presumptive ectodermal areas from the presumptive mesoderm. Note also the dotted line that starts at the dorsal lip and extends around the embryo. That is the line that marks the region of invagination.

Figure 80 is a slice of a 30-hour gastrula through the meridian that includes the animal and vegetal poles and the dorsal lip. Most of the notochord cells have curled around the dorsal lip of the blastopore and form the roof of the tiny archenteron. When gastrulation has ended and the neural folds have begun to form, in the 47-hour embryo, the embryonic layers are assuming their final positions (figure 81). The ectoderm covers the entire outer surface. The dorsal portion is in the position where it will form the central nervous system and the rest will become the epidermis. The endoderm lines the large archenteron. The

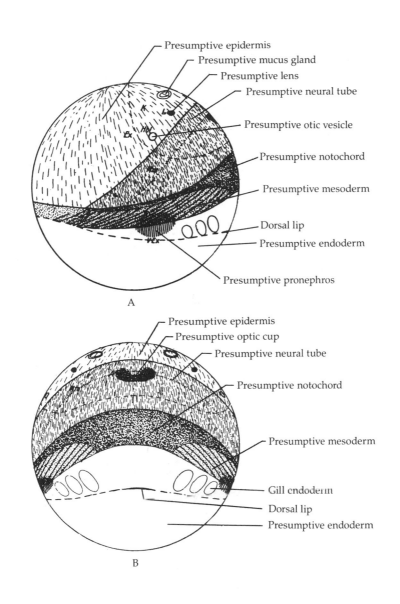

Presumptive epidermis
Presumptive mucus gland
Presumptive lens
Presumptive neural tube

Presumptive otic vesicle

Presumptive notochord

Presumptive mesoderm

Dorsal lip
Presumptive endoderm

Presumptive pronephros

A

Presumptive epidermis
Presumptive optic cup
Presumptive neural tube

Presumptive notochord

Presumptive mesoderm

Gill endoderm
Dorsal lip
Presumptive endoderm

B

79   *Vogt's fate map for* Bombinator *(1929).*

mesoderm is forming the notochord on the roof of the archenteron and is spreading around the body. A cross section of an embryo of this age is shown in figure 82. The neural folds can be seen on the dorsal side as they can in the whole embryo of figure 74. A few hours later the folds have closed, as shown in the whole embryos of figures 74 and 75 and the sectioned embryos of figures 81 and 82.

The distribution of the internal structures of an 80-hour embryo are shown in figure 83. The cross section of the head shows that the neural tube has enlarged to form the brain, and from its ventrolateral walls the optic cups have grown out. The optic cups will form the retina, the light-sensitive portion of the eye. The epidermis adjacent to the optic cup forms the lens and the cornea. The notochord does not appear in this very anterior section. Figure 81 shows why.

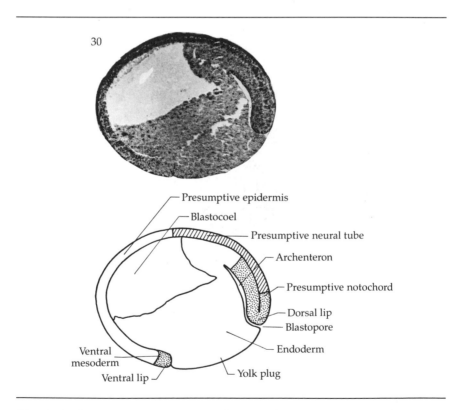

80 *Sagittal section and interpretative diagram of a 30-hour frog gastrula.*

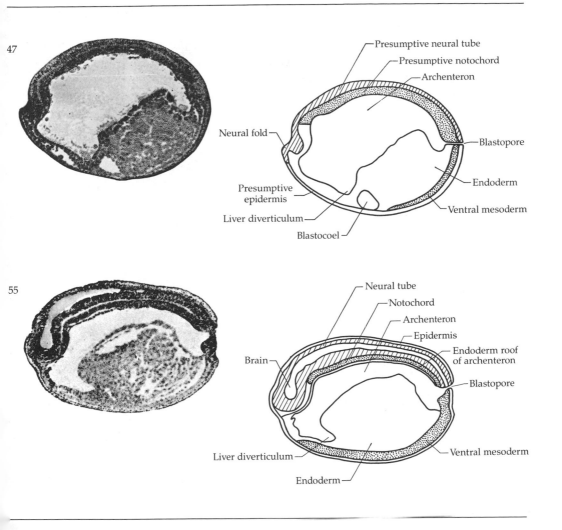

81 Sagittal sections (left) *and interpretative diagrams* (right) *of a 47-hour and a 55-hour neurula.*

A section made in the heart region of the same embryo shows additional structures. The neural tube at this level is the hindbrain, which will form the medulla. Otic vesicles have developed from the outer ectoderm and will differentiate into the inner ear. The heart is forming as a delicate tube beneath the archenteron. The cavity surrounding it is the pericardium, which is part of the coelom.

The cross section of the middle of the body shows the first stage in the development of the excretory system—the pronephros. The mesoderm on either side of the nerve tube and notochord has differentiated into the myotomes or somites, which will form the voluntary muscles and parts of the skeleton. The more ventral mesoderm will eventually split along its length, and the cavity so formed will be the coelom.

This brief survey of early development of the amphibian embryo will provide a basis for understanding the experiments that, beginning in the 1850s, sought to explain differentiation. Now that we have surveyed *what* happens, we can try to understand *how* it happens.

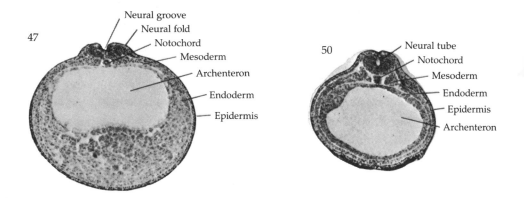

82 Cross sections of a mid and late neurula. The 50-hour embryo shows the very large archenteron, which will become the alimentary canal and associated organs such as the liver and pancreas, the dorsal nerve tube, and the notochord. The three embryonic layers are clear. The ectoderm is represented by the neural tube and the epidermis; the mesoderm by the notochord and a band of cells that surround the embryo just under the epidermis. This layer will form muscles, the excretory and circulatory systems, and the skeleton. The endoderm is the thick wall of the archenteron.

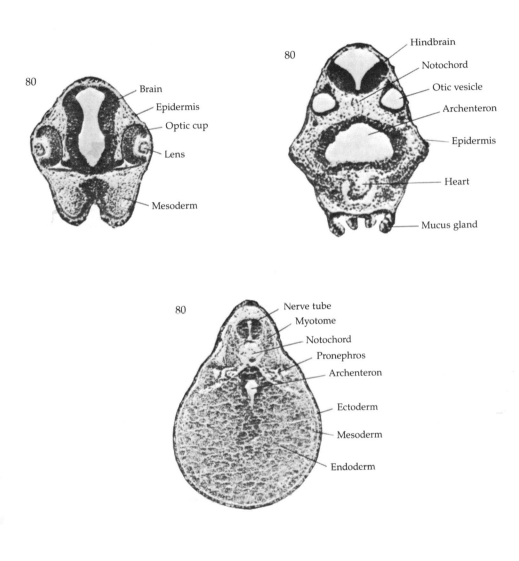

80

Brain
Epidermis
Optic cup

Lens

Mesoderm

80

Hindbrain
Notochord
Otic vesicle
Archenteron

Epidermis

Heart

Mucus gland

80

Nerve tube
Myotome
Notochord
Pronephros
Archenteron

Ectoderm

Mesoderm

Endoderm

83 Cross sections of a tailbud embryo in the optic cup, otic vesicles, and trunk regions. The optic cup and the lens will form the eye and the otic vesicles part of the ear. The embryo's first kidney, the pronephros, is shown in the cross section of the trunk.

chapter 21

The Dawn of Analytical Embryology

George Newport (1802–1854) has been credited with performing the first experiment on embryos. He found that the point of sperm entry determines the axis of the developing embryo. Newport was primarily interested in fertilization and the factors influencing it. First he studied the ovarian eggs of frogs, noted the breakdown of the germinal vesicle, and described the passage of the ova through the body cavity into the oviducts and their storage in the uterus. He found that the jelly layers deposited while the ova passed through the oviducts are necessary for fertilization. He stripped semen from the males and carried out many experiments on the relation of sperm concentration and motility to fertilization, and he tried various temperatures and chemical solutions to find their effects on fertilization.

At first he did not accept other reports or his own observations as indicating that the sperm actually enters the ovum. Eventually he did so, and he is now generally regarded as the first person to offer conclusive proof of this fundamental event.

The fertilized egg of a frog is an enormous sphere compared with a single sperm. Therefore the chance of observing sperm penetration is slight, since one would be searching a huge surface (as it would appear under a microscope) for a tiny contact point. Newport successfully solved this problem by controlling the point of sperm entry. He prepared a sperm suspension and then dipped the point of a pin into it. The pin was then touched gently to the jelly membranes as he looked through the microscope and became convinced that the sperm did penetrate the egg.

Newport also constructed little glass cells that would hold the developing embryo in place. The ability to immobilize embryos permitted Newport to make some very important observations about the polarity that developed in the later embryo.

> *Obs.* 1.—I took an egg that had just divided for the first time, and placed it in a glass cell only sufficiently large to contain it when the jelly was fully expanded, and filled the cell with water. The dorsal surface turned uppermost, as usual, consequently I had under my eye the whole surface; and could watch the changes with the microscope. I marked the plate of glass supporting the cell with a line parallel to the primary cleft of the yelk, and indicated the position of the ends of the sulcus [cleavage furrow] by other marks. The whole was placed in a temperature of 60° Fahr.
>
> At the time of the closing-in of the dorsal laminae [neural folds], I found the correspondence between the axis of the embryo and the line of the first cleft to be exact . . .

The observations were repeated:

> March 14. In each of the eight instances the axis of the body is more or less precisely in the line of the sulcus: thus in five it was in the exact line, in one about five degrees to the left, in another about three degrees to the left, and in the remaining one more to the left of the given line. (1854, pp. 241–242)

Now comes the remarkable experiment showing that the axis of the embryo could be controlled.

> *On the power of the Spermatozoon to influence in artificial impregnation the direction of the first cleft of the Yelk.*
>
> In connection with the influence of the spermatozoon on the egg, I determined to try whether the artificial application of that body to different parts of the egg's surface could affect the position of the first cleft of the yelk.
>
> Similar experiments were repeated four other times, and the results showed that the first cleft of the yelk is in a line with the point of the egg artificially impregnated, and that the head of the young frog is turned toward the same point. (1854, pp. 242–243)

In 1854 Newport became ill after a collecting trip in a swampy area near London and died. Not only had he established a causal relationship between the entrance point of the sperm, the plane of first cleavage, and the primary axis of the embryo, but he could control the relationship. That ability to control development in such a basic way made possible the experimental analysis of differentiation. One could begin to ask meaningful questions and have some hope of being able to answer them.

It is important to note that a breakthrough in experimental science frequently comes as a result of observations having little to do with the problem being explored. Newport was initially mainly interested in fertilization. The tube-cells that he constructed to observe sperm penetration held the embryo in a fixed position, but the same experimental setup proved to be valuable in another way—to make possible his observations associating the entrance point of the sperm, first cleavage, and the embryonic axis.

Experimental, analytical embryology was under way. Well, not quite. Newport's remarkable discoveries were not to be extended for several decades. Darwin was shortly to capture the interest of embryologists, and experimentation was to receive scant attention in the next decade. In the 1870s when the experimental analysis of development began to attract more investigators, there was a need to define the problem. It was stated well by E. B. Wilson. Although the following quotation was written in 1900, he expressed a point of view that would have been much the same two decades earlier.

> Every discussion of inheritance and development must take as its point of departure the fact that the germ is a single cell similar in its essential nature to any one of the tissue-cells of which the body is composed. That a cell can carry with it the sum total of the heritage of the species, that it can in the course of a few days or weeks give rise to a mollusk or a man, is the greatest marvel of biological science. In attempting to analyze the problems that it involves, we must from the onset hold fast to the fact, on which Huxley insisted, that the wonderful formative energy of the germ is not impressed upon it from without, but is inherent in the egg as a heritage from the parental life of which it was originally a part. The development of the embryo is nothing new. It involves no breach of continuity, and is but a continuation of the vital processes going on in the parental body. What gives development its marvelous character is the rapidity with which it proceeds and the diversity of the results attained in a span so brief.
>
> But when we have grasped this cardinal fact, we have but focussed our instruments for a study of the real problem. *How* do the adult characteristics lie latent in the germ-cells; and how do they become patent as development proceeds? This is the final question that looms in the background of every investigation of the cell. In approaching it we may well make a frank confession of ignorance; for in spite of all that the microscope has revealed, we have not penetrated the mystery, and inheritance and development

still remain in their fundamental aspects as great a riddle as they were to the Greeks . . . The real problem of development is *the orderly sequence and correlation of . . . phenomena toward a typical result.* We cannot escape the conclusion that this is the outcome of the organization of the germ-cells; but the nature of that which, for lack of a better term, we call "organization," is and doubtless long will remain almost wholly in the dark. (pp. 396–397)

Yet something could be said about that organization. Since the egg is part of the parent, as Wilson emphasized, its organization must be a part of the organization of the parent. The egg has, therefore, "something" of the parents. That inherent something is encased in a single-celled zygote, and the problem becomes to discover the mechanisms that convert the zygote to adult.

His, Roux, and Mosaic Development

In 1874 William His (1831–1904) proposed the hypothesis of germinal localization, which stimulated much interest and experimentation. He worked mainly with chick embryos, and his problem was the eternal one: "If the body of the chick is not preformed in the germ, what is?" He suggested that, if the parts were not preformed, whatever is responsible for them is present at the beginning of development.

It is clear, on the one hand, that every point in the embryonic region of the blastoderm must represent a later organ or part of an organ, and, on the other hand, that every organ developed from the blastoderm has its preformed primordium in a definitely located region of the flat germ-disc . . . The material of the primordium is already present in the flat germ-disc, but it is not yet morphologically marked off and hence not directly recognizable. But by following the development backwards we may determine the location of every such primordium even at a period when the morphological differentiation is incomplete or before it occurs; logically, indeed, we must extend this process back to the fertilized or even the unfertilized egg. According to this principle, the germ-disc contains the primordia of the organs spread out in a flat plate, and, conversely, every point of the germ-disc reappears in a later organ; I call this *the principle of organ-forming primordial-regions.* (Wilson 1900, p. 398)

(In this translation by E. B. Wilson of His's paper, the term "germ" is used in two ways, one meaning embryonic area on the surface of the yolk, as in "germ-disc," the other referring to the substances necessary for the formation of organs. For these latter I have substituted "primordia" for the sake of clarity.)

Today it may be hard to understand why His's hypothesis was thought important. Would not one expect that the parts of the older embryo and adult come from the substance of the zygote? What other possible source could there be? However, His was saying something else—that the organization of the egg consists of the localization of the factors, unknown but presumably material, that are responsible for the development of the parts of the embryo and adult. Thus the zygote was not to be regarded as a totally unorganized bit of protoplasm but as having some *substances*—not force or immaterial organizing principle—that were the *sine qua non* for differentiation. His was suggesting that by careful observation one could prepare a fate map of the chick embryo much as Vogt was to do a half-century later for the amphibian embryo.

Although His spoke of the "principle" of organ-forming germ regions, "hypothesis" would have been a better term—he suggested, he did not prove. Nevertheless, his hypothesis was a useful way to think of the egg's organization, and it suggested experimental approaches to Roux and others.

Analytical embryology, or *Entwicklungsmechanik*, became a full-fledged program of experimentation in the hands of the German biologist Wilhelm Roux (1850–1924). He was a gifted, vigorous, outspoken, and dedicated scientist who was prominent even in the Germany of his famous teacher, Ernst Haeckel. Roux's main hypotheses were to require much modification and many of his experiments proved to be defective, but with brilliance and perseverance he raised the questions that brought experimental embryology into full flower. He initiated and for years was the editor of the first important journal devoted to analytical embryology, *Wilhelm Roux' Archiv für Entwicklungsmechanik*, which began in 1894–1895 and continues to this day.

Together with his compatriot August Weismann, Roux developed the first important hypothesis of differentiation from which deductions could be made and then tested by observation and experiment. The Roux–Weismann hypothesis, usually called "theory," was based mainly on the observations, experiments, and interpretations of Roux plus theoretical elaboration by Weismann.

Roux's key paper for the discussion that follows was published in 1888. Roux posed several fundamental questions:

The following investigation represents an effort to solve the problem of self-differentiation—to determine whether, and if so how far, the fertilized egg is able to develop independently as a whole and in its individual parts. Or whether, on the contrary, normal development can take place only through direct formative influences of the environment on the fertilized egg or through the differentiating interactions of the parts of the egg separated from one another by cleavage. (Willier and Oppenheimer, 1964, p. 4)

His first question, whether or not the development of an egg requires specific stimuli from the environment, may seem strange to us today. It was not strange in the 1880s. Botanists had been describing the many diverse effects of the environment on the growth and differentiation of plants. Light had a pronounced effect on the production of chlorophyll, the rate of growth, the pattern of growth, leaf retention or loss, and seemingly just about everything plants did. Gravity, temperature, wind, moisture, and soil chemistry all had their effects on plant growth and development. Roux sought to determine if frog embryos were similarly affected by these environmental factors by rotating the embryos constantly so that gravity, light, heat, and magnetic forces would not be able to exert an influence from a constant direction. The embryos developed perfectly normally.

We can conclude from this that the typical structures of the developing egg and embryo do not need any formative influence by such external agencies for their formation, and that in this sense the morphological development of the fertilized egg may be considered as self-differentiation. (p. 4)

Having answered his question for the embryo as a whole, Roux sought to ask the same question for its parts. The very fact that he was able to ask such a question at all depended not only on his work but also the work of those who had preceded him or who were his contemporaries. We must never forget this most important aspect of scientific work. The questions that can be asked at any time relate to the state of the field, which means that others have prepared the groundwork for the scientist's research. For example, in the 1880s there were exciting new discoveries in cell biology, especially about chromosomes. In addition, a huge amount of information was available about development, notably His's postulation that differentiation depended on the presence of determinants for the structures of the embryo. This information was general and could not suggest to Roux what he should do the next morning when he went to his laboratory. However, some very specific facts that he had learned suggested the possibility of a truly impressive experiment which, if successful, would throw great light on the age-old problem of the causes of differentiation.

Roux reported that he had discovered some fascinating rules involving the early development of frog embryos. The first of these was that the plane of first cleavage coincides with the median plane of neurulae and later embryos. There was even the possibility of this being a rule of broad applicability because others had found it to be true for such different embryos as those of bony fish and ascidians. Newport had discovered this long before, as Roux noted. But no notice had been taken of Newport's discovery in 1854 because no one had the remotest idea of how to profit by it. The field was "not ready."

Roux confirmed, for the most part, Newport's other discovery of the relation of the point of sperm entrance to the plane of first cleavage and the future polarity of the frog embryo. But Roux found another relationship: Shortly after fertilization a broad crescent in the lower part of the animal hemisphere, opposite the point of sperm entry, loses some of its dark pigment and becomes the gray crescent. The gray crescent persists at most for a few cleavages. By keeping embryos in a fixed position, Roux found that the dorsal lip of the blastopore appears where the gray crescent had been.

There seemed, therefore, to be these relations. (1) The sperm enters the ovum. (2) The gray crescent forms 180° from the sperm's entrance point. (3) The plane of first cleavage is in the meridian of the entrance point of the sperm and the animal pole. (4) The plane of first cleavage bisects the gray crescent. (5) The dorsal lip forms where the gray crescent had been. (6) The anterior-posterior axis of the embryo forms in relation to the dorsal lip: when the neural folds form, the blastopore will be at their posterior end. Thus, the plane of first cleavage divides the embryo into a right and left half.

Roux saw the possibility of testing His's hypothesis, for if primordia are absolutely necessary for the formation of the parts of the embryo, this deduction follows logically:

If some of the primordia can be destroyed, and the embryo still be able to develop to some extent, the structures normally determined by those primordia must be absent.

Since the primordia were hypothetical structures, it was impossible to identify and then manipulate them. Roux sought to achieve that end, however, in an indirect way. This involved a subsidiary hypothesis and this deduction:

If the plane of first cleavage divides the embryo into a right and left half, each half must contain the primordia for that specific half. Therefore, the destruction of one cell of the two-cell stage would also destroy the primordia for half of the body.

After trying various methods, Roux destroyed one cell of the 2-cell stage with a hot needle.

> I heated the needle by holding it against a brass sphere for a heat supply, heating the sphere as necessary. In this case only a single puncture was made, but the needle was ordinarily left in the egg until an obvious light brown discoloration of the egg substance appeared in its vicinity . . . I now had better results; they were as follows. In about 20% of the operated eggs only the undamaged cell survived the operation, while the majority were completely destroyed and a very few, where the needle had possibly already become too cold, developed normally. I thus developed and preserved over a hundred eggs with one of their halves destroyed, and, of these, 80 were sectioned completely. (p. 9)

In the 20 percent where the untreated cell survived, various results could be expected:

> For example, abnormal processes might intervene which would lead to bizarre structures. Or the single half of the egg, which, after all, according to many authors, is a complete cell with a nucleus completely equivalent in quality to the first segmentation nucleus, might develop into a correspondingly small individual . . . But . . . an even more amazing thing happened; the one cell developed in many cases into a half-embryo generally normal in structure, with small variations occurring only in the region of the immediate neighborhood of the treated half of the egg. (p. 12)

Figure 84 shows some of the results. In embryo *A* the left blastomere had been killed but the right blastomere lived and formed a half blastula. In embryo *B* the right blastomere had been killed and was later sloughed off; the living side produced an embryo with a single neural fold and with the mesodermal layer extending from the notochord around the left side of the embryo only. There is what may be described as half an archenteron, though it is hard to recognize half a hole. What was one to conclude?

> In general we can infer from these results that each of the first two blastomeres is able to develop independently of the other and therefore does develop independently under normal circumstances . . . All this provides a new confirmation of the insight we had already achieved earlier that developmental processes may not be considered a result of the interaction of all parts, or indeed even of all the nuclear parts of the egg. We have, instead of such differentiating interactions, the self-differentiation of the first blastomeres and of the complex of their derivatives into a definite part of the embryo . . . The development of the frog gastrula and of the embryo initially produced from it is, from the second cleavage on, a mosaic of at least four vertical pieces developing independently. (pp. 25–28)

Roux was formulating a hypothesis which came to be called *mosaic development*. Figure 85 is a schematic representation of Roux's interpretations. It shows the segregation of the determinants that produces "a mosaic of at least four vertical pieces developing independently."

These results can be taken as a dramatic test of the deductions, and Roux's hypothesis that determinants are localized is made more probable—as is, of course, His's hypothesis that there are determinants for differentiation.

It is hard to overemphasize the importance of Roux's 1888 hypothesis for the rapidly developing field of experimental embryology. However, important ideas in science must be tested in a variety of ways and by

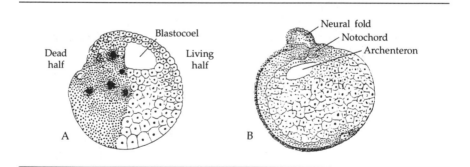

84 *Wilhelm Roux's (1888) drawing of half-embryos obtained after killing one cell of the two-cell stage of a frog embryo. In A the dead half remains. In B it has been sloughed off.*

other scientists before they can be accepted. Such requirements help to eliminate faulty hypotheses and faulty experiments. Roux's ideas, which were center stage for at least a decade, eventually had to be drastically altered.

Even Roux knew there were problems. Although the hypothesis appeared to be fully confirmed by the development of the half embryos up to the neurula stages, some of the half embryos gradually formed a whole embryo—surely a most discouraging phenomenon. Roux called this *postgeneration*.

The simplest interpretation of postgeneration would be that there had been no destruction of the determinants for one side. It had been assumed that each cell of the 2-cell stage had the determinants *only* for that half. Certainly development to the neurula stage seemed to indicate that the determinants for the operated side had been destroyed. If so, they could not "come to life" and produce a whole embryo. Yet they must have been preserved in some way since postgeneration would have been impossible without them.

Roux developed a subsidiary hypothesis to account for postgenera-

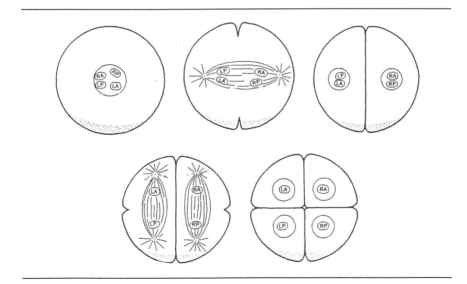

85   *Schematic representation of Roux's revised hypothesis for the segregation of determinants during cleavage. LA = left anterior determinants; RA = right anterior determinants; LP = left posterior determinants; RP = right posterior determinants.*

tion, but the fact that he had to do so greatly weakened his original hypothesis. He held, in effect, that there were two sorts of determinants. The main sort consisted of the determinants that were divided *qualitatively* during cell division and specified the organs and parts of the embryo. In addition, another sort of determinant was held in reserve. It was divided *quantitatively* and kept intact a complete set of determinants. Later in development, if a part was lost, this reserve set made it possible for there to be a complete regeneration. This *deus ex machina* solution was without much merit. It was proposed to save the original hypothesis, and it seems impossible that experiments could be devised to test it. Roux himself reported some experiments that made the original hypothesis questionable when he found that whole embryos would develop, without the need for any postgeneration, from single cells of the 2-cell stage.

Driesch and Regulative Development

Four years after Roux's seminal paper, Hans Driesch (1892) sought to confirm or deny Roux's hypothesis of mosaic development using sea urchin embryos. He, too, started with the hypothesis of His's organ-forming primordial areas and, as had Roux, sought to separate the cells of the 2-cell stage. Instead of killing one of the cells, Driesch put about 100 embryos in a small tube with a little sea water and shook the tube violently for about 5 minutes or more. Some of the 2-cell embryos were found to have broken through their membranes and the cells had separated.

How would they develop? The first check could be made of the cleavage stages. In normal embryos the planes of the first two cleavages are vertical and the third horizontal, as in the frog. The result is 8 cells of almost equal size. The fourth cleavage, however, is very different. The 4 animal hemisphere cells divide about equally. The 4 vegetal cells divide very unequally to give 4 large macromeres and 4 very small micromeres.

Would the isolated blastomeres of the 2-cell stage produce micromeres and, if so, at which division? If they behaved as whole normal embryos starting to develop, they would form micromeres at their fourth division. If they behaved as though they were still part of a whole embryo, the micromeres would be formed at their third division (the isolated cells would, of course, have already gone through the first division and then been shaken apart). Driesch found that the first two cleavages were equal, giving 4 cells looking just like a half of a normal embryo. The third cleavage of the isolated cells produce micromeres and the result was an embryo that was identical with half of a normal 16-cell embryo.

So far, the isolated cells were behaving just as His and Roux would have predicted. In fact, they continued to exhibit mosaic development, forming a half blastula, that is, one that resembled a cup in being open on one side.

That was the evening of the first day and, having observed the experimental embryos all day, Driesch went to bed. What would the morrow bring when he knew that normal embryos would gastrulate and develop into pluteus larvae? When he checked, he found that some had formed pluteus larvae that were normal, except for being small. Apparently Driesch was astonished but not overjoyed with this discovery. It seemed to him "almost a step backward along a path considered well established."

How could he account for these results? Driesch suggested that, after all, frogs are not sea urchins. Since that answer did not seem adequate, he thought that maybe the difference was that Roux had not really "isolated" blastomeres at the 2-cell stage. He had killed one cell, but it remained in contact with the one living one. Possibly the dead one was having an inhibitory effect. His guess was correct, for in 1910 Mc-Clendon removed one cell of the 2-cell stage of frog embryos by sucking it out with a tiny pipette, and found that normal, though small, larvae would result.

Driesch's results were extremely difficult to deal with. When the cells of the 2-cell stage remained together, each would produce half of an older embryo. Yet when the cells were separated from one another, each formed normal pluteus larvae. One had to assume that there is some overall control exerted by the whole embryo over its constituent parts—that is, the embryo is not a complete mosaic of independent self-differentiating parts. Thus, Driesch hypothesized that there must be some harmonizing control by the entire embryo of its equipotential cells. The sea urchin embryo was a "harmonious equipotential system."

But this was found not to be true of some other invertebrate embryos, which seemed to have a purely mosaic pattern of development. The ctenophores are beautiful, medusalike marine invertebrates with bodies of glasslike transparency. They move slowly through the ocean water propelled by 8 rows of comb plates. Several investigators isolated the blastomeres of ctenophore eggs, among them Driesch and Morgan

(1895). They used *Beroe ovata* in which the first three cleavages are vertical, producing an 8-cell stage with the blastomeres almost in a flat plane. Morgan summarized their results:

> When the first two blastomeres are separated from each other by a sharp needle or cut apart by a pair of small scissors, each continues to cleave as a half, *i.e.* as though it were still in contact with its fellow-blastomere. When the organs appear in the larva, only half the full number of rows of swimming-paddles appear. Each row, however, has its full complement of paddles . . .
>
> The isolated one-fourth blastomere [that is, one blastomere from the four cell stage] segments also as a part of a whole, and develops in some cases into a one-fourth larva, having only *two rows* of paddles (*i.e.* one-fourth the normal number) . . . The three-fourth embryos [three cells of the four-cell stage] develop six rows of paddles . . . (1897, pp. 129–130)

The fact that the isolated blastomeres went on to produce larvae with the number of comb plates that each would have produced if it had remained part of an entire embryo seemed to indicate strictly mosaic development. This deeply impressed embryologists, as did later experiments showing an isolated cell from the 8-cell stage would produce a larva with one row of comb plates. That's about as mosaic as an embryo could get.

Thus sea urchin and ctenophore embryos exhibit two fundamentally different patterns of development—mosaic and regulative. The first is a pattern of independently developing parts, while the latter is a pattern of parts that could regulate and form more than they were normally destined to do.

Regulative development was a disturbing notion. What could be the controlling mechanism that restrained the individual cells of the 2-cell sea urchin embryo and molded them into parts of a single organism but released those restraints if the same cells were isolated, allowing each part to form an entire larva? It had all seemed so clear and intellectually satisfying if development were, as Roux suggested, fixed from the onset. It had been equally satisfying, long before, when it was accepted that the embryo was preformed in the ovum (or sperm) and equally when finally it was shown convincingly that development is epigenetic. Driesch puzzled about the implications of his discovery, that backward step as he saw it, and he eventually abandoned experimental science and devoted full time to philosophy.

Novelty in Development

The concepts of preformation and mosaic development avoided the central problem of development: How can novelty arise? The concepts of epigenesis and regulative development must come to grips with that central problem.

In the same year that Driesch published his paper on sea urchins, E. B. Wilson, working at the Stazione Zoologica, a marine biological laboratory at Naples, attempted to solve the problem of regulative versus mosaic development. He repeated Driesch's experiments with eggs of amphioxus, using the same basic technique as Driesch—vigorously shaking the cleaving egg until the individual cells fell apart. This is what he found:

> An isolated 1/2 blastomere [that is, one cell of the two-cell stage] undergoes a cleavage identical with, or approximating to, that of a normal embryo. It produces a normally-formed blastula and gastrula of half the normal size, and finally may give rise to a half-sized dwarf larva exactly agreeing, except in size, with the normal larva up to the period when the first gill-slit is formed . . .
>
> An isolated 1/4 blastomere may undergo a cleavage nearly or quite identical with that of a normal ovum, but often varies more or less widely from it . . . The [larval] stage, with a notochord, is rarely attained and no normally constituted ones were observed . . .
>
> The 1/8 blastomeres are of two sizes (micromeres and macromeres) which, as far as could be determined, do not differ essentially in mode of development. The isolated blastomere segments in a form approaching that of a complete ovum . . . but the gastrula stage is never attained. (1893, pp. 587–589)

Those results did not fit totally with the hypothesis of either regulative or mosaic development—cleavage of the isolated cells might be close to that of an entire embryo, but "their power of development progressively diminishes as cleavage advances" until a 1/8 blastomere cannot gastrulate.

Wilson explained the results by hypothesizing that every cell has the same genetic material, and,

> as the ontogeny advances the [genetic material] of the cells undergoes gradual and progressive *physiological* modification (brought about by the interaction of the various parts of the embryo), without, however losing any of its elements. The isolation of a blastomere restores it in a measure

to the condition of the original ovum and the [genetic material], therefore, tends to return to the condition of the original germ-plasm and thus to cause a repetition of the development from the beginning.

But as development continues the genetic material becomes progressively modified.

By the 8-celled stage [in amphioxus] it is incapable of returning to the original state, and the normal type of cleavage is no longer repeated . . . The specialization of the [genetic material], like that of the cell as a whole, appears to be a cumulative process that results in a more and more fixed mode of action . . . The independent, self-determining power of the cell, therefore, steadily increases as the cleavage advances. In other words: *the ontogeny assumes more and more of the character of a mosaic-work as it goes forward. In the earlier stages the morphological value of a cell may be determined by its location. In later stages this is less strictly true and in the end the cell may become more or less completely independent of its location, its substance having become finally and permanently changed.* (pp. 606–610)

Wilson pushes the analysis back to the beginning of development by suggesting that we regard

ontogeny as a connected series of interactions between the blastomeres in which each step conditions that which succeeds. The character of the whole series depends on the first step, and this in turn upon the constitution of the original ovum . . . The entire series of events is primarily determined by the organization of the undivided ovum that forms its first term, and, as such, conditions every succeeding term. (pp. 613–614)

Cell Lineage

Few observations speak so forcefully for the importance of the organization of the ovum as those on cell lineage, which was one of the main contributions of the American school of embryology in the 1890s and early 1900s. Cell lineage traces the products of cell division, beginning with the uncleaved egg, to the point where the rudiments of the embryonic organs have become distinct. Such studies sought to answer the question whether or not there is a fixed lineage of cells.

Individual cells can be traced only if it is possible to recognize them —they must differ in some way from one another, either in size, coloration, or position. As embryologists coursed up and down the animal kingdom looking for suitable embryos to study, they found many, especially those of marine invertebrates, with distinctive patterns of coloration and cleavage and with different sizes of cells. Some embryos were even transparent, allowing one to observe cells of the interior. Nature was providing naturally stained eggs that could serve the same purpose as Vogt's vitally stained embryos. The patterning of pigmentation of the eggs was found not to be a random affair but part of a basic organization. The planes of cleavage were constant in relation to the pigmented area and, in many cases, the differently colored regions of the egg seemed to have a fixed relation to the germ layers and to the structures they would form.

This visible organization of some eggs at the very beginning of development made it difficult to regard a just-fertilized ovum as an amorphous mass of protoplasm awaiting the directing influences of either idioplasm, determinants, gemmules, nuclei, chromosomes, or whatever. One could not deny organization when it was so striking and constant in what it was and did.

There was, however, an opposing view that considered the uncleaved ovum to be *isotropic*, that is, with no axial organization and with all parts of the cytoplasm equivalent. This hypothesis appealed to many investigators who were impressed by Driesch's experiments on sea urchins and some other experiments in which two eggs were fused and found to produce a single embryo.

One of the first painstaking studies of cell lineage was that of Charles Whitman (1878) on the embryos of a leach. The first two cleavages produce 4 cells of equal size, which Whitman called *a, b, c,* and *x*. At the next division these divide to give 4 very small cells and 4 large ones. The 4 small cells are the progenitors of the ectoderm. The cleavages then become irregular. The cell derivatives of *x* could be followed and were found to give rise to the mesoderm and the nervous system. In fact, Whitman found that entire organ systems could be traced back to their origin in pairs of cells. One pair gave rise to the mesoderm bands, another pair to ventral nerve cord, another to the trunk nephridia, and so on.

Whitman related his observations to the explanatory hypotheses of the day:

> In the fecundated egg slumbers potentially the future embryo. While we cannot say that the embryo is predelineated, we can say that it is predetermined. The "Histogenetic sundering" of embryonic elements begins with the cleavage, and every step in the process bears a definite and invariable relation to antecedent and subsequent steps . . . It is, therefore, not surprising to find certain important histological differentiations and

fundamental structural relations anticipated in the early phases of cleavage, and foreshadowed even before cleavage begins.

The egg is, in a certain sense, a quarry out of which, without waste, a complicated structure is to be built up; but more than this, in so far as it is the architect of its own destiny. (pp. 263–264)

Whitman expressed a point of view that His had proposed in 1874 and Roux held a few years later: the parts of the future embryo existed as primordia from the very beginning and the course of development is determined, not regulative.

In 1892, in the same year that Driesch published his studies of regulative development, E. B. Wilson published a magnificent study of cell lineage in the embryos of the marine polychaete worm *Nereis* in which he described the following findings:

(1) First cleavage cuts across what will become the future longitudinal axis of the embryo, dividing the egg into a small anterior cell, called *AB*, and a large posterior cell, called *CD* (figure 86 A).

(2) The second cleavage coincides with the median plane of the future body and it produces 4 large macromeres: *AB* dividing into *A* and *B* and *CD* into *C* and *D* (figure 86 B, C).

(3) Third cleavage (figure 86 D) is horizontal and unequal. Each large macromere gives off a small micromere. This first quartet of micromeres Wilson designated as a^1, b^1, c^1, and d^1. Each micromere does not come off directly above a macromere but in a slightly clockwise direction. This pattern is known as *spiral cleavage*.

(4) The fourth division (figure 86 E) is also unequal and horizontal. This time the spindles of the macromeres slant in the opposite direction and a second quartet of micromeres comes off in a counterclockwise direction. At the same time the first quartet of micromeres divides.

(5) At the fifth division (figure 86 F), the third quartet of micromeres comes off the macromeres in a clockwise direction. These first three quartets of micromeres form the entire ectoderm.

(6) At the next division, which is no longer synchronous throughout the embryo, the *D* macromere divides into a large cell, still called *D*, and a smaller cell, d^4 (the bottom embryo in figure 86; this has been simplified by omitting the divisions of the micromeres). That d^4 was to become famous, because localized in

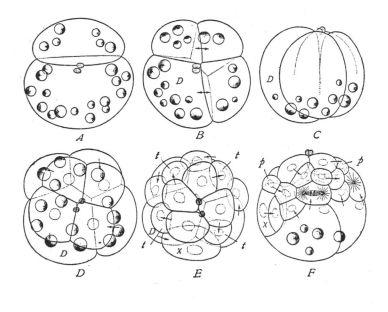

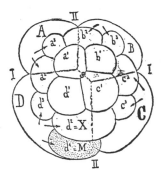

86 *E. B. Wilson's study of early cleavage in* Nereis. *All are polar views except C and F, which are side views. A is the two-cell stage; the circles are oil drops. B is the four-cell stage. C is a side view of the four-cell stage. D is the eight-cell stage; the first quartet of micromeres has come off clockwise. E is the 16-cell stage; the blastomeres marked* t *will form part of the prototroch;* X *will form the nerve cord and some other structures. F is a side view of the 29-cell stage. The lower figure is a simplified drawing showing the micromeres as they come off in quartets but omitting their subsequent divisions; the mesoblast is* d^4 *or* M.

that small cell was the entire material that would form meso-dermal structures. The remaining micromeres form ectoderm and the macromeres the endoderm.

Working out the complete cell lineage of *Nereis* was a difficult problem. The eggs are tiny, 0.12 to 0.14 mm in diameter, and Wilson's optical equipment could not match that available today. The time of breeding was most inconvenient—after dusk. The adult males and females swarm at the surface of the ocean water, where they can be netted and then placed in separate dishes. Back in the laboratory when males and females were put together, spawning would occur. Observations on the embryos began about 9 p.m. and continued throughout the night.

Most embryologists in the 1890s, even the experimentalists, were still influenced by the Haeckelian paradigm, and Wilson sought to relate his study of *Nereis* to studies on other embryos with spiral cleavage. At the very same time that he was working at Woods Hole, E. G. Conklin was there studying cell lineage in a mollusk, the limpet *Crepidula* (1897). The two of them made the astonishing discovery that the details of early cleavage in the annelid worm *Nereis* and the mollusk *Crepidula* were nearly identical. In both the three quartets of micromeres came off in the typical pattern of spiral cleavage: the first quartet clockwise, the second counter-clockwise, and the third clockwise. But the truly startling discovery was that both formed a d^4 cell from which all mesodermal structures are derived in later development. Thus it was hard not to conclude that annelids and mollusks, phyla that differ so widely in the structure of their adults, retain some "ancestral reminiscences" (as Wilson in 1898 called them) in the details of their early development.

One of the more remarkable cases of visible organization in the uncleaved ovum and the early cleavage stages was provided by Conklin (1905), using the eggs of an ascidian *Cynthia* (now *Styela*).

> The very first lot of the living eggs of *Cynthia* which I examined showed a most remarkable phenomenon and one which modified the whole course and purpose of my work; for there on many of the unsegmented eggs, which were of a slaty-gray color, was a brilliant orange-yellow spot, which in other eggs appeared in the form of a crescent or band. Further observation showed that this crescent became divided into two equal parts at the first cleavage and that it could be followed through the later cleavages and even into the tadpole stage. I therefore, for a considerable portion of the summer, devoted myself to the study of the living eggs of *Cynthia*.

And no wonder. Conklin had struck embryological gold, and he was the careful and capable person worthy to develop the strike. He followed the changes from ovarian egg to fully formed larva.

The mature oocyte has a large transparent germinal vesicle. The interior consists of a mass of gray yolk and the periphery contains a yellow pigment. When the germinal vesicle ruptures at the onset of meiosis, it liberates a quantity of clear material. At fertilization the sperm enters near the vegetal pole, and this starts a dramatic rearrangement of the cytoplasm.

Conklin discovered that at the close of first cleavage these distinctively colored regions of the embryo have a precise relationship with the structures that would form subsequently. The fate of the yellow crescent is to form muscles and mesenchyme, the gray yolky cytoplasm forms endoderm, and the clear cytoplasm of the animal hemisphere forms ectodermal structures. Conklin could even distinguish the area that would form the neural plate and the notochord.

The striking aspect of these observations is not that the positions of the structures-to-be are already fixed at the very beginning of development but that because the pigments correspond to the boundaries of the germ-layers, this allows the embryologist to trace them through early development.

These studies of Whitman, Wilson, Conklin, and many others on cell lineage demonstrated that the mature ovum is a complex and highly organized structure. All one can really conclude from these studies, however, is that in the course of normal development identifiable regions of the very early embryo develop into specific structures of the older embryo, as His had proposed long before. One cannot say that those regions can form *only* those structures of the older embryo. Neither can we say that the structures of the older embryo can be formed only by those delineated parts of the early embryo.

Thus a careful distinction must be made between *fate* and *capacity*. Fate means what an area of a younger embryo will form in a later embryo. Capacity means what the cells of that area of the younger embryo *are able to do* under a variety of experimental conditions.

The fate and capacity of a region of an early embryo may be the same if the region is *irreversibly determined*. That is, it can self-differentiate into the specific later structure without influences from other parts of the embryo. Alternatively, that region of the early embryo might, under different conditions, have the capacity to produce much more than its normal fate would suggest, that is, it would have the capacity to regulate. A frequent synonym for capacity is competence.

Thus the distinction between mosaic development and regulative development, which has been applied to the whole embryo, can also be applied to its parts. Problems of this sort, and especially the determination of capacity, can be solved only by experimentation.

Nucleus or Cytoplasm?

The repertoire of techniques available to experimental embryologists at the turn of the century was limited and very crude. One could push hot needles into cells to kill them or one could shake them apart. It was found that when cleavage stages of some marine invertebrates were placed in sea water without calcium ions, the blastomeres separated, which meant that the isolation of cells was made much easier. Simple hand centrifuges enabled one to stratify the more fluid parts of uncleaved eggs. It was discovered that some embryos could be cut with a scalpel—and survive.

Since there were not many experimental techniques available, embryologists adopted a strategy common in biology: search for organisms that differ from those that have already been studied in the hope of finding a new pattern of development—an experiment that nature had done—that might provide new information and new insights.

One interesting variant that nature provides is the presence of polar lobes in the early cleavage stages of many invertebrate embryos. Polar lobes are nonnucleated structures that push out from cells and then flow back into one of the daughter cells. They appear to be a mechanism for redistributing cytoplasmic materials in the early cleavage stages.

Polar lobes are found in the embryos of the mollusk *Dentalium*. Figure 87, from Wilson's classic study (1904*a*), shows events up to the 4-cell stage. When the eggs are shed from the ovary, they are divided into three zones: a clear cytoplasm at the animal pole, a central reddish portion, and another clear area at the vegetal pole.

Before first cleavage the first polar lobe forms at the vegetal pole, as shown in figure 87. It contains essentially all of the clear cytoplasm of the vegetal hemisphere. The first cleavage plane is such that the first polar lobe is attached to only one blastomere, called *CD*. The first polar lobe is then withdrawn into *CD*. As a result, *CD* is larger than *AB*. *AB* has clear cytoplasm only at the animal hemisphere, but *CD* has it not only there but also in the vegetal hemisphere—the contents of the first polar lobe.

A second polar lobe forms from the vegetal hemisphere of *CD*, and at its completion the contents of the second polar lobe are incorporated in *D*. At third cleavage *D* forms the third polar lobe, which then flows back into *D*. Before this cleavage starts the clear cytoplasm near the animal pole moves clockwise and, when the cells divide, it becomes incorporated into the first quartet of micromeres.

From fertilized egg to the trochophore larva requires one day. The trochophore is top-shaped, with an apical tuft of long, stiff cilia and an equatorial band—the prototroch—of three rows of motile cilia, a ciliated pre-trochal region, and a nonciliated posttrochal region (figure 88, embryo 29).

Wilson sought to learn the significance of the polar lobes by removing them with a scalpel and observing subsequent development. When he cut off the first polar lobe, the second polar lobe failed to form. Otherwise the cleavages were normal. Excised polar lobes do not develop. After 24 hours, however, the larva was a disaster (embryo 32). It had three rows of prototrochal cilia that were larger than normal. The pre-trochal region is present—it can be identified by its covering with short cilia. The apical tuft is absent and so is the entire posttrochal region—the embryo ends at the prototroch. Embryo 29 is an unoperated control of the same age shown for comparison.

Embryo 36 had its second polar lobe removed. It also formed a larva with an exaggerated prototroch and no posttrochal region. It does, however, possess a normal apical tuft. Thus the first polar lobe has

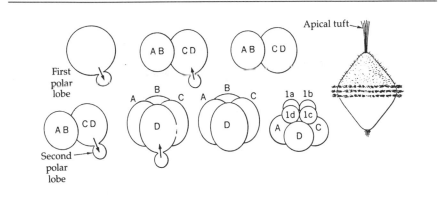

87 *Early cleavage stages and the trochophore larva in the mollusk* Dentalium.

something necessary for the apical tuft, but the second polar lobe does not. Both are necessary for the posttrochal region.

Wilson was much impressed with the importance of the polar lobes especially since "the amount of material removed with the polar lobe . . . is wholly disproportionate to the effect produced. The polar lobe includes less than one-fifth the volume of the egg; yet its removal does not merely cause a structural effect of like extent, but inhibits the whole process of growth and differentiation in the post-trochal region" (pp. 56–57).

Figure 89 shows the results of Wilson's experiments in isolating blastomeres. When the blastomeres were cut apart at the 2-cell stage, the results were strikingly different. Embryos 45 and 46 are isolates from the same 2-celled embryo. Embryo 45 developed from the *CD* blastomere and has an apical tuft, a ciliated pretrochal region, a prototroch of normal size, and the nonciliated posttrochal region. Embryo 46 developed from the *AB* blastomere and is the same as embryos from which the first polar lobe is removed (figure 88, embryo 32). This is not surprising since in a normal embryo the first polar lobe goes into the *CD* blastomere.

Wilson then isolated blastomeres at the 4-cell stage. Embryos 47 and 48 (figure 89) are both from a separated *CD* blastomere. Embryo 47,

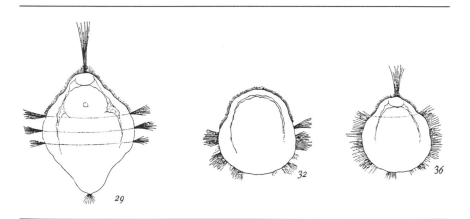

88 *Wilson's (1904) polar lobe elimination experiments. Embryo 29 is a normal trochophore larva. Embryo 32 is a larva that had its first polar lobe removed. Embryo 36 is a larva that had its second polar lobe removed.*

from the isolated D blastomere, is fairly normal, having both an apical tuft and a posttrochal region. Embryo 48, from the isolated C blastomere, is very abnormal. Its capacity for development is about equal to the isolated AB blastomere (embryo 46) or to an embryo from which the first polar lobe has been removed (figure 88, embryo 32).

Finally when he isolated the micromeres after the third cleavage, an important new bit of information was obtained. Embryo 49 (figure 89) developed from the $1d$ cell and embryo 50 from $1c$ of the same embryo. The $1d$ cell produced an embryo with an apical tuft but $1c$ did not.

Putting all these data together, Wilson concluded that the substances in the egg that are necessary for the posttrochal region to develop are originally in the clear cytoplasm of the vegetal hemisphere of the uncleaved egg. They are then successively located in the first polar lobe, the CD blastomere, the second polar lobe, and finally in the D blastomere.

Similarly, the materials necessary for the apical tuft are first in the

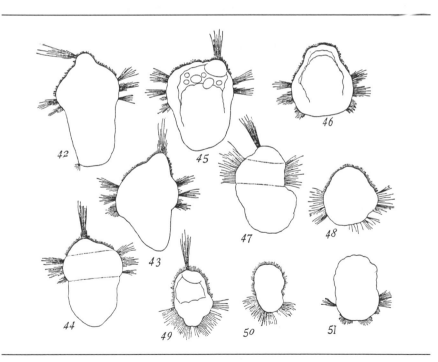

89   Wilson's experiments in the isolation of blastomeres.

vegetal hemisphere, then successively in the first polar lobe, *CD* blastomere, *D* blastomere, and then the 1*d* micromere.

Since the polar lobes do not contain a nucleus, the substances responsible for the apical tuft and posttrochal region *must be cytoplasmic and they must be present in the egg before fertilization*. Does this conclusion mean that the determinants are unrelated to genes? Almost certainly not. The hypothesis that would eventually be developed later is that genes control the synthesis of the determinants while the ovum matures in the ovary. Essentially this conclusion was reached by Wilson nearly a century ago.

> My observations demonstrate conclusively, I think, both the mosaic character of cleavage in these eggs, and the definite prelocalization of some of the most important morphogenic factors in the unsegmented egg. The *Dentalium* egg shows, even before it breaks loose from its attachment in the ovary, and long before even the initial changes of maturation, a visible definite topographical grouping of the cytoplasmic materials. This is proved by the experiments to stand in definite causal relation to the subsequent differentiation of the embryo in such wise that the removal of a particular cytoplasmic area [he had also cut off parts of eggs] of the unsegmented egg results in definite defects in the resulting embryo that are not restored by regenerative or other regulative processes within the time limits of the experiment [there was none of Roux's postgeneration] . . . (p. 55)
> The conclusion is therefore unavoidable that the specification of the blastomeres in these eggs is due to their reception, not of a particular kind of chromatin, but of a particular kind of cytoplasm; and that the unsegmented egg contains such different kinds of cytoplasm in a definite topographical arrangement. (p. 56)

But his final conclusion is that all is ultimately under nuclear control:

> It therefore appears possible, not to say probable, that every cytoplasmic differentiation, whether manifested earlier or later, has been determined by a process in which the nucleus is directly concerned, and that the regional specifications of the egg-substance are all essentially of secondary origin. (p. 64)

Wilson's experiments on *Dentalium* were done at the Naples Zoological Station between February and August of 1903, a period in his life when his long-held view of the importance of the nucleus, and specifically of the chromosomes, in inheritance—including development, of course—was prominent in his mind. His close friend Th. Boveri had recently published his experiments on dispermic sea urchin embryos, which showed that normal development depends on a balanced set of chromosomes. But more importantly, his student W. S. Sutton had just published his remarkable papers linking chromosomes and Mendelian inheritance. In a few years Wilson was to essentially abandon embryological work and devote his full energies to establishing the cytological basis of genetics. At the same time Thomas Hunt Morgan, his colleague at Columbia University, would soon be making genetics an exact science.

Fin de Siècle

If we ask "What were the Big Questions?" that concerned experimental embryologists during the last decades of the nineteenth century and the first one of the twentieth, we will find that few were new. The dominant question was whether early development could be best described as mosaic or regulative. That was no more than a variation of the age-old debate over preformation versus epigenesis. Studies of the organization of the mature ovum, the pattern of early cleavages, cell lineages, and the isolation of blastomeres were all designed to ascertain the degree to which the parts of the ovum are determined (irrevocably committed to a specific developmental pattern) or regulative (having the capacity to do more than their normal fate would indicate).

These attempts to dissect the fundamental phenomena of development sought to understand differentiation better. The questions could not be answered by watching normal embryos develop. In a normal embryo the capacity of the parts of an embryo and what the parts actually do are identical. The fate (prospective significance) of a part and the capacity (prospective potency) of that part, however, may differ widely. Capacity must be determined by subjecting the part to various abnormal situations.

Thus one cannot ask "What is the capacity of a single blastomere of the 2-cell stage of a sea urchin embryo?" and obtain an answer by watching normal development. All one could determine would be that half of an embryo produces half of a larva. However, if we isolate that blastomere, we learn that it has the capacity to do all that an entire embryo can do.

Tentative answers to these questions were obtained for the major groups of animals by the turn of the century. It was possible to describe early embryos as being mosaic, regulative, or some mixture of these basically different patterns. Mollusks, ctenophores, polyclads, and annelids were thought to be strongly mosaic. The amphibians and echinoderms were thought to be intermediate, and *Amphioxus* was regarded as the most regulative in the early cleavage stages. It is highly probable that *Homo sapiens* is a regulative species at least up to the end of second cleavage, since identical twins or identical triplets are derived from a single fertilized egg. These characterizations of any species related to the early cleavage stages only, since it was generally understood that eventually all embryos reached a mosaic stage where the parts would self-differentiate.

Before trying to make sense of this diversity, one needs two critical bits of information that, although of the greatest importance, are not usually emphasized. The first, as we have seen, is that even the most regulative embryos tend to become mosaic at some stage in their development. The second is that in the regulative species even though one blastomere of the 2 or even the 4-cell stage can produce a normal larva, we are not justified in concluding that *any* half an egg can produce a whole embryo. Here nature might be misleading the egg shakers. In all eggs that had been studied it was realized that the unfertilized egg is organized to some degree. There was often a difference in pigmentation of animal and vegetal poles, frequently there was a gradient in the quantity of yolk granules in cells and, wherever it was possible to test, the polar bodies formed in a specific area of the ovum. There is a third important fact that will be developed later, namely, that in even the most strictly mosaic species, their mosaicism is a transitory state— the annelid worms, for example, have remarkable powers of regeneration when they are adults, even though their eggs are highly mosaic.

All one can conclude from the development of isolated blastomeres of the 2-cell stage is that a half embryo cut by cleavage along the animal–vegetal axis will develop in a certain way. We cannot conclude that any half embryo—such as one derived from an egg that cleaved horizontally to give an animal hemisphere cell and a vegetal hemisphere cell—will develop normally. That notion occurred to those shaking the eggs apart. Driesch wondered if the *Echinus* egg had cleaved horizontally instead of vertically, would he have obtained the same result? He suspected the answer would be no, and there were some data supporting that hunch. It was to remain for Hörstadius, a half century later, to provide the answer in some most elegant experiments.

In spite of all the variations among the embryos and disputes among the scientists—some held firmly to the hypothesis that regulative development was the rule, others held firmly to the hypothesis of mosaic development—it did seem possible to provide a conceptual scheme to cover all embryos. By 1900 Wilson had developed such a scheme:

> The cytoplasm of the ovum possesses a definite primordial organization which exists from the beginning of its existence even though invisible, and is revealed to observation through polar differentiation, bilateral symmetry, and other obvious characters in the unsegmented egg . . . [These] promorphological features of the egg are as truly a result of development as the characters coming into view at later stages. They are gradually established during the preembryonic stages, and the egg, when ready for fertilization, has already accomplished part of its task by laying the basis for what is to come. (pp. 384, 386)
>
> In *Amphioxus* the differentiation of the cytoplasmic substance is at first very slight, or readily alterable, so that the isolated blastomere, as a rule, reverts at once to the condition of the entire ovum . . . In the snail and ctenophore we have the opposite extreme to *Amphioxus*, the cytoplasmic conditions having been so firmly established that they cannot be readjusted, and the development must, from the onset, proceed within the limits thus set up . . .
>
> [Thus] we reach the following conception. The primary determining cause of development lies in the nucleus, which operates by setting up a continuous series of specific metabolic changes in the cytoplasm. This process begins during ovarian growth, establishing the external form of the egg, its primary polarity, and the distribution of substances within it. The cytoplasmic differentiations thus set up form as it were a framework within which the subsequent operations take place in a course which is more or less firmly fixed in different cases. (pp. 424–425)

The data available to Wilson supported the hypothesis that all eggs and even some embryos begin as highly regulative and then gradually become mosaic. The various species differ in the time when ovulation and fertilization occurs in relation to this transition from the regulative mode to the mosaic. That time comes early in *Amphioxus* and late in mollusks and ctenophores.

When Wilson and others were reaching these conclusions, the next paradigm of experimental embryology was being formulated. It would be concerned not so much with the development of isolated parts of embryos as with the interactions among the parts. We will consider two examples: the work of Hörstadius on the sea urchin and that of the Spemann school on amphibian organizers.

chapter 22

Interactions during Development

D uring the 1920s and 1930s Sven Hörstadius, a Swedish experimental embryologist, performed a remarkable series of experiments on the embryos of the sea urchin *Paracentrotus lividus*. He was skilled at operations on these minute embryos, and his results, and their interpretation, are one of the main contributions to developmental biology in the first half of the twentieth century.

Mature *Paracentrotus* ova have a pigmented equatorial band that serves as a convenient landmark. As is true with so many species, the first two cleavages are meridianal and the third is equatorial (figure 90). The resulting 8 cells are of approximate equal size. The fourth cleavage, giving 16 cells, is vertical in the animal hemisphere, the result being a single layer of 8 cells. The cleavage plane in the vegetal hemisphere is horizontal and unequal—resulting in 4 large macromeres and 4 small micromeres. We will note the fifth cleavage only for what happens to the layer of 8 cells in the animal hemisphere. They divide horizontally into two layers—called an_1 and an_2. Figure 90 is also a fate map for *Paracentrotus*. The boundaries of the cells, and their corresponding regions in the uncleaved egg are shown by different symbols so they can be traced throughout cleavage and up to the pluteus larval stage.

Experiments involving the separation and recombination of different layers of the embryo led to the hypothesis that the egg and early embryo may have concentration gradients of animalizing and vegetalizing materials. The animalizing substances are assumed to be necessary for the development of structures normally formed from the ectodermal areas. The vegetalizing substances are assumed to be necessary for the formation of those parts normally derived from the presumptive mesoderm and endoderm. It appeared that normal development is not so much a question of the parts involved but whether or not there is a proper balance of these two hypothetical substances.

The hypothetical animalizing substances are assumed to be in highest concentration at the animal pole and lowest at the vegetal pole. For convenience let us assume that they have a concentration of 5 in an_1, 4

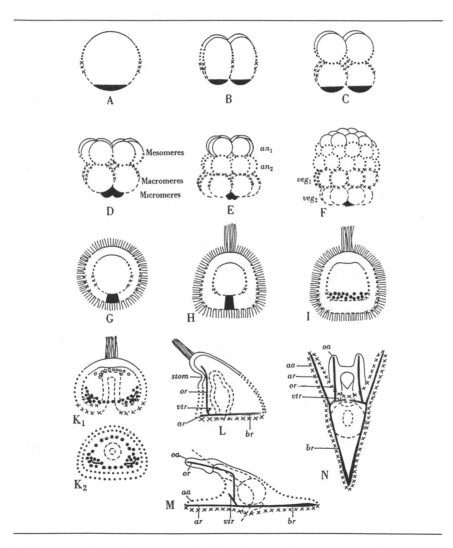

90 Sven Hörstadius's (1939) illustration of normal development of the sea urchin embryo. Equivalent regions of the embryo are shown by specific symbols.

in an_2, 3 in veg_1, 2 in veg_2, and 1 in the micromeres. The vegetalizing substances are assumed to have the highest concentrations, let us say 5, in the micromeres and then decreasing one number per layer until they have a value of 1 in an_1. Let us also assume that normal development is possible only when the concentrations of the animalizing and vegetalizing substances are nearly equal in the whole embryo or fragment produced experimentally. Thus in a normal embryo, if we sum the values from an_1 to the micromeres, there will be a total of 15 animalizing units ($5 + 4 + 3 + 2 + 1 = 15$) and the total will be the same for the vegetalizing substances ($1 + 2 + 3 + 4 + 5 = 15$).

The hypothesis that development is related to these substances can be tested by cutting the embryo horizontally between an_2 and veg_1. The animal hemisphere half would have $5 + 4 = 9$ animalizing units and $1 + 2 = 3$ vegetalizing units. That ratio of 9 animalizing to 3 vegetalizing units is far from equal. The vegetal hemisphere half would have 6 animalizing and 12 vegetalizing units.

The results of such an experiment are shown in figure 91. The upper row illustrates the blastulae derived from the animal hemisphere halves (consisting of $an_1 + an_2$—both presumptive ectoderm). When the blastula stage is reached, the apical organ, instead of being of normal size (figure 90, H), may be expanded to cover nearly the entire embryo. The embryos A_1 through A_4 show the range of results, the majority being like A_1 or A_2.

The bottom row shows the development of the lower half ($veg_1 + veg_2 +$ the micromeres); that is, one layer of ectoderm, one mainly of endoderm, and the micromeres, which form the primary mesenchyme and the skeleton. These plutei usually have an enlarged gut, usually poorly developed arms or none at all, and often no mouth. Earlier they usually lacked the apical organ. These results seemed to support the hypothesis of normal development depending on a balance of animalizing and vegetalizing influences.

Hörstadius's hypothesis was tested in many other ways. He developed the techniques to separate the individual layers at either the 32- or 64-cell stage and combine them at will (figure 92). These are some of the results (with our hypothetical values for the animalizing and vegetalizing materials in parentheses).

(1) $an_1 + an_2 =$ blastula with large apical tuft; almost never any gastrulation (9 animalizing and 3 vegetalizing units). (Figure 92 A.)

(2) $an_1 + an_2 + veg_1$ = apical tufts normal but almost never any gastrulation. These three layers consist of the entire ectoderm (12 animalizing and 6 vegetalizing units). (Figure 92 B.)

(3) $an_1 + an_2 + veg_2$ = normal apical tuft. Reasonably normal pluteus larva (11 animalizing and 7 vegetalizing units). (Figure 92 D.)

(4) $an_1 + an_2 + veg_1$ + micromeres = normal development (13 animalizing and 11 vegetalizing units). (Figure 92 E.)

(5) $an_1 + an_2$ + micromeres = normal development (10 animalizing and 8 vegetalizing units). (Figure 92 F.)

Thus there is normal development when the ratio of animalizing to vegetalizing substances are close: 13/11, and 10/8; an intermediate condition when the ratio is 11/7; and abnormal development when the ratios differ markedly: 9/3 and 12/6.

Note that in experiment 3 the addition of a single vegetal hemisphere

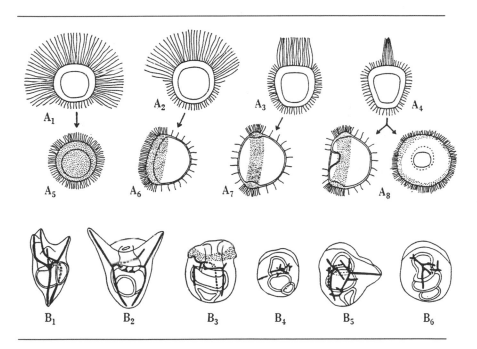

91 *The development of isolated animal (A) and vegetal (B) hemispheres of sea urchin embryos (Hörstadius 1939).*

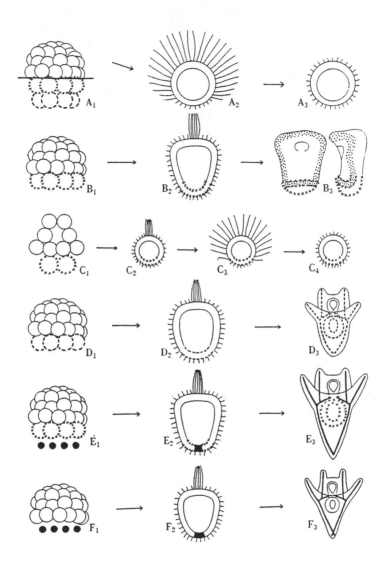

92   *The development of combinations of cell layers of sea urchin embryos (Hörstadius 1939).*

layer, veg_2 is enough to balance the animalizing influences. A fairly normal pluteus is obtained even though a third of the presumptive ectoderm (veg_1), and the micromeres that normally form the primary mesenchyme and the skeleton, are missing. Other cells, however, are able to alter their normal fates and produce the structures that the missing layers would have formed in a normal embryo.

Experiment 4 shows a similar result. The layer that would normally form the endoderm, veg_2, has been removed. The embryo remaining consists only of the presumptive ectoderm and the presumptive primary mesenchyme. Nevertheless, an archenteron is formed.

These results, plus many more not listed, lead to many important conclusions:

(1) When the blastomeres of the 2-cell stage are isolated each will have the entire range of substances that are localized along the animal pole-vegetal pole (A-V) axis. However, if the half-embryo is obtained by an equatorial cut, isolating the animal hemisphere and the vegetal hemisphere, as in the experiments just described, development is abnormal.

(2) Thus the experiments on the isolation of halves show that the sea urchin embryo is of the regulative type when the separation is along a meridian plane (A-V axis) but largely mosaic when the separation is along an equatorial plane.

(3) Although development can be explained by assuming concentration gradients of substances distributed along the A-V axis, these substances are not localized in specific areas. Any one of the five tiers of cells—an_1, an_2, veg_1, veg_2, and the micromeres—can be eliminated and a normal larva result. *Thus the development of a part depends on the entire embryo.* That is, the development of the part is regulated in such a manner that the end result is as normal as the entire fragment will permit.

There are many morals to be learned from the research on sea urchin embryos. Probably the most important is that the "facts" of science are to be accepted only for the precise phenomena they are assumed to describe. Sea urchin embryos were the model for regulative development, a "fact" based on the development of halves obtained by the separation of blastomeres along the meridian extending from animal pole to vegetal pole. This "fact" is replaced by a better "fact" when experiments produce halves by isolation of animal and vegetal hemispheres. Thus, we no longer can describe sea urchin embryos as "regulative" or "mosaic" but must specify which conditions and which parts are being discussed.

Driesch was not wrong; his statements were merely incomplete. Since the questions he asked were fundamental to our understanding of development, others sought to repeat his experiments. When they used his techniques, they usually obtained his results. Hörstadius was able to ask the question in a different way and obtain a different answer that expanded our understanding of early development. The test of a single deduction rarely establishes a hypothesis as "true beyond all reasonable doubt."

Amphibian Organizers

In the early 1920s a new paradigm began to attract notice. This was the line of work started by the German embryologist Hans Spemann (1869–1941), who sought to discover how the parts of an amphibian embryo influence one another. This led to the hypothesis that one part of an embryo, the *organizer*, can influence the differentiation of another part, the *reacting tissue*.

The cells of an amphibian embryo in the late blastula stage appear to be essentially the same. To be sure there is a gradient of increasing size, with the smallest cells at the animal pole and the largest at the vegetal pole. There is also a gradient in the concentration of yolk granules, with less in the cells at the animal pole and more in those at the vegetal pole. The cells of the animal hemisphere are packed with melanin granules whereas those of the vegetal hemisphere are relatively pigment free. Apart from these differences there is nothing to suggest the widely divergent destinies of the cells of different regions. Thus, the conversion of the single-celled zygote into the many-celled late blastula is brought about by cleavage with little differentiation of the cells: they just get smaller.

The first system to form in an amphibian embryo is the nervous system, so it is not surprising that it engaged the attention of embryologists. Although in the interior of the adult, it appears on the outside of the early embryo. At the end of gastrulation a flattened area, the neural plate, becomes visible—extending anteriorly from the closed blastopore. Neural folds appear at the edges of the neural plate, move to the center, and fuse along their crests, forming a tube that lies under the outer epidermis (figures 74, 75, 82).

When embryos of these stages are examined in sectioned material, we find that, by the time the neural plate is forming, gastrulation movements have brought a sheet of presumptive notochord cells into a position below the neural plate. Repeated observation would show that these events always occur in normal development—as Vogt's fate map indicates. The neural tube forms in a constant relation to the positions of the blastopore, archenteron, and polarity of the embryo. These constant relations must be important because, if something always happens in the same way, it is assumed that it is a fixed phenomenon, and possibly with cause–effect relationships.

Thus our problem is to understand how, at the end of gastrulation, those presumptive ectodermal cells that are in the area above the roof of the archenteron become the neural tube, whereas the remaining presumptive ectodermal cells, which look identical, become the epidermal covering of the body. In our own case the difference is quite spectacular. One set of our presumptive ectodermal cells becomes so changed that it can think about the epidermis; the epidermal cells can never think about the brain at all.

So we ask: "Why these different fates?" The answer can only come from experimentation, but, as usual, there is that awesome problem of knowing what to do—that is, how to ask a question that is answerable. We could start by asking, "Is the part mosaic or regulative?" Two alternative hypotheses to explain how presumptive neural tube cells of the early gastrula become the neural tube suggest themselves:

Hypothesis 1. The presumptive neural tube cells of an early gastrula possess an inherent capacity to form neural tissue. They are determined, that is, they have within themselves all that is necessary to differentiate into a neural tube.

Hypothesis 2. The presumptive neural tube cells of an early gastrula do not possess an inherent capacity to form neural tissue. That is, they are still in a regulative stage and influences from outside the presumptive neural tube area are necessary for them to differentiate into a neural tube.

These alternative hypotheses can be tested by the same experiments. Thus, if the presumptive neural tube cells are already determined and possess within themselves all that is necessary for the differentiation of a neural tube, this deduction follows logically:

The presumptive neural tube cells should be able to differentiate into a neural tube if they are separated from the remainder of the embryo.

We now have to devise experimental means of verifying or denying the deduction. One such experiment was performed by Johannes Holtfreter (1901–1992), a student of Spemann. Pieces of the blastocoel roof of an early gastrula were cut out and cultured in a dilute salt solution. No external source of food is required since each cell has many yolk granules. Such *explants* remain alive for days—many more than are necessary for the control embryos to gastrulate and form neural tissue. Explants were taken from two areas—the presumptive neural tube area and the presumptive epidermis. The results of many experiments were the same: neither type of explant differentiated as neural tissue. Both formed only simple epidermallike cells (figure 93, top).

If these results can be accepted as an adequate test of the deduction, we must conclude that hypothesis 1 has not been supported. But hy-

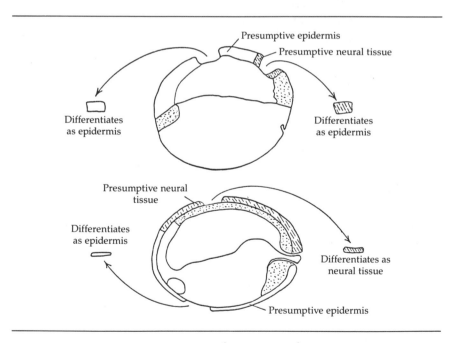

93 *Johannes Holtfreter's experiments explanting presumptive neural tissue and presumptive epidermis in an early* (above) *and late* (below) *gastrula.*

pothesis 2 has been. The experiment can be criticized, of course, as having injured the excised piece of the blastocoel roof. This possibility can be partially ruled out since self-differentiation by other explants is possible, as we will soon see.

The evidence from this first experiment suggests that the presumptive neural tube cells have not been determined by the onset of gastrulation, since they are unable to self-differentiate. Nevertheless, they must become determined within a day because at that time they begin to form a neural tube.

Holtfreter now did the experiment at the end of gastrulation but before there was any indication of the neural plate. The results were dramatically different: neural tissue was formed by presumptive neural tissue but not by presumptive epidermis (figure 93, bottom).

Holtfreter found, quite by accident, another way to test his hypotheses. In some experiments designed for an entirely different problem, early gastrulae were placed in water to which extra salts had been added. Then the jelly membranes surrounding the gastrula were removed and the embryos were rotated so the animal hemisphere was down. Under these conditions gastrulation movements were abnormal. The presumptive ectoderm cells did not move over the vegetal hemisphere but tended to pull away from the rest of the embryo. The result was a dumbbell-shaped embryo known as an exogastrula. In extreme cases the presumptive ectodermal cells formed an irregular mass connected by only a thin strand of cells with the presumptive endodermal and mesodermal cells (figure 94).

Development of the two parts was very different. The presumptive endoderm and presumptive mesoderm differentiated into heart, muscle, parts of the alimentary canal, and other organs normally formed from these two layers. These layers were able to self-differentiate. In marked contrast, the presumptive ectoderm remained essentially undifferentiated. There was no trace of a nerve tube.

Exogastrulae are strange not only in this separation of presumptive ectoderm from the other regions but in the abnormal movements of the other two presumptive regions. The embryo turns inside out. As a result the mesoderm is inside the endoderm and the *lining* of the archenteron faces outward (figure 94).

Once again, the data indicate that the presumptive neural tissue is undetermined at the onset of gastrulation. We know, however, that it is determined by the end of gastrulation. Thus some change must occur in the interval between the early gastrula and the late gastrula. This change, however, does not occur in the presumptive neural tube tissue during the time it is an explant or part of an exogastrula—its cells do not become determined. We might suspect, therefore, that the change is due to influences from other parts of the embryo, and this would mean almost certainly influences from either the presumptive mesoderm or presumptive endoderm, or both. That fits our second hypothesis, which implies that the presumptive ectoderm is completely undetermined at the onset of gastrulation and that some outside influence results in part of it being determined to become neural tissue. Since only part of the presumptive ectoderm becomes neural tissue, the stimulus from outside must be localized. If this is the case, the following deduction can be made:

If the relative positions of the animal hemisphere, which contains the presumptive ectoderm, and the rest of the embryo are altered, the position of the neural tube should be altered accordingly.

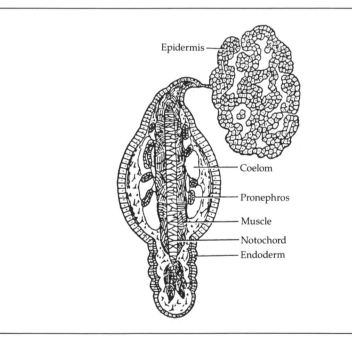

Epidermis

Coelom

Pronephros

Muscle

Notochord

Endoderm

94 *Differentiation of the parts of an exogastrula (Holtfreter 1933b).*

An experimental test of this deduction was made by Spemann. He cut off the upper part of the animal hemisphere of an early gastrula, rotated it 180°, and stuck it back on the lower portion of the embryo. The two parts healed and the embryo went on to form a normal larva, not in relation to the presumptive regions of the animal hemisphere but of the ventral part.

Figure 95 shows the experiment. The upper figures are normal, un-operated embryos. As Vogt had established, the presumptive notochord area (stippled in the figure) is above the dorsal lip and the presumptive neural tube above that. The lower two figures show the operation. The animal hemisphere was cut along the dashed line and then rotated. As a consequence, the presumptive neural tube area is now 180° from its normal position and the presumptive epidermis is adjacent to the presumptive notochord. The operated embryo continues to develop, but the neural folds form *in relation to the dorsal lip of the blastopore*. This means that the presumptive epidermis formed the nerve tube and the presumptive neural tube cells formed epidermis.

This experiment shows that the differentiation of the presumptive

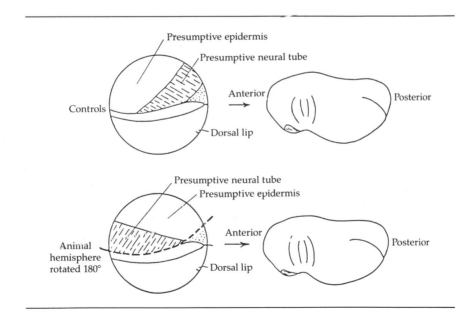

95 *Schematic representation of Hans Spemann's experiments involving the rotation of the animal hemisphere.*

ectoderm is greatly influenced by the ventral part of the embryo. But what part? The constant relation of the dorsal lip of the blastopore to the position of the neural plate and neural tube, both in normal development and in the just-mentioned experiment on rotation of the animal hemisphere, suggests that the dorsal lip might be involved. The dorsal lip is the place where the presumptive notochord cells turn in, forming the archenteron roof and lying beneath the presumptive neural plate.

These experiments and their analysis suggest a variation on hypothesis 2.

Hypothesis 2a. The presumptive neural plate cells of an early gastrula do not possess an inherent capacity to form neural tissue. Instead, the presumptive neural plate cells become determined as a result of stimulation by the presumptive notochordal cells of the archenteron roof.

If this hypothesis is accepted as true, the following deduction can be a test of it:

If the dorsal lip cells are removed from a donor embryo and grafted into a host embryo, and if they are able to invaginate, a nerve tube should be produced from the overlying presumptive ectoderm of the host.

This difficult experiment was performed in 1924 by Hilda Mangold, when she was a student of Spemann. It is one of the classics of embryology, winning a Nobel Prize for Spemann in 1935 (Hilda Mangold had died shortly after the experiments were performed).

In order to recognize the origin of the cells, the embryos of two species of salamander were used; in one species the embryos are nearly white and in the other they are brownish. A small piece of tissue was removed from the dorsal lip region of the donor embryo and then transplanted to a site 180° from the host's dorsal lip (figure 96).

The host, therefore, had two dorsal lips—its own and the donor's. Invagination occurred at both. Because of the difference in pigmentation of the two species, it could be established that the dorsal lip cells of the donor invaginated. At the time the host's neural folds were forming (the primary embryo), neural folds also appeared above the region where the donor dorsal lip cells had invaginated (the secondary embryo).

An important question now confronts us. Is the secondary embryo formed from the donor tissue, host tissue, or both? Again, we can tell

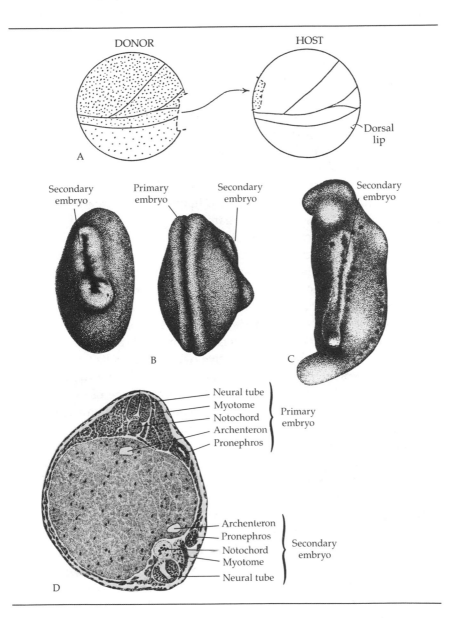

DONOR

HOST

Dorsal
lip

A

Secondary
embryo

Primary
embryo

Secondary
embryo

Secondary
embryo

B

C

Neural tube
Myotome
Notochord
Archenteron
Pronephros

⎫
⎬
⎭ Primary
embryo

Archenteron
Pronephros
Notochord
Myotome
Neural tube

⎫
⎬
⎭ Secondary
embryo

D

96 *The dorsal lip transplantation experiment of H. Spemann and Hilda Mangold.*
A is a diagram of the operation—see figure 79 for full labels. B and C show the
secondary embryos. D is a cross section showing the structure of the primary and
secondary embryos (B, C, and D are modified from Spemann and Mangold, 1924).

because of the difference in pigmentation of host and donor cells. The answer is both. The donor tissue forms the archenteron roof of the secondary embryo, which later becomes the notochord. It also forms other structures, mainly mesodermal. The neural tube, however, is formed almost entirely from host cells. Thus host cells that normally would have formed epidermis now form a nerve tube.

Thus Spemann and Mangold showed that the presumptive notochordal cells that invaginate at the dorsal lip and form the roof of the archenteron have a profound effect on development. They spoke of these cells as the *organizer* and their action on the undetermined ectodermal cells as *induction*.

The experiments so far described suggest that, in normal embryos, the neural tube is formed under the influence of the organizer. At the beginning of gastrulation, the organizer region consists of the cells above the dorsal lip corresponding roughly to the presumptive notochordal region of Vogt's fate map (figure 79). This region invaginates to form the roof of the archenteron. The roof of the archenteron then induces the overlying ectoderm to form a neural tube. Without this inductive influence these cells will form only simple epidermis.

We now have a theoretical basis to interpret the experimental results from explantation of tissues, exogastrulation, and the rotation of the animal hemisphere. When presumptive neural plate cells from an early gastrula are explanted, they will never be stimulated by the organizer and hence cannot form neural tissue. The same is true of exogastrulae —the presumptive ectoderm is never in contact with the organizer. In this case, however, Holtfreter made some most interesting observations. He found that by varying the culture conditions he could obtain partial exogastrulae. In these instances the presumptive ectoderm that was in contact with the presumptive mesoderm and endoderm was induced to form neural tissue.

The experiment on rotating the roof of the blastocoel has a similar explanation. The original presumptive neural tube area was moved to a position where the archenteron roof would not make contact with it —and it remained as epidermis. The presumptive epidermis, however, came to be situated over the archenteron roof, and it was induced to form a neural tube.

Secondary Organizers

Other organizers were soon found in addition to the primary organizer associated with the tissue of the dorsal lip, which later forms the archenteron roof. Secondary organizers were discovered for the mouth, heart, eye, lens, otic vesicle, olfactory organs, pronephros, and many other structures. In fact there is much evidence to suggest that the primary axial organization of the early embryo is controlled by the invaginated material that forms the walls of the archenteron.

The formation of the optic cup and lens of the eye can serve as an example. Vogt's fate map shows the amphibian eye as having a dual origin (figure 79). The bottom diagram shows the presumptive optic cups in the middle of the presumptive neural tube area. The lenses, however, are the small ovals above, to the right and left, in the presumptive epidermis area. They are shown and labeled in the upper diagram.

The complete eye has the lens centered, which is of course necessary for normal vision. An off-centered lens would be useless. When we remember the complicated movements of the presumptive areas during gastrulation and neurulation, one can only marvel that the processes are so precise that the lens always ends up exactly where it should. There is a reason.

The optic cups are induced by the archenteron roof. This would seem to indicate that the primary organizer has multiple specific influences. That is, it (or they) induces not only the overlying ectoderm to form a neural tube but also other structures as well.

Shortly after the closure of the neural folds, the optic cups begin to grow laterally from the floor of the brain (figure 83, 80 hours). When the optic cup reaches the epidermis, a lens begins to form from the inner layer of the epidermis adjacent to the optic cup. Subsequently the outer layer of the epidermis, still full of pigment granules and quite opaque in figure 83, begins to clear and form the cornea.

The interrelations of these events can be tested experimentally. When the optic cups are beginning to form, a slit is made in the head epidermis and the optic cup on one side is cut off. Then the epidermis is pushed back in position and it heals in a few minutes. The embryo is allowed to develop for two days and then fixed and prepared for microscopic examination. The optic cup on the unoperated side—we have an experimental and control animal in a single individual—is found to have produced a normal eye with lens. On the operated side, however, the brain is found to have healed and there is neither optic cup nor lens. The brain cells, therefore, could not regulate to replace the excised optic cup. Thus, in the absence of an optic cup, lens differentiation does not occur. This result suggests that the optic cup is an organizer for the lens.

The next experiment supports that conclusion. An optic cup is removed when it starts forming and placed under the epidermis of the trunk region. The wound heals and, at the time a lens would normally form, the trunk epidermis adjacent to the transplanted optic cup forms a lens. Thus it seems true beyond all reasonable doubt that the optic cup induces the overlying epidermis to form a lens. That trunk epidermis would normally have continued to differentiate as epidermis, but this experiment shows that it still has the ability—or competence, in the language of embryologists—to do more than its fate suggests.

That eye back in the flank, which may appear to be structurally normal, is nonfunctional. It does not enable the tadpole to see where it has been or, at least, who is sneaking up behind it. The transplanted eye never makes the proper nerve connections. We "see" with our brains, not with our eyes.

The Reacting Tissue

These descriptions of the induction of neural tubes and lenses have emphasized the role of the inducing agent. This may have given the impression that the reacting tissue is passively molded by the organizer. Not so. The ability of tissue to respond to organizers is limited in several ways.

Age is one limitation. The experiments described before showed that any portion of the presumptive ectoderm of an early gastrula can be induced to form a neural tube, but this specific competence is short lived. At or about the stage when the neural folds close, the presumptive epidermis is no longer capable of being induced by the archenteron-roof organizer. However, it is still competent to respond to other organizers—the optic cup, for example, and form a lens.

Tissue specificity is another limitation. Explantation is a test of the degree to which a tissue has been determined at the time of explantation and hence the extent to which it can self-differentiate without further influences from the embryo. Such tests show that the presumptive ectoderm of an early gastrula has not been irreversibly determined. If similar explantation experiments are done with the presumptive notochord and adjacent mesodermal regions of an early gastrula, another result is obtained. Both kinds of explants, although too small to produce organs, differentiate into notochordal, neural, and some other tissue types. These cells, therefore, are partially determined. They can form differentiated tissues, but they are not completely determined—or they would form only what their fate suggests. There are problems with endodermal explants, as the cells tend to fall apart, but indirect evidence suggests that the presumptive endoderm of a gastrula is probably fully determined.

The ability of tissues to respond—their competence—can be tested in other ways. When small pieces of an early gastrula are transplanted to various parts of the body of an older embryo, such as a neurula, one discovers another important property of the reacting tissue. If pieces of presumptive ectoderm are transplanted, they are found to participate in the formation of whatever structure is present in the region where they are placed (figure 97). If transplanted to the heart region, heart tissue is formed; to the liver region, liver; to the kidney region, kidneys; to the brain region, brain. The same is true of the presumptive chorda-mesoderm as well.

The presumptive ectoderm and presumptive chorda-mesoderm, therefore, do not exhibit germ-layer specificity. Seemingly the cells of those regions do not know to which germ layers they belong. Figure 98 summarizes the embryological state of the parts of an early gastrula.

Genetic specificity is another limitation of the reacting tissue. The dorsal lip transplantation experiments of Spemann and Mangold involved embryos of *Triton taeniatus*, which are pigmented, and *Triton cristatus*, which are pale. This pigmentation difference can even be detected in histological preparations. Thus when a *cristatus* dorsal lip was transplanted to *taeniatus*, it was possible to say that the *taeniatus* host ectoderm formed the neural tube.

But which kind of neural tube? Was it a *taeniatus* neural tube or a *cristatus* neural tube? That is, does the structure of the induced neural tube conform to the host or the donor species? That question cannot be answered, since the neural tubes of the two species are identical in shape and overall appearance. What is required is a system where the induced structure is recognizably different in host and donor embryos.

Again nature supplied the material. The mouth regions of frog and salamander larvae differ greatly. The frog larval mouth is bordered by black, horny jaws and rows of tiny teeth (these are formed by the ectoderm and are not homologous with the true jaws and teeth). The salamander larva lacks both ectodermal jaws and teeth; its mouth is just a hole in the head.

It is possible to interchange the ectoderm of the region where the mouth will form in frog and salamander embryos. Thus we have the prospect of answering the question: "If a mouth region is induced, will it be characteristic of the host or of the donor species?" The results of such experiments are clear-cut. The frog ectoderm on the salamander embryo, though induced by the salamander mouth-region organizer, forms a frog mouth, with horny jaws and teeth. In the reciprocal experiment the salamander ectoderm on a frog host produces a salamander mouth.

Other experiments of this sort have been tried and a general rule emerges: the tissue responds in accordance with its specific genetic constitution. Competent tissues can react to organizers, but they must

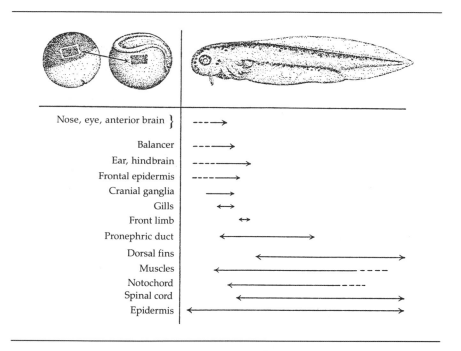

97 Holtfreter's experiments (1933) testing the competence of the presumptive ectoderm of an early gastrula. Gastrula tissue of a salamander was transplanted to various sites in an older embryo, where it formed structures appropriate to the location in the host. The lines with arrows show the structures, listed at the left, that were induced in the donor tissue. Thus a line drawn directly down from the two spots just anterior to the larva's gills shows that the transplanted ectoderm can form olfactory organs, eyes, forebrains, midbrains, balancers, ears, hindbrains, frontal epidermis, neural crest, gills, pronephric ducts, muscles, epidermis, and connective tissue when placed at that site on the host.

do so their own way. One is left with the impression that organizers are general stimuli and that the end result of their action is modulated by the genetic limitations of the reacting tissue. In normal embryos, of course, there is no problem—both the organizing tissue and the reacting tissue are from the same individual and hence have the same genes. It

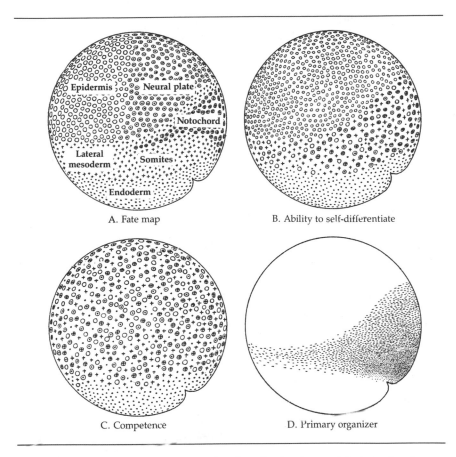

A. Fate map

B. Ability to self-differentiate

C. Competence

D. Primary organizer

98 Holtfreter's (1936) experiments showing the developmental state of early gastrula cells. A shows the fate, that is, the presumptive regions (compare with figure 79); the symbols repeat in B and C. B shows the ability of explanted bits of tissue from various parts of the gastrula to self-differentiate when explanted. C shows the capacity, or competence, of cells to form structures when transplanted to various regions of older embryos. D shows the distribution and relative potency of the primary organizer.

is only under experimental conditions that unite inducing and reacting tissues of different genetic types that we uncover this principle of the limitation of the reacting tissue's response.

This extensive and detailed work on the experimental embryology of amphibians made important contributions to our understanding of recapitulation. For example, one can easily understand why that "useless" structure, the notochord, forms in the amphibian larva only to be replaced by the vertebral column in the adult. The experimental evidence points to the vital role of the notochordal area in the induction of the central nervous system. Other experiments have shown that it plays the same role in other vertebrates. The notochord, therefore, although of transitory importance as a skeletal element, is part of the basic organization of the vertebrate embryo. It is present in all vertebrates because it is necessary if the embryo is to get past the gastrula stage.

There is a similar group of experiments for the pronephros, which is also recapitulated in all vertebrate embryos. It is functional in the amphibian larva but is replaced by the mesonephros, which is the functional kidney of the tadpole and adult frog. Chick embryos start with a pronephros, but it is never functional. Their functional embryonic kidney is the mesonephros; and their adult kidney is the metanephros. Why bother with that useless pronephros? It turns out it isn't useless at all—when the pronephric duct is cut in either amphibian or chick, the mesonephros fails to develop. Like the notochord, the pronephros plays a vital, though brief, role in embryonic development.

The discovery of the inductive role of some recapitulated structures allows us to reevaluate the concept that was so puzzling, and so important, to nineteenth-century embryologists and morphologists. They made two fundamental errors: first, it was assumed that structures such as the notochord and the pronephros are "useless" in more complex vertebrates, and second, that development and evolution were so demanding that inefficiencies such as the production of "useless" structures would be rapidly eliminated by natural selection.

Our final conclusion is that some structures *are* recapitulated, and for good reasons. The notochordal area might not be a necessary part of the skeletal system, but it is necessary as an organizer. This is largely the recapitulation as envisioned by von Baer—the sharing of a common pattern of development by the diverse organisms of a natural group. That being the case, it is inevitable that to some degree ontogeny gives the appearance of recapitulating phylogeny but, to an even greater degree, ontogeny recapitulates ontogeny.

The Chemical Nature of the Organizer

Clearly the dorsal lip organizer is of great importance in development, so not surprisingly there was great eagerness to learn its chemical nature. Questions of this sort were asked in the 1930s when endocrinologists were discovering more and more hormones and were able to purify some of them. Could the organizer be a hormonelike substance? One could hypothesize that the roof of the archenteron might secrete such a hormonelike substance that caused the overlying ectoderm to form a nerve tube.

The organizer was found to be widely distributed. The structures in other vertebrates—fish, reptiles, birds, and mammals—equivalent to the dorsal lip of amphibians act as organizers when tested on amphibian embryos. That is, they induce a secondary embryo. This was exciting, but an even more exciting discovery was that pieces of the dorsal lip, or of the archenteron roof, could be killed by heat or chemical means and still induce undetermined ectoderm to form neural tissue. This was so important because it implied that the organizer was a stable chemical substance, and that meant it might be extracted and purified.

But soon things started to get out of hand, or at least out of theory. Not only would dead dorsal lips induce but so would dead tissue from any part of an amphibian gastrula. Earlier experiments to determine the extent of tissue that could serve as an organizer had shown that such ability is restricted largely to the presumptive notochordal region and the presumptive endoderm above the dorsal lip in living embryos (figure 98). There was no organizing ability in living presumptive ectoderm, but when killed there was.

It was also discovered that tissues of many invertebrates, none of which possess a notochord or dorsal nerve tube, would also induce when killed. What was equally baffling was the finding that dead adult tissues, such as kidney or liver, could induce. And the list became ever more bizarre: silica, kaolin, methylene blue, steroids, egg albumin, and polycyclic hydrocarbons were all found to have inductive power.

Some investigators suggested that these substances are not really organizers but are acting as toxic substances that somehow stimulate amphibian embryonic cells to form neural tissue. Although this is not a satisfying explanation, there was no other.

There is no question that some tissues having no obvious relation to archenteron roofs are potent organizers. The liver of adult mice or guinea pigs, especially if treated with alcohol, can induce head structures in amphibian embryos. Guinea pig kidney, on the other hand, is a potent inducer of trunk structures.

The problem seems insoluble. There is simply no way to identify specifically the substance in the archenteron roof that causes the overlying ectoderm to form a neural tube if such a wide variety of other substances have the same effect. If one is searching for a substance, there must be some way of identifying it. The original test was the ability of the archenteron roof to induce neural tissue in competent ectoderm. But since essentially any tissue when killed will induce, one is left with no way of screening for the *real* organizer substance. Half a century later the search still goes on, aided by the powerful techniques of molecular biology.

Putting It All Together

The research done during the nineteenth and twentieth centuries permits us to understand development as a consequence of the interactions of genes and cytoplasm. This is a theory of development at the level of cells and tissues that, today, is being rapidly extended to the molecular level. These are some of the main steps in developing this theory.

Sexual reproduction by multicellular organisms results in the formation of a single-celled zygote. Complex mechanisms are required to convert that zygote into the multicellular adult with its diversity of differentiated cell types. The principle events in development are an *increase in the number of cells*, the *rearrangement of cells*, and, finally their *differentiation* and association as tissues and organs.

Mitotic cell division is the universal way that individuals increase the number of their cells. The embryo's cells become rearranged, usually drastically, and in their final positions become the primordia of the future structures. But the prime problem of developmental biology is differentiation, and we can think of it in terms of these four axioms.

(1) Cells are the biological units of structure and function.
(2) Genes control the cellular syntheses and thereby cell structure and function.
(3) The gene-controlled cytoplasms exert feedback control of gene activity.
(4) Individual organisms are integrated systems that have overall control of their separate parts.

Axioms 1 and 2 are so well established that no more need be said about them. Axioms 3 and 4 require explanation so far as their relations to development are concerned.

Axiom 3 maintains that the mature ovum is highly structured and has localized cortical and cytoplasmic determinants that largely control early development. This statement is based on observations and experiments on many different species, but it was largely ignored during the first half of the twentieth century when genes were thought to have almost independent control over the events in cells. Yet there was evidence that what genes do is influenced by the cytoplasm in which they function.

A more balanced view regards genes and cytoplasm as interacting entities. The specificity of cells, organs, individuals, and species depends ultimately on the information encoded in their DNA. But the products of gene action—cellular substances and activities—have feedback control of the genes themselves. This cytoplasmic control is of enormous importance in early development.

Evidence for the importance of the cytoplasm accumulated in the late nineteenth century. A case that greatly influenced contemporary thought was Boveri's discovery that the chromosomes of the round-worm *Ascaris* destined for the cells of the somatic tissues differ greatly from those destined for the gametes. The difference is caused by substances in the cytoplasm. The nuclei that become incorporated in the cytoplasm of one small part of the early embryo, and they alone, will differentiate as germ cells.

Another similar example: a recognizably different cytoplasm, called the pole plasm, is present in a specific portion of the eggs of some insects. Nuclei that enter this zone are incorporated into cells that become gametes. If nuclei are prevented from entering the pole plasm, the embryos develop into adults that produce no gametes. It is possible to manipulate the nuclei and establish the fact that any nucleus forced to enter the pole plasm will become part of a gamete.

A more recent and dramatic example was provided by Gurdon and Brown (1965). They studied the production of ribosomal RNA in the early development of frog embryos. Essentially none is produced before gastrulation, but thereafter the rate of synthesis increases rapidly. One can interpret this to mean that the rRNA genes are "turned off" before gastrulation and "turned on" thereafter. Using the techniques of nuclear transfer, Gurdon and Brown removed an rRNA synthesizing nucleus from a neurula and injected it into an enucleated uncleaved ovum.

Development began, and the question was, "Will the nucleus continue to synthesize rRNA or will its genes be turned off and synthesize none?"

One group of experimental embryos was allowed to develop to blastulae and then the amount of their rRNA measured. None had been synthesized. Another group of embryos was allowed to develop to the neurula stage and then their rRNA production measured. These had resumed rRNA production.

Thus the neurula nuclei had, when returned to the cytoplasm of an early embryo, behaved as a nucleus of a normal early embryo. Then at the normal time—the cytoplasm's normal time—the transplanted nucleus began to produce rRNA. We could say that those turned-on rRNA genes of the neurula were turned off by the cytoplasm of the early embryo and turned on again at the normal time.

In mature ova and early embryonic cells the molecules responsible for the basic organization are situated mainly in the cortex. When we recall that early development is striking in the constancy of its events, it is not surprising that organization is built into the relatively stable cortex compared with the more fluid cytoplasm. There is evidence of some organization in the more fluid cytoplasm as well. Wilson's observations on the determinants of the apical tuft of *Dentalium* suggest most strongly that the determinants are located first near the vegetal pole and then, after a few cleavages, in cells near the animal pole. It would be hard for such shifts to occur in the cortex.

Most of the data, however, indicate that the determinants are to be found in the cortex. Those strikingly different pigmented areas of the cortex of *Crepidula* and other ova are so closely associated with the formation of specific embryonic structures that one suspects that they are at least markers, and may be the determinants in some cases.

The surprising results obtained by centrifuging eggs pointed to the importance of the cortex. Fertilized eggs can be centrifuged until the cytoplasm is divided into layers of materials differing in density—all the yolk granules at the bottom and all the oil drops at the top, for example. Nevertheless such embryos develop normally or almost so. However, when greater centrifugal force is used, enough to disrupt the pattern of the cortex, abnormalities are observed.

Cytoplasmic localization was well known to the grand masters of embryology, but this was not always understood by others. After 1900 the rapid rise of genetics, compared with the measured tread of embryology, left many biologists with the opinion that cells, especially those of embryos, were somewhat leaky bags of assorted molecules awaiting instructions from the genes. New discoveries found the genes doing more and more things, and soon nothing seemed to be left for the cytoplasm. Thus what was clear to E. B. Wilson and others by the early 1890s ceased to be part of a general theory of development.

But now the question becomes, "What is the cause of cytoplasmic differentiations?" The evidence suggests that the organization of the ovum is largely determined by stimuli from without. It appears that the mature ova of all animal species are organized to a considerable degree by the time of ovulation. This basic organization is established under the influence of maternal genes. While the ova are in the ovary they are not isolated. They are cells of the mother's body, formed from preexisting maternal cells, and supplied with the requisites for life. Since ova are cells of the adult female, we should find it no more of a problem to accept that they are organized than to accept that the mother's neurons or kidney cells are highly organized.

If we accept that during oogenesis the ovum becomes differentiated, a difficult technical problem emerges for the developmental biologist— ovarian eggs cannot be manipulated with the same ease as early embryos of amphibians, sea urchins, or ctenophores. Hence clues had to be sought through correlations between the organization of the oocyte and external conditions. Many were found. For example, in some species of marine invertebrates the basic polarity—the animal pole–vegetal pole axis—is determined by the position of the egg in the ovary. Another example comes from insects, many of which have elongate ova, and the long axis often parallels the main axis of the female's body.

The egg that seems to have the least organization at the time of ovulation is that of the marine alga *Fucus*, studied by D. M. Whittaker. However, a protuberance appears on the undivided egg, and at first cleavage the egg divides into two unequal cells. The fate of these two cells is established at this time. The larger cell becomes the thallus and the one with the protuberance becomes the rhizoid. So far as Whittaker could tell, the formation of the protuberance, which determines the pattern of future development, is due to some external influence. He suspected this when he noticed that in groups of cells the protuberance formed inward (figure 99). This suggested that possibly the concentration of some substance produced by the cells was the stimulus. In addition, tests of various environmental components, such as pH, light, and temperature, showed that the site of the protuberance could be controlled at will.

By experimental means, therefore, a fundamental step in differentiation could be controlled. Genes that during the first mitotic division happen to be allocated to the cell with the protuberance will participate in the formation of the rhizoid. Those entering the other cell at the first mitotic division will participate in the formation of the thallus. It is clear, therefore, that what genes do can be affected by the cytoplasm in which they function.

Another classic example of cytoplasmic influence is organizer action. Chemical substances formed by the archenteron roof organizer diffuse to the overlying ectoderm and induce it to form the neural tube, optic cups, divisions of the brain, etc.

One of the first examples of external stimuli affecting the organization of the early embryo relates to the origin of bilaterality. This is the reason that those observations of Newport, Roux, and others on the entrance point of the sperm were so electrifying. Here was a stimulus that not only determined the position of a gray crescent but also marked the

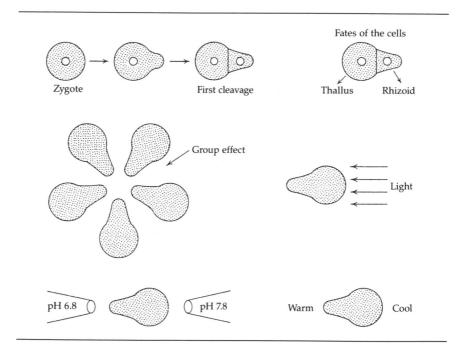

99 *The early development of* Fucus *and some of the factors that influence the formation of the protuberance. Thus external factors can control the primary polarity of the embryo and thus the course of future development.*

plane of first cleavage, the position where the dorsal lip was to appear, and finally the anterior–posterior axis of the embryo and adult.

Thus the evidence suggests strongly that each cell receives a complete set of genes, different genes are expressed in different ways in different embryonic cells, and this expression is controlled in part by cytoplasmic molecules of both cortical and noncortical regions.

The Roux–Weismann hypothesis of qualitative nuclear division was short-lived, and most of the grand masters accepted the hypothesis that all cells receive the same set of genes—and what they did with them in very early development depended largely on the cytoplasm. This hypothesis was hard to prove because one could not at that time study the genetics of somatic cells. Nevertheless the indirect evidence was fairly good. The data of regeneration of planarians and hydroids seemed to indicate that all cells retained the full genetic capability of the species. The cells remaining were able to restore the parts removed.

The isolation of cells of regulative eggs showed that a full set of genes goes to each cell, at least for the first few cleavages. When differences among the chromosomes of somatic cells were recognized, it became possible to trace them throughout successive mitotic divisions and find that all somatic cells have the same set of chromosomes. This individuality of the chromosomes is evidence that all cells are genetically equivalent.

None of these earlier investigations was wholly convincing, and it was not until Briggs and King (1952) and later Gurdon (1962) perfected methods for transferring nuclei from older embryos and from differentiated cells to enucleated ova that better data became available (figure 100). Normal development occurs in varying percentages of the cases, depending on the source of the nuclei.

That even some nuclei from differentiated cells have the ability to support normal development is taken as evidence that any nucleus can do so. However, one cannot conclude from such results that all somatic nuclei are undifferentiated. One can only conclude that they are not irreversibly differentiated. It is beyond argument that the nuclei of differentiated cells are differentiated. The fact that a frog erythrocyte synthesizes hemoglobin but not pepsinogen, and a stomach cell synthesizes pepsinogen but not hemoglobin, shows that different genes are active in the two cell types.

Embryos are integrated systems with the whole having overall control of the parts. There are innumerable examples of the whole embryo or adult organism controlling its parts, and the mechanisms are varied.

Holtfreter's experiments on the transplantation of pieces of early gastrula ectoderm to older embryos and finding that each developed according to its surroundings is a fine example (figure 97). Hörstadius's results with combinations of different cell layers in *Paracentrotus* find their explanation less in the specific layers involved than in the portions of the gradients they contain. Each blastomere of the two-cell stage of *Amphioxus* or *Echinus* has the capacity to produce a whole embryo, but that capacity is restrained in an entire embryo. The fate of an embryonic cell reflects its position in the embryo rather than its innate capacity.

Some of the more dramatic examples of the control of the whole over its parts come from experiments on regeneration. Consider the case of a planarian flatworm cut in half—across the long axis of its body. Each half will undergo an extensive reorganization and produce a complete planarian. The events in regeneration can be explained by assuming the presence of a gradient. The "high" point of the gradient is the head end, and there is a gradual decline in its effect until we reach the tail.

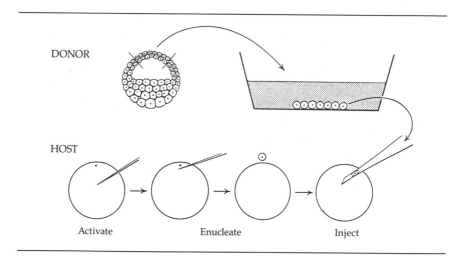

DONOR

HOST

Activate   Enucleate   Inject

100   *The Robert Briggs and Thomas King method of transferring nuclei. A piece of an older donor embryo, in this case the roof of the blastocoel, is removed and placed in a solution that causes the cells to disassociate. A donor egg, fresh from the uterus, is then activated by a jab with a glass needle and its nucleus removed by flicking it out with a glass needle. A single cell donor is drawn into a pipette and then injected into the host embryo. No sperm are involved and the host then develops with the injected donor nucleus.*

When the body is cut crosswise, the anterior half will regenerate a tail at its hind end. The posterior half will regenerate a head at its front end. The cells at the posterior end of the anterior half and those at the anterior end of the posterior half were adjacent before the cut so they should have been as similar to one another as is possible to imagine. Nevertheless their fates in regeneration are entirely different. The conclusion is inescapable that the regenerating whole is controlling the events in its parts.

That last statement reflects a truly extraordinary biological phenomenon. Let us consider some of the implications. A planarian when cut begins to regenerate and stops when its body is complete. What stops this regeneration? Why does it not continue as a cancerous growth forever? Each fragment must have the complete information on "How to make a whole planarian" and also a mechanism to shut off regeneration when that has been done. In the case of planarians, the marvel is not only that the lost part is restored but that each fragment is entirely reformed. That is, the entire structure of each fragment is altered so that at the end of regeneration a perfect, though small, planarian results.

Thus we do have a conceptual framework that will account for the general features of embryonic development. Some of the basic concepts cannot be rigorously defined—the control of the whole over the parts, for example—but there is no doubt whatsoever that such control exists. The framework allows us to comprehend what has been discovered and serves as a basis for further analysis at the level of cell and organism. It can, as well, serve to extend the analysis to the molecular level.

Conclusion

As the twentieth century stands down, we have a depth of understanding of that most distinctive and puzzling feature of life—its ability to self-replicate—that is satisfying to almost everyone, scientist and nonscientist alike. The answer to that puzzle is fairly simple: the basic units of life are cells, and self-replication involves the production of new cells. We have general answers not only for how cells can replicate but also how cells differentiate in the development of individuals and how the basis of evolution is built into the mechanisms for cell reproduction through errors in the replication of DNA.

Is the science of biology finished? Hardly. The concepts that inform our understanding, now mainly at the level of populations, whole organisms, and cells—holistic biology, if you will—are being revolutionized by discoveries at the molecular level. Events within cells, once a veritable black box, are now being explained in terms of specific chemical reactions. The contributions of molecular biology to evolution are to be found in the ability to make detailed comparisons of the genetic makeup of individuals of different species and, therefore, to better understand relationships and phylogenies. It is even proving possible to extract DNA from the fossil remains of organisms that lived many millions of years ago and compare it with the DNA of living creatures. And for genetics itself, molecular probes are providing details about the architecture of genes and increasingly exact data about how the information built into the structure of genes is used to produce the substances, and to control the activities, within cells. This knowledge is providing a far better understanding of development. Now it is possible to specify when genes are turned on and off and thus to relate gene products to developmental events.

Some look upon this new knowledge from molecular biology with dread because soon we will be able to manipulate life in ways never before possible. Already genes can be transferred from one individual to another, and this raises profound ethical problems that many might prefer not to consider. But knowledge in itself cannot be bad: it has the potential for being either threat or opportunity. Which it becomes depends on who uses it and for what purpose. In short, human beings remain in control.

The answers to the fundamental questions of biology are so extraordinary that it is of interest to consider the methods by which they were obtained. In the last century these methods have given us a world dominated by science and its operational companion, technology. It is important, therefore, to inquire why science has become so powerful and pervasive in our lives. What is unique about science as a way of knowing compared with other ways of knowing?

Science is both knowledge of the natural world expressed in naturalistic terms and the procedures for obtaining that knowledge. The scientific way of knowing is not the only way of knowing and, for most of human history, it has not been the dominant mode. What it is can be best understood by first describing what it is not. From the earliest of times the dominant mode has been to "explain" natural things and processes as consequences of supernatural forces: sickness is due to divine wrath; all animals and plants were created by a deity; drought and sickness can be ended by prayer; lightning is a thunderbolt of Zeus; the sun is Apollo and his fiery chariot. That is, the events of the natural world are controlled by a putative supernatural world that is formless, unknowable, immune from discovery, and, in the view of many, nonexistent. The relationships of the natural and supernatural worlds need be neither constant nor predictable. The gods may have their own reasons and behave as they wish. Thus, if the cause of a natural event is the whim of a deity, the event is neither predictable nor fully understandable.

Science deals not with the gods above but with the worlds below. It does not refute the gods; it merely ignores them in its explanations of the natural world. The basic assumption of the sciences is that nature is, in principle, knowable and its phenomena are assumed to have constant cause–effect relationships. If oxygen and hydrogen unite today to form water, it is assumed that they will tomorrow. In discovering these cause–effect relationships, scientists strive to admit only those data obtained through observation and experimentation.

Explanations in science consist of relating natural phenomena one to another. A mountain is more than a pile of dirt. It is a consequence of an uplifting of the earth's crust and its wearing down by erosion. It may be composed of rocks formed from molten lava or sedimentary rocks formed at the bottom of some ancient sea. Usually it will have a thin skin of soil, formed by life itself, that supports many species of plants and animals. If the mountain is high, the climate will vary with the altitude and this will be reflected in differences in the species present.

To take another example, chromosomes became of interest when their behavior in cell division and their relationship to inheritance were discovered and established beyond all reasonable doubt. That phrase "beyond all reasonable doubt" is the goal of any scientific statement. It suggests a tentativeness because experience has shown that the science of today will be replaced by a better science tomorrow. Darwin's understanding of inheritance was full of doubt, Mendel's was much improved, and Sutton's was much better than that of any of his predecessors. Nevertheless, it was far less adequate than is ours today.

Science is a way of knowing by accumulating data from observations and experiments, seeking relationships of the data with other natural phenomena, and excluding supernatural explanations and personal wishes. It has proved a powerful procedure for understanding nature.

But why did it take so long to discover that cell replication is the basis of all self-replication? The answer to that question is that the scientific problems that can be solved depend on the state of society at the time. There was no way that Aristotle, in the fourth century BC, could have answered these questions. The answer was hidden in an invisible world not to be revealed until microscopes were invented in the late sixteenth century AD. But of even greater importance, how would Aristotle have begun an investigation? He could not have known how to ask a question—pose a hypothesis—that could be tested.

In fact, the clues that eventually led to an answer seemed totally unrelated to inheritance. The discovery and early studies of cells were not theory-driven—they were just isolated descriptions of aspects of the natural world. Mendel's experiments, on the other hand, *were* theory driven, and they sought to provide a clearer picture of what happens in inheritance. He did not seek the physical basis of inheritance, but Mendel's laws made predictions for what the physical basis must be. It remained for Sutton to show that the behavior of chromosomes met those demands.

So science can tell us much about the natural world. Its methods are simple. They require disciplined minds capable of accurately recording observations, using the data from those observations to develop a tentative explanation (a hypothesis), testing the necessary deductions from that hypothesis, and relating the conclusions that are true beyond all reasonable doubt to the existing body of scientific information. Not only does the testing of hypotheses make science a self-correcting enterprise but so does the practice of one scientist testing the conclusions of another. The result is that science is the most powerful mechanism we have for obtaining confirmable information about the natural world.

But powerful as it is, we must never forget that decisions affecting human beings must be made by human beings; they do not emerge from science. Nevertheless, when human beings select some goal, not infrequently the data and procedures of science can help to achieve that goal. Alternatively, science might suggest that the goal is not obtainable—human beings are part of nature, after all, so human life will always be constrained by the basic laws of nature, which the gods cannot annul.

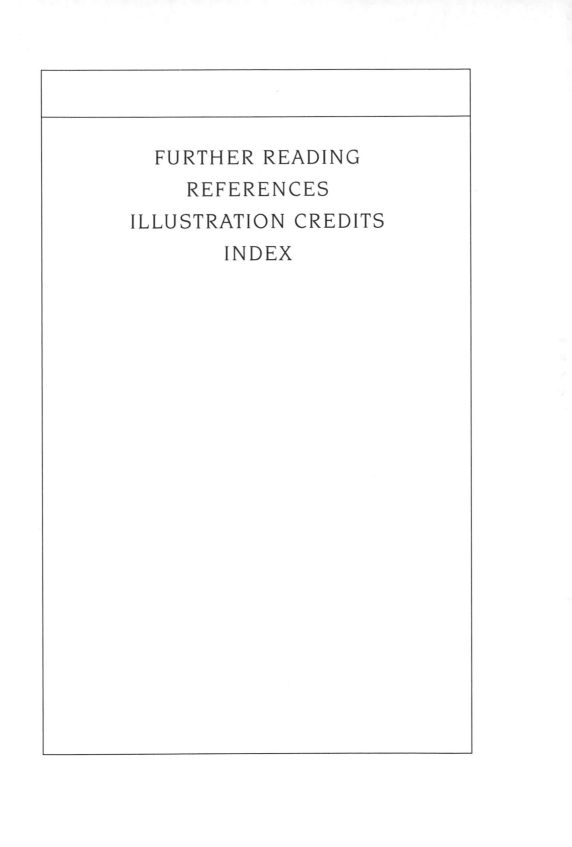

FURTHER READING
REFERENCES
ILLUSTRATION CREDITS
INDEX

Further Reading

Much of the material in this book first appeared in the *American Zoologist*. For those who wish to consult the original versions, which are longer and provide many more references, the citations are: "Understanding Nature," *American Zoologist* 28 (1988): 449–584; "Evolutionary Biology," *American Zoologist* 24 (1984): 467–534 and 30 (1990): 723–858; "Genetics," *American Zoologist* 26 (1986): 583–747; and "Developmental Biology," *American Zoologist* 27 (1987): 415–573.

Short and authoritative biographies of nearly all of the historically important scientists discussed are to be found in the multivolume *Dictionary of Scientific Biography* (New York: Scribners).

A few general references now follow for those who wish more information about the subjects that have been covered.

PART ONE: UNDERSTANDING NATURE

Antecedents of Scientific Thought

Animism, Totemism, Shaminism: Campbell 1988; Eliade 1987; Frazer 1926, 1934; Redfield 1953.

The Paleolithic View: Bataille 1955; Breuil 1952; Campbell 1988; Gowlett 1984; Highwater 1981; Lévi-Strauss 1966; Pfeiffer 1982, 1985; Ruspoli 1987; Wenke 1990.

Mesopotamia and Egypt: Breasted 1909; Burns et al. 1986; Frankfort 1977; Hawkes 1973; Hawkes and Woolley 1963; Kramer 1959, 1972; Kramer et al. 1972; McNeill 1963.

The Greek View of Nature

Downey 1962; Edelstein 1967; Farrington 1949; Mayr 1982; Pellegrin 1986; Preus 1975; Sarton 1952, 1954; Siegerist 1951, 1961; Singer 1950.

The Judeo-Christian Worldview

Artz 1965; Crombie 1959; Haskins 1927; Ley 1968; McMullen 1970; B. Russell 1945; H. O. Taylor 1951; A. D. White 1898; T. H. White 1954.

The Revival of Science

Bradbury 1968; Cohen 1985; Deyrup-Olsen 1988; Dobell 1960; 'Espinasse 1962; Graubard 1964; A. R. Hall 1983; Lindberg and Westman 1990: McMurrich 1930; O'Malley 1964; B. Russell 1945; A. E. Taylor 1927.

Geology before Darwin

Cole 1975; Gillispie 1959; J. Greene 1959; Mayr 1982; Rudwick 1985.

PART TWO: THE GROWTH OF EVOLUTIONARY THOUGHT

General References to Nineteenth-Century Evolutionary Thought

Dawkins 1986; Eisley 1958; Ghiselin 1969; J. Greene 1959, Irvine 1955; Mayr 1982, 1991; Romanes 1892–1897; Tax 1960.

General References to Twentieth-Century Evolutionary Thought

Dawkins 1986; Dobzhansky 1970; Dobzhansky et al. 1977; Futuyma 1986; J. Huxley 1942; Mayr 1963, 1982; Simpson 1949, 1953; Stebbins 1950; Wright 1968–1978.

PART THREE: CLASSICAL GENETICS

Genetics before Morgan

J. R. Baker 1988; Bateson 1902; Bennett 1965; Dunn 1951, 1965; Gabriel and Fogel 1955; Hughes 1959; Iltis 1932; Lock 1906; Mayr 1982; Moore 1972a, 1972b; Olby 1966; Stern and Sherwood 1966; Stubbe 1972; Sturtevant 1965; Zirkle 1935.

Classical Genetics

Carlson 1967; Dunn 1951, 1965; Mayr 1982; Moore 1972a, 1972b; Morgan 1919, 1922, 1926; Stubbe 1972; Sturtevant 1965; Sturtevant and Beadle 1939; Wilson 1914.

DNA and Molecular Genetics

Alberts et al. 1989; Judson 1979; Lewin 1990; Schrödinger 1945; Watson 1987.

PART FOUR: THE ENIGMA OF DEVELOPMENT

Historical and Descriptive Embryology

Adelmann 1942, 1946; Cole 1930; Farley 1977, 1982; Gasking 1967; Kumé and Dan 1988; Needham 1959; Oppenheimer 1967; Roe 1981; Willier and Oppenheimer 1974.

Experimental Embryology

Alberts et al. 1989; Berrill 1971; Child 1941; Davidson 1986; DePomerai 1985; Horder et al. 1986; J. Huxley and de Beer 1934; Needham 1942; Slack 1983; Spemann 1938; Waddington 1956; Willier, Weiss, and Hamburger 1955.

References

Adelmann, Howard B. 1942. *The Embryological Treatises of Hieronymus Fabricius of Aquapendente.* 2 vols. Ithaca: Cornell University Press.

———— 1966. *Marcello Malpighi and the Evolution of Embryology.* 5 vols. Ithaca: Cornell University Press.

Agassiz, Louis. 1859. *An Essay on Classification.* London: Longmans. Reprinted 1962 by Harvard University Press, Cambridge.

Alberts, Bruce, et al. 1989. *Molecular Biology of the Cell.* 2nd ed. New York: Garland.

Allen, Garland E. 1978. *Thomas Hunt Morgan: The Man and His Science.* Princeton: Princeton University Press.

Alvarez, L. W., et al. 1980. "Extraterrestrial Cause for the Cretaceous-Tertiary Extinctions." *Science* 208:1095–1108.

*Aristotle. [1910]. *The Works of Aristotle.* Vol. 4. *Historia Animalium.* Oxford: Clarendon Press. (Thompson trans.)

———— [1912]. *The Works of Aristotle.* Vol. 5. *De Partibus Animalium. De Motu. De Incessu Animalium. De Generatione Animalium.* Oxford: Clarendon Press. (Ogle, Farquharson, and Platt trans.)

———— [1928]. *The Works of Aristotle.* Vol. 8. *Metaphysica.* Oxford: Clarendon Press. (Ross trans.)

———— [1943]. *Generation of Animals.* Loeb Classical Library. Cambridge: Harvard University Press. (Peck trans.)

———— [1955]. *Organon.* Loeb Classical Library. Cambridge: Harvard University Press. (Cooke and Tredennick trans.)

———— [1955]. *Parts of Animals. Movement of Animals. Progression of Animals.* Loeb Classical Library. Cambridge: Harvard University Press. (Peck and Forster trans.)

———— [1957]. *On the Soul. Parva Naturalia. On Breath.* Loeb Classical Library. Cambridge: Harvard University Press. (Hett trans.)

———— [1965–1970]. *Historia Animalium.* 2 vols. Loeb Classical Library. Cambridge: Harvard University Press. (Peck trans.)

*The citations to Aristotle in the text refer not to pages but to paragraphs or even sentences, a system introduced in Bekker's standard Greek editions.

——— [1972]. *Aristotle's De Partibus Animalium I and De Generatione Animalium I.* Oxford: Clarendon Press. (Balme trans.)

Artz, Frederick B. 1965. *The Mind of the Middle Ages: A.D. 200–1500.* 3rd. ed. New York: Knopf.

Augustine, Saint. [1844]. *De Genesi ad Litteran.* (See J. H. Taylor 1982.)

Avery, O. T., C. M. MacLeod, and M. McCarty. 1944. "Studies on the Nature of the Substance Inducing Transformation of Pneumococcal Types." *Jour. Exper. Medicine* 79:137–158.

Avicenna (=Ibn Sina). [1930]. *A Treatise on the Canon of Medicine of Avicenna, Incorporating a Translation of the First Book by O. Cameron.* London: Luzak.

Bacon, Sir Francis. 1857–1874. *The Works of Francis Bacon.* Edited by James Spedding, R. L. Ellis, and D. D. Heath. 14 vols. London: Longman. Reprinted 1968 by Garrett Press, New York.

——— 1937. *The Great Instauration.* Edited by Gail Kennedy. Garden City: Doubleday, Doran.

——— 1960. *New Organon and Related Writings.* Edited by Fulton Anderson. Indianapolis: Bobbs Merrill.

Bailey, L. H. 1895. *Plant-breeding.* New York: Macmillan.

Baker, H. 1756. "A Supplement to the Account of a Distempered Skin, Published in the 424th Number of the Philosophical Transactions." *Phil. Trans.* 49, part 1:21–24.

Baker, John R. 1988. *The Cell Theory: A Restatement, History, and Critique.* New York: Garland.

Barlow, Nora, ed. 1958. *The Autobiography of Charles Darwin, 1809–1882; With Original Omissions Restored; edited with appendix and notes by his granddaughter.* London: Collins.

Bataille, Georges. 1955. *Prehistoric Painting: Lascaux or the Birth of Art.* Cleveland: World.

Bateson, W. 1900. "Hybridisation and Cross-breeding as a Method of Scientific Investigation." *Jour. Royal Horticultural Soc.* 24:59–66.

——— 1902. *Mendel's Principles of Heredity: A Defence.* Cambridge: Cambridge University Press.

——— 1909. *Mendel's Principles of Heredity.* Cambridge: Cambridge University Press.

——— 1913. (Above with additions.)

Bateson, William, and E. R. Saunders. 1902. *Reports to the Evolution Committee of the Royal Society.* Report I. London: Harrison.

Belon, Pierre. 1555. *L'Histoire de la Nature des Oyseaux.* Paris: Benoist Prevost.

Bennett, J. H., ed. 1965. *Experiments in Plant Hybridization: Mendel's Original Paper in English Translation with Commentary and Assessment by the Late Sir Ronald Fisher Together with a Reprint of W. Bateson's Biographical Notice of Mendel.* Edinburgh: Oliver and Boyd.

Berrill, N. J. 1971. *Developmental Biology.* New York: McGraw-Hill.

Bourgeosis, J., et al. 1988. "A Tsunami Deposit at the Cretaceous-Tertiary Boundary in Texas." *Science* 241:567–570.

Boveri, Th. 1888. "Die Befruchtung und Teilung des Eies von Ascaris megalocephala." *Jenaische Zeit.* 21:423–515.

———— 1902. "Über mehrpolige Mitosen als Mittel zur Analyse des Zellkerns." *Verh. der Phys. Med. Ges. Wurzburg* 35:67–90. English translation in Willier and Oppenheimer 1964.

———— 1907. *Zell-studien VI. Die Entwicklung dispermer Seeigel-eier.* Jena: Fisher.

Boyle, R. 1666. "A Way of Preserving Birds Taken Out of the Egge, and Other Small Faetus's." *Phil. Trans. Royal Soc. London* 1:199–200.

Bradbury, Savile. 1968. *The Microscope Past and Present.* New York: Pergamon.

Breasted, James H. 1909. *A History of Egypt from the Earliest Times to the Persian Conquest.* New York: Scribner's.

Breuil, Henri. 1952. *Four Hundred Centuries of Cave Art.* Montignac, France: Centre d'Études et de Documentation Préhistoriques.

Bridges, C. B. 1916. "Non-disjunction as Proof of the Chromosome Theory of Heredity." *Genetics* 1:1–52, 107–163.

———— 1921. "Triploid Intersexes in Drosophila melanogaster." *Science* 54:252–254.

———— 1939. "Cytological and Genetic Basis of Sex." In E. Allen, ed., *Sex and Internal Secretions.* 2nd ed. Baltimore: Williams and Wilkins.

Briggs, R., and T. J. King. 1952. "Transplantation of Living Nuclei from Blastula Cells into Enucleated Frog's Eggs." *Proc. Nat. Acad. Sci.* 38:455–463.

Brown, R. 1833. "The Organs and Mode of Fecundation in Orchideae and Asclepideae." *Trans. Linnaean Society (of London)* 16:710–713. (Reprinted in Gabriel and Fogel 1955.)

Burns, Edward M., et al. 1986. *World Civilizations.* 2 vols. 7th ed. Norton: New York.

Campbell, Joseph. 1970. *The Masks of God. Primitive Mythology.* New York: Viking.

———— 1988. *Historical Atlas of World Mythology.* Vol. 1. *The Way of Animal Powers.* Part 1. *Primitive Hunters and Gatherers.* Part 2. *Mythologies of the Great Hunt.* New York: Harper & Row.

Carlson, Elof Axel. 1967. *Gene Theory.* Belmont: Dickenson.

Child, Charles M. 1941. *Patterns and Problems of Development.* Chicago: University of Chicago Press.

Cohen, I. Bernard. 1985. *Revolution in Science.* Cambridge: Harvard University Press.

Colbert, E. H. 1980. *Evolution of the Vertebrates.* New York: John Wiley.

Cole, F. J. 1930. *Early Theories of Sexual Generation.* Oxford: Oxford University Press.

———— 1975. *A History of Comparative Anatomy from Aristotle to the Eighteenth Century.* New York: Dover.

Conklin, E. G. 1897. "The Embryology of Crepidula, a Contribution to the Cell-lineage and Early Development of Some Marine Gasteropods." *Jour. Morph.* 13:1–226.

———— 1905. "The Organization and Cell-Lineage of the Ascidian Egg." *Jour. Acad. Nat. Sci. Philadelphia.* 2nd ser. 13:1–119.

Cooke, James. 1762. *A New Theory of Generation, According to the Best and Latest Discoveries in Anatomy, Further Improved and Fully Displayed.* London: Buckland, Dilly, Keith, and Johnson.

Cornford, F. M. 1930. "Anaxagoras' Theory of Matter. I, II." *Classical Quarterly* 24:14–30, 83–95.

Correns, C. 1900. "G. Mendel's Regel über das Verhalten der Nachkommenshaft der Rassenbastarde." *Berichte Deutschen Botanischen Gesellschaft* 18:158–168. (English translation in Stern and Sherwood 1966.)

Creighton, H. B., and B. McClintock. 1931. "A Correlation of Cytological and Genetical Crossing-over in *Zea mays*." *Proc. Nat. Acad. Sci.* 17:485–497.

Crombie, Alistair C. 1959. *Medieval and Early Modern Science*. Vol. 1. *Science in the Middle Ages: V-XIII Centuries*. Vol. 2. *Science in the Later Middle Ages and Early Modern Times: XIII-XVII Centuries*. Garden City: Doubleday Anchor.

Cuvier, Georges. 1831. *Discourse on the Revolution of the Surface of the Globe*. Philadelphia: Carey and Lee.

Darlington, C. D. 1960. "Chromosomes and the Theory of Heredity." *Nature* 187:892–895.

Darwin, Charles. 1859. *On the Origin of Species*. London: John Murray. 2nd ed., 1859. 3rd ed., 1861. 4th ed., 1866. 5th ed., 1869. 6th ed., 1872. All by John Murray, London. Harvard University Press, Cambridge, published a facsimile of the first edition, with an introduction by Ernst Mayr, in 1964.

——— 1868. *The Variation of Animals and Plants under Domestication*. 2 vols. London: John Murray.

——— 1871. "Pangenesis." *Nature* 3:502–503.

Davidson, Eric H. 1986. *Gene Activity in Early Development*. 3rd ed. New York: Academic Press.

Dawkins, Richard. 1986. *The Blind Watchmaker*. New York: Norton.

De Graaf, René. 1672. *De Mulierum Organis Generationi . . .* Leiden.

DePomerai, David. 1985. *From Gene to Animal: An Introduction to the Molecular Biology of Animal Development*. New York: Cambridge University Press.

De Solla Price, Derek. 1975. *Science since Babylon*. New Haven: Yale University Press.

De Vries, H. 1900. "Das Spaltungsgesetz der Bastarde." *Ber. Deutsch. Bot. Gesell.* 18:83–90. (English translation in Stern and Sherwood 1966.)

Deyrup-Olsen, Ingrith. 1988. "Five Examples Illustrating the Study of Form and Function." *Amer. Zool.* 28:585–617.

Dobell, Clifford. 1960. *Antony van Leeuwenhoek and His "Little Animals."* New York: Dover.

Dobzhansky, Th. 1970. *Genetics of the Evolutionary Process*. New York: Columbia University Press.

——— 1973. "Nothing in Biology Makes Sense Except in the Light of Evolution." *American Biology Teacher* 35:125–129.

Dobzhansky, T., F. J. Ayala, G. L. Stebbins, and J. W. Valentine. 1977. *Evolution*. San Francisco: W. H. Freeman.

Downey, Glanville. 1962. *Aristotle: Dean of Early Science*. New York: Franklin Watts.

Driesch, H. 1892. "Entwicklungsmechanische Studien. I. Der Werth der beiden ersten Furchungzellen in der Echinodermenentwicklung." *Zeit. wissenschaft. Zool.* 53:160–178.

Driesch, H., and T. H. Morgan. 1895. "Zur Analyse der ersten Entwicklungsstadien des Ctenophoreneises. I. Von der Entwicklung einzelner Ctenophorenblastomeren. *Arch. Entwicklungsmech. Org.* 2:204–215.

Dunn, Leslie C. 1965. *A Short History of Genetics: The Development of Some of the Main Lines of Thought, 1864–1939.* New York: McGraw-Hill.

———— ed. 1951. *Genetics in the 20th Century: Essays on the Progress of Genetics during Its First 50 Years.* New York: Macmillan.

Edelstein, Ludwig. 1967. *Ancient Medicine.* Baltimore: Johns Hopkins University Press.

Eiseley, Loren. 1958. *Darwin's Century: Evolution and the Men Who Discovered It.* Garden City: Doubleday.

Eliade, Mircea, ed. 1987. *The Encyclopedia of Religion.* 15 vols. New York: Macmillan.

'Espinasse, Margaret. 1962. *Robert Hooke.* Berkeley: University of California Press.

Fabricius. See Adelmann 1942.

Farley, John. 1977. *The Spontaneous Generation Controversy from Descartes to Oparin.* Baltimore: Johns Hopkins University Press.

———— 1982. *Gametes and Spores: Ideas about Sexual Reproduction, 1750–1914.* Baltimore: Johns Hopkins University Press.

Farrington, Benjamin. 1949. *Greek Science.* Vol 1. *Thales to Aristotle.* Vol. 2. *Theophrastus to Galen.* Harmondsworth: Penguin.

Fisher, R. A. 1936. "Has Mendel's Work Been Rediscovered?" *Annals of Science* 1:115–137. (Reprinted in Stern and Sherwood 1966.)

Fleming, D. 1955. "Galen on the Motions of the Blood in the Heart and Lungs." *Isis* 46:14–21.

Flemming, Walther. 1882. *Zellsubstanz, Kern, und Zelltheilung.* Leipzig: Vogel.

Focke, W. O. 1881. *Die Pflanzen-mischlinge.* Berlin: Borntraeger.

Frankfort, Henri, et al. 1977. *The Intellectual Adventure of Ancient Man.* Chicago: University of Chicago Press.

Frazer, Sir James G. [1934]. *Totemism and Exogamy: A Treatise on Certain Early Forms of Superstition and Society.* 4 vols. Reprinted 1968 by Dawsons of Pall Mall, London.

Futuyma, Douglas J. 1986. *Evolutionary Biology.* 2nd ed. Sunderland: Sinauer Associates.

Gabriel, Mordecai L., and Seymour Fogel. 1955. *Great Experiments in Biology.* Englewood Cliffs: Prentice-Hall.

Galen. [1916]. *Galen on the Natural Faculties.* Translated by A. J. Brock. Loeb Classical Library. Cambridge: Harvard University Press. Also in the *Great Books of the Western World* published by Encyclopaedia Britannica, Chicago.

———— [1968]. *On the Usefulness of the Parts of the Body.* 2 vols. Ithaca: Cornell University Press.

Galton, F. 1871*a*. "Experiments in Pangenesis." *Proc. Royal Soc. London (Biol.)* 19:393–404.

——— 1871*b*. "Pangenesis." *Nature* 4:5–6.

Garboe, Axel. 1958. *The Earliest Geological Treatise (1667) by Nicolaus Steno.* New York: St. Martin's Press.

Gasking, Elizabeth B. 1967. *Investigations into Generation 1651–1828.* Baltimore: Johns Hopkins University Press.

Gesner, Konrad. 1551–1587. *Historiae Animalium.* 5 vols. Frankfort: Tigvri.

Ghiselin, Michael T. 1969. *The Triumph of the Darwinian Method.* Berkeley: University of California Press.

Gillispie, Charles C. 1959. *Genesis and Geology: The Impact of Scientific Discoveries upon Religious Beliefs in the Decades before Darwin.* New York: Harper & Row.

Gould, Stephen J. 1977. *Ever since Darwin.* New York: Norton.

Gowlett, John. 1984. *Ascent to Civilization.* New York: Knopf.

Graubard, Mark. 1964. *Circulation and Respiration: The Evolution of an Idea.* New York: Harcourt, Brace & World.

Greene, John C. 1959. *The Death of Adam: Evolution and Its Impact on Western Thought.* Ames: Iowa State University Press.

Gurdon, J. B. 1962. "Adult Frogs Derived from Nuclei of Single Somatic Cells." *Dev. Biol.* 4:256–273.

Gurdon, J. B., and D. D. Brown. 1965. "Cytoplasmic Regulation of RNA Synthesis and Nucleolus Formation in Developing Embryos of *Xenopus laevis.*" *Jour. Mol. Biol.* 12:27–35.

Guthrie, William K. C. 1962. *A History of Greek Philosophy.* Vol. 1. Cambridge: Cambridge University Press.

Haeckel, Ernst. 1866. *Generelle Morphologie der Organismen.* Berlin: Reimer.

——— 1868. *Natürliche Schöpfungsgeschichte.* Berlin: Reimer. For translation see 1876.

——— 1874. *Anthropogenie: Keimes- und Stammesgeschichte des Menschen.* Leipzig: Engelmann. For translation see 1905.

——— 1876. *The History of Creation.* 2 vols. New York: Appleton.

——— 1905. *The Evolution of Man.* New York: Eckler.

Hall, Alfred Rupert. 1983. *The Revolution in Science, 1500–1750.* New York: Longman.

Hamilton, Edith. 1942. *The Greek Way.* New York: Norton.

Hardin, Garrett. 1985. *Filters against Folly.* New York: Viking.

Hartsoeker, Nicolaas. 1694. *Essai de Dioptrique.* Paris: Anisson.

Harvey, William. [1628]. *Exercitatio Anatomica de Motu Cordis et Sanguinis in Animalibus.* Translation by C. D. Leake. 3rd ed. Springfield: Charles C. Thomas, 1941.

——— 1651. *Anatomical Exercises on the Generation of Animals.* Reprinted 1965 in *The Works of William Harvey* by Johnson Reprint Corp., New York.

——— [1982]. *De Generatione Animalium.* Translation by G. Witteridge. Oxford: Blackwell.

Haskins, Charles. 1924. *Studies in Medieval Science.* 2nd ed. Cambridge: Harvard University Press.

Hawkes, Jacquetta. 1973. *The First Great Civilizations: Life in Mesopotamia, the Indus Valley, and Egypt.* New York: Knopf.

Hawkes, Jacquetta, and Sir Leonard Woolley. 1963. *History of Mankind.* Vol. 1. *Prehistory and the Beginnings of Civilization.* New York: Harper & Row.

Henking, H. 1891. "Untersuchungen über die ersten Entwicklungsvorgänge in den Eiern der Insekten." *Zeit. wiss. Zool.* 51:685–736.

Herodotus. [1954]. *The Histories.* Baltimore: Penguin.

Hershey, A. D., and M. Chase. 1952. "Independent Functions of Viral Protein and Nucleic Acid in Growth of Bacteriophage." *Jour. General Physiology* 36:39–56.

Hertwig, O. 1876. "Beiträge zur Kenntniss der Bildung, Befruchtung und Theilung des thierischen Eies." *Morphologisches Jahrbuch* 1:343–434.

Highwater, Jamake. 1981. *The Primal Mind.* New York: Harper & Row.

Hippocrates. [1923–1931]. *Hippocrates.* Translated by W. H. S. Jones. Loeb Classical Library. Cambridge: Harvard University Press.

———— [1950]. *Medical Works. A New Translation from the Original Greek Made Especially for English Readers by the Collaboration of John Chadwick and W. N. Mann.* Oxford: Blackwell.

Holtfreter, J. 1933a. "Der Einfluss von Wirtsalter und verschiedenen Organbezirken auf die Differenzierung von angelagertem Gastrulaektoderm." *Roux' Arch. Entwicklungsmechanik* 127:619–775.

———— 1933b. "Organisierungsstufen nach regionaler Kombination von Entomesoderm mit Ektoderm." *Biol. Zentralblatt* 53:404–431.

———— 1936. "Regionale Induktionen in xenoplastisch Zusammengesetzen Explantaten." *Roux' Arch. Entwicklungsmechanik* 134:466–549.

Hooke, Robert. 1665. *Micrographia: or Some Physiological Descriptions of Minute Bodies Made by Magnifying Glasses with Observations and Inquiries Thereupon.* London: Martyn and Allestry. Reprinted 1961 by Dover, New York.

———— 1674. *Animadversions on the First Part of the Machina Coelestes of . . . Johannes Helvelius.* London: John Martyn.

———— 1705. *The Posthumous Works of Robert Hooke . . .* Ed. Richard Waller. London: Smith and Waller. Reprinted 1969 by Johnson Reprint Corp., New York.

Horder, Timothy J., J. A. Witkowski, and C. C. Wyle, eds. 1986. *A History of Embryology.* Cambridge: Cambridge University Press.

Hörstadius, S. 1939. "The Mechanics of Sea Urchin Development, Studied by Operative Methods." *Biol. Rev.* 14:132–179.

Hughes, Arthur. 1959. *A History of Cytology.* New York: Abelard-Schüman.

Hutton, James. 1788. "Theory of the Earth, Or an Investigation of the Laws Observable in the Composition, Dissolution, and Restoration of Land upon the Globe." (Read March 7 and April 4, 1785.) *Trans. Royal Society Edinburgh* 1:209–304.

Huxley, Julian. 1942. *Evolution: The Modern Synthesis.* New York: Harper Bros.

Huxley, Julian S., and G. R. de Beer. 1934. *The Elements of Experimental Embryology.* Cambridge: Cambridge University Press.

Huxley, Thomas H. 1868. "Remarks upon Archaeopteryx lithographica." *Proc. Roy. Soc. London* 16:243–248.

Iltis, Hugo. 1932. *Life of Mendel.* New York: Norton. Reprinted 1966 by Haffner, New York.

Irvine, W. 1955. *Apes, Angels, and Victorians: The Story of Darwin, Huxley, and Evolution*. New York: McGraw-Hill.

Janssens, F. A. 1909. "La Théorie da la Chiasmatypie." *La Cellule* 25:387–406.

Jenkin, F. 1867. "The Origin of Species." *North British Review* 46:277–318.

Judson, Horace F. 1979. *The Eighth Day of Creation: Makers of the Revolution in Biology*. New York: Simon and Schuster.

Kerr, R. A. 1988. "Was There a Prelude to the Dinosaur's Demise?" *Science* 239:729–730.

Kimura, M. 1983. *The Neutral Theory of Molecular Evolution*. New York: Cambridge University Press.

Kölliker, Rudolf A. von. 1841. *Beiträge zur Kenntniss der Geschlectverhältnisse und der Samenflüssigkeit wirbellose Tiere*. Berlin.

Kottek, S. S. 1981. "Embryology in Talmudic and Midrashic Literature." *Jour. History Biology* 14:299–315.

Kramer, Samuel N. 1959. *History Begins at Sumer*. Garden City: Doubleday.

——— 1972. *Sumerian Mythology*. Philadelphia: University of Pennsylvania Press.

Kramer, Samuel N., et al. 1969. *Cradle of Civilization*. New York: Time-Life Books.

Kuhn, Thomas S. 1970. *The Structure of Scientific Revolutions*. 2nd ed. Chicago: University of Chicago Press.

Kumé, Matazo, and Katsuma Dan. 1988. *Invertebrate Embryology*. New York: Garland.

Leeuwenhoek, Antony van. 1798–1807. *The Selected Works of Antony van Leeuwenhoek Containing His Microscopical Discoveries in Many of the Works of Nature*. London: H. Fry. Reprinted 1977 by Arno Press, New York.

Lévi-Strauss, Claude. 1966. *The Savage Mind*. Chicago: University of Chicago Press.

Lewin, Benjamin. 1990. *Genes IV*. New York: Oxford University Press.

Lurker, Manfred. 1980. *The Gods and Symbols of Ancient Egypt*. New York: Thames and Hudson.

Lyell, C. 1826. "Art. IX.—Transactions of the Geological Society of London. vol. i, 2d. Series. London, 1824." *Quart. Rev.* 34:507–540.

——— 1830–1833. *Principles of Geology: Being an Attempt to Explain the Former Changes of the Earth's Surface by Reference to Causes Now in Operation*. 3 vols. London: Murray.

——— 1854. *A Manual of Elementary Geology: Or the Ancient Changes of the Earth and Its Inhabitants, as Illustrated by Geological Monuments*. 4th ed. New York: Appleton.

MacCurdy, Edward. 1938. *The Notebooks of Leonardo da Vinci*. London: Reynal & Hitchcock.

Malthus, T. R. 1798. *An Essay on the Principle of Population as It Affects the Future Improvement of Society*. London: Johnson.

Mayr, E. 1963. *Animal Species and Evolution*. Cambridge: Harvard University Press.

——— 1982. *The Growth of Biological Thought: Diversity, Evolution, and Inheritance*. Cambridge: Harvard University Press.

———— 1991. *One Long Argument: Charles Darwin and the Genesis of Modern Evolutionary Thought*. Cambridge: Harvard University Press.

McClendon, J. F. 1910. "The Development of Isolated Blastomeres of the Frog's Egg." *Amer. Jour. Anatomy* 10:425–430.

McClung, C. E. 1901. "Notes on the Accessory Chromosome." *Anat. Anz.* 20:220–226.

McKenzie, John L. 1965. *Dictionary of the Bible*. New York: Macmillan.

McMurrich, James P. 1930. *Leonardo da Vinci: The Anatomist (1452–1519)*. Baltimore: Williams and Wilkins.

McNeill, William H. 1986. *The Rise of the West: A History of the Human Community*. 6th ed. Chicago: University of Chicago Press.

Medawar, P. B. 1981. "Review of Gould's *The Panda's Thumb*." *New York Review*, Feb. 19, p. 35.

Mendel, Gregor. 1865. "Versuche über Pflanzen-hybriden." *Verhandlungen des naturforschenden Vereines im Brunn* 4:3–47. (For English translation see Stern and Sherwood 1966; Bateson 1909; Bennett 1965.)

Montgomery, T. H., Jr. 1901. "A Study of the Chromosomes of the Germ-Cells of Metazoa." *Trans. Amer. Philosophical Soc.* 20:154–236.

Moore, John A. 1972a. *Heredity and Development*. 2nd ed. New York: Oxford University Press.

———— 1972b. *Readings in Heredity and Development*. New York: Oxford University Press.

Morgan, Thomas H. 1897. *The Development of the Frog's Egg: An Introduction to Experimental Embryology*. New York: Macmillan.

———— 1909. "What are 'Factors' in Mendelian Explanations?" *Amer. Breeders Assoc.* 5:365–368.

———— 1910. "Sex Limited Inheritance in Drosophila." *Science* 32:120–122.

———— 1911a. "An Attempt to Analyze the Constitution of the Chromosomes on the Basis of Sex-limited Inheritance in Drosophila." *Jour. Exp. Zool.* 11:365–413.

———— 1911b. "Random Segregation Versus Coupling in Mendelian Inheritance." *Science* 34:384.

———— 1919. *The Physical Basis of Heredity*. Philadelphia: Lippincott.

———— 1922. "Croonian Lecture: On the Mechanism of Heredity." *Proc. Royal Soc. B.* 94:162–197. (Reprinted in Moore 1972b.)

———— 1926. *The Theory of the Gene*. New Haven: Yale University Press. Reprinted 1964 by Hafner, New York.

Morgan, T. H., A. H. Sturtevant, H. J. Muller, and C. B. Bridges. 1915. *The Mechanism of Mendelian Heredity*. New York: Holt.

Muller, H. J. 1959. "One Hundred Years without Darwin Are Enough." *School Science & Math.*, April, pp. 304–316.

Munro, Dana C. 1903. *Essays on the Crusades*. Burlington.

Needham, Joseph. 1942. *Biochemistry and Morphogenesis*. London: Cambridge University Press. Revised edition 1966.

———— 1959. *A History of Embryology*. 2nd ed. New York: Abelard-Schuman.

Newell, N. D. 1963. "Crisis in the History of Life." *Sci. Amer.*, Feb., pp. 76–92.

Newport, G. 1854. "Researches on the Impregnation of the Ovum in the Amphibia; And on the Early Stages of the Development of the Embryo. (Third series.)" *Phil. Trans. Royal Society* 144:229–244. (Reprinted in Gabriel and Fogel 1955.)

Nicolson, Marjorie. 1956. *Science and Imagination.* Ithaca: Cornell University Press.

——— 1972. *The Microscope and English Imagination.* Northampton: Smith College.

Nirenberg, M., and J. H. Matthaei. 1961. "The Dependence of Cell-free Protein Synthesis in *E. coli* upon Naturally Occurring or Synthetic Polyribonucleotides." *Proc. Nat. Acad. Sci.* 47:1588–1602.

Olby, Robert C. 1966. *Origins of Mendelism.* New York: Schocken Books.

O'Malley, Charles D. 1964. *Andreas Vesalius of Brussels 1514–1564.* Berkeley: University of California Press.

——— 1956. "On the Genesis of the Ovum of Mammals and of Man by Karl Ernst von Baer." *Isis* 47:117–153. English translation of von Baer 1827.

Oppenheimer, J. M. 1963. "K. E. von Baer's Beginning Insights into Causal-Analytical Relationships." *Devel. Biol.* 7:11–21.

——— 1967. *Essays in the History of Embryology and Biology.* Cambridge: MIT Press.

Owen, R. 1963. "On the Archaeopteryx of von Meyer . . ." *Phil. Trans. Roy. Soc. London.* 153:33–47.

Paley, William. [1802]. *Natural Theology; or, Evidences of the Existence and Attributes of the Deity, Collected from the Appearances of Nature.* My quotations are from the 13th edition of 1811 published by J. Faulder, London.

Pellegrin, Pierre. 1986. *Aristotle's Classification of Animals.* Berkeley: University of California Press.

Pfeiffer, John E. 1982. *The Creative Explosion: An Inquiry into the Origins of Art and Religion.* New York: Harper & Row.

——— 1985. *The Emergence of Humankind.* 4th ed. New York: Harper & Row.

Playfair, John. 1802. *Illustrations of the Huttonian Theory of the Earth.* Edinburgh: William Creech.

Pliny the Elder. [1938–1962]. *Natural History.* 10 vols. Translated by H. Rackman et al. Loeb Classical Library. Cambridge: Harvard University Press. Vols. 2 and 3 discuss animals and vols. 4–7 plants.

Preus, Anthony. 1975. *Science and Philosophy in Aristotle's Biological Works.* New York: Georg Olms.

Punnett, R. C. 1928. "Ovists or Animalculists." *Amer. Nat.* 62:481-507.

Raup, D. M. 1979. "Size of the Permo-Triassic Bottleneck and Its Evolutionary Implications." *Science* 206:217–218.

Ray, John. 1691. *The Wisdom of God Manifested in the Works of Creation.* London: S. Smith.

Redfield, Robert. 1953. *The Primitive World and Its Transformations.* Ithaca: Cornell University Press.

Roe, S. A. 1981. *Matter, Life and Generation: 18th Century Embryology and the Haller-Wolff Debate.* Cambridge: Cambridge University Press.

Romanes, G. J. 1892–1897. *Darwin and after Darwin.* 3 vols. Chicago: Open City Publishing Co.

Roux, W. 1888. "Beiträge zur Entwicklungsmechanik des Embryo . . ." *Virchows Arch.* 114:113–153, 289–291. (Translated in Willier and Oppenheimer 1964.)

Rudwick, Martin J. S. 1985. *The Meaning of Fossils: Episodes in the History of Palaeontology.* 2nd ed. Chicago: University of Chicago Press.

Ruspoli, Mario. 1987. *The Cave of Lascaux: The Final Photographs.* New York: Abrams.

Russell, Bertrand. 1945. *A History of Western Philosophy.* New York: Simon and Schuster.

Sarton, George. 1952. *A History of Science: Ancient Science through the Golden Age of Greece.* Cambridge: Harvard University Press.

———— 1954. *Galen of Pergamon.* Lawrence: University of Kansas Press.

Schleiden, Matthias J. 1838. "Beiträge zur Phytogenesis." *Muller's Arch.,* pp. 137–176. (Translated in "Sources of Science" series no. 40. Johnson Reprint Corp, New York.)

Schneider, A. 1873. "Untersuchungen über Platyhelminthen." *Oberhessischen Gesellschaft für Natur- und Heilkunden* 14:69–140.

Schrödinger, Erwin. 1945. *What Is Life? The Physical Aspects of the Living Cell.* New York: Macmillan.

Schwann, Theodor. 1839. *Mikroskopische Untersuchungen über die Uebereinstimmung in der Struktur und dem Wachsthum der Thiere und Pflanzen.* Berlin: Sander'schen Buchhandlung (G. E. Reimer).

Shadwell, Thomas. [1676]. *The Virtuoso.* Reprinted 1966 by University of Nebraska Press, Lincoln.

Simpson, G. G. 1949. *The Meaning of Evolution.* New Haven: Yale University Press. Revised 1967.

———— 1953. *The Major Features of Evolution.* New York: Columbia University Press.

Singer, Charles. 1950. *A History of Biology.* Revised ed. New York: Schuman.

Sinnott, Edmund, and L. C. Dunn. 1925. *Principles of Genetics.* New York: McGraw-Hill.

Slack, J. M. W. 1983. *From Egg to Embryo: Determinate Events in Early Development.* New York: Cambridge University Press.

Spallanzani, Lazzaro. 1784. *Dissertations Relative to the Natural History of Animals and Vegetables.* London: Murray.

Spemann, Hans. 1938. *Embryonic Development and Induction.* New Haven: Yale University Press.

Spemann, H., and H. Mangold. 1924. "Über Induktion von Embryonalanlagen durch Implantation artfremder Organisatoren." *Roux' Arch. Entwicklungsmechanik* 100:599–638.

Stebbins, G. L. 1950. *Variation and Evolution in Plants.* New York: Columbia University Press.

Steno, Nicolaus. [1669]. *The Prodromus of Nicolaus Steno's Dissertation Concerning a Solid Body Enclosed by a Process of Nature.* Reprinted 1968 by Hafner, New York.

Stern, C. 1931. "Zytologisch-genetische Untersuchungen als Beweise für die Morganische Theorie des Faktorenaustauschs." *Biologisches Zentralblatt* 51:547–587.

——— 1957. "The Problem of Complete Y-linkage in Man." *Amer. J. Human Genetics* 9:147–165.

Stern, Curt, and E. R. Sherwood. 1966. *The Origins of Genetics: A Mendel Source Book.* San Francisco: W. H. Freeman.

Stubbe, Hans. 1972. *History of Genetics from Prehistoric Times to the Rediscovery of Mendel's Laws.* Cambridge: MIT Press.

Sturtevant, A. H. 1913. "The Linear Arrangement of Six Sex-linked Factors in Drosophila, as Shown by their Mode of Association." *Jour. Exp. Zool.* 14:43–59.

——— 1965. *A History of Genetics.* New York: Harper & Row.

Sturtevant, Alfred H., and George W. Beadle. 1939. *An Introduction to Genetics.* Philadelphia: Saunders. Reprinted by Dover, New York.

Sutton, W. S. 1902. "On the Morphology of the Chromosome Group in Brachystola magna." *Biol. Bull.* 4:24–39.

——— 1903. "The Chromosomes in Heredity." *Biol. Bull.* 4:231–251.

——— 1917. *Walter Stanborough Sutton. April 5, 1887, November 10, 1916.* Published by his family.

Tax, S., ed. 1960. *Evolution after Darwin.* 3 vols. Chicago: University of Chicago Press.

Taylor, Alfred E. 1927. *Francis Bacon.* New York: Oxford University Press.

Taylor, Henry O. 1951. *The Medieval Mind: A History of the Development of Thought and Emotion in the Middle Ages.* 4th ed. Cambridge: Harvard University Press.

Taylor, John H. 1982. *The Literal Meaning of Genesis: St. Augustine.* New York: Newman Press.

Theophrastus. [1916]. *Theophrastus: Enquiry into Plants.* Translated by Sir Arthur Hort. 2 vols. Loeb Classical Library. Cambridge: Harvard University Press.

Topsell, Edward 1607. *The History of Four-footed Beasts and Serpents and Insects.* 3 vols. Reprinted 1967 by Da Capo Press, New York.

Tylor, Edward B. 1881. *Anthropology: An Introduction to the Study of Man and Civilization.* New York: Appleton.

Tyndall, John. 1863. *Heat Considered as a Mode of Motion.* New York: Appleton.

Vesalius, Andreas. [1543]. *De Humani Corporis Fabrica Libri Septem.* Basileae. Reprinted 1964 by Culture et Civilization, Brussels.

Virchow, Rudolf. 1863. *Cellular Pathology as Based upon Physiological and Pathological Histology.* New York: Dewitt. Reprinted 1971 by Dover, New York.

Vogt, W. 1925. "Gestaltungsanalyse am Amphibienkeim mit örtlicher Vitalfärbung. Vorwort über Wege und Zeile. I." *Roux' Arch. Entwicklungsmechanik* 106:610.

——— 1929. Ibid. II. *Roux' Arch. Entwicklungsmechanik* 120:385–706.

von Baer, Karl E. 1827. *De Ovi Mammalium et Hominis Genesi.* Lipsiae: Leopold Vossi. Reprinted 1966 by Culture et Civilisation, Brussels. English translation by O'Malley 1956.

Vorzimmer, Peter J. 1970. *Charles Darwin: The Years of Controversy.* Philadelphia: Temple University Press.

Waddington, Conrad H. 1956. *Principles of Embryology.* New York: Macmillan.

Watson, James D. 1987. *The Molecular Biology of the Gene.* 4th ed. Menlo Park: Benjamin-Cummings.

Watson, J. D., and F. H. C. Crick. 1953*a*. "Molecular Structure of Nucleic Acid." *Nature* 171:737–738.

——— 1953*b*. "Genetical Implications of the Structure of Deoxyribonucleic Acid." *Nature* 171:964–967.

Weismann, August. 1889. *Essays upon Heredity and Kindred Biological Problems.* Oxford: Clarendon Press.

Weldon, W. F. R. 1902. "Mendel's Laws of Alternative Inheritance." *Biometrika* 1:228–254.

Wenke, Robert J. 1990. *Patterns in Prehistory: Humankind's First Three Million Years.* New York: Oxford University Press.

White, Andrew D. 1898. *A History of the Warfare of Science with Theology in Christendom.* 2 vols. New York: Appleton.

White, T. H. 1954. *The Bestiary: A Book of Beasts Being a Translation from a Latin Bestiary of the Twelfth Century.* New York: Putnam's.

Whitman, C. O. 1878. "Embryology of Clepsine." *Quart. Jour. Micr. Sci.* 18:215–315.

Whittaker, D. M. 1940. "Physical Factors of Growth." *Growth Suppl.* 1940:75–90.

Whittington, H. B. 1985. *The Burgess Shale.* New Haven: Yale University Press.

Wightman, William Persehouse Delisle. 1951. *The Growth of Scientific Ideas.* New Haven: Yale University Press.

Willier, Benjamin H., P. A. Weiss, and V. Hamburger, eds. 1955. *Analysis of Development.* Philadelphia: Saunders.

Willier, Benjamin H., and Jane M. Oppenheimer. 1964. *Foundations of Experimental Embryology.* Englewood Cliffs: Prentice-Hall. Reprinted 1974, with a new introduction, by Hafner Press (Macmillan), New York.

Wilson, E. B. 1892. "The Cell-lineage of Nereis. A Contribution to the Cytogeny of the Annelid Body." *Jour. Morph.* 6:361–480.

——— 1893. "Amphioxus and the Mosaic Theory of Development." *Jour. Morph.* 8:579–639.

——— 1895. *An Atlas of the Fertilization and Karyokinesis of the Ovum.* New York: Columbia University Press

——— 1898. "Considerations on Cell-lineage and Ancestral Reminiscence." *Annals New York Acad. Sci.* 11, no. 1:1–27.

——— 1900. *The Cell in Development and Inheritance.* New York: Macmillan. 1st ed. 1896.

——— 1904. "Experimental Studies on Germinal Localization. I. The Germ-Regions in the Egg of Dentalium." *Jour. Exp. Zool.* 1:1–72.

——— 1911. "Studies on Chromosomes. VII. A Review of the Chromosomes of Nezara; With Some More General Considerations." *Jour. Morp.* 22:71–110.

———— 1914. "Croonian Lecture: The Bearing of Cytological Research on Heredity." *Proc. Roy. Soc. London* B 88:333–352. (Reprinted in Moore 1972*b*.)

———— 1925. *The Cell in Development and Heredity.* 3rd ed. New York: Macmillan. Reprinted with corrections, 1928.

Wolbach, W. S., R. S. Lewis, and E. Anders. 1985. "Cretaceous Extinctions: Evidence for Wildfires and Search for Meteoritic Material." *Science* 230:167–170.

Wolff, Caspar F. 1759. *Theoria Generationis.* Halae ad Salam.

Wright, S. 1968–1978. *Evolution and the Genetics of Populations: A Treatise.* Chicago: University of Chicago Press.

Wythe, J. H. 1880. *The Microscopist: A Manual of Microscopy.* Philadelphia: Lindsay and Blakiston.

Zirkle, Conway. 1935. *The Beginnings of Plant Hybridization.* Philadelphia: University of Pennsylvania Press.

Illustration Credits

Unless otherwise indicated, illustrations were provided by the author.

1, 2. Sketches by Abbé Henri Breuil from Henri Breuil and Hugo Obermaier, *The Cave of Altamira at Santillana del Mar, Spain* (Topografía de Archivos, 1935).

4. From George Smith, *The Chaldean Account of Creation* (1876).

5. Copyright British Museum.

6. From George Perrot and Charles Chipiez, *A History of Art in Ancient Egypt* (1883).

8. From Edward Topsell, *A History of Four-Footed Beasts and Serpents and Insects* (1607; reprinted in 1967 by Da Capo Press, New York).

9. Courtesy Arnold Arboretum, Houghton Deposit, Harvard University.

10–14. Andreas Vesalius, *De Humani Corporis Fabrica* (1543; reprinted in 1950 in *Illustrations from the Works of Andreas Vesalius*, Dover Books, New York).

15. From Robert Hooke, *Micrographia* (1665).

16. From Robert Hooke, *The Posthumous Works of Robert Hooke* (1705; reprinted in 1969 by Johnson Reprint Corp., New York).

17, 18. Photos by Michael A. Murphy.

21. Plaster cast of Berlin specimen in the Humbolt Museum. Photo courtesy The Natural History Museum, London.

22. Drawing by Laszlo Meszoly.

30. J. W. Schopf, "Paleobiology of the Archean," in J. W. Schopf and C. Klein, eds., *The Proterozoic Biosphere: A Multidisciplinary Study* (New York: Cambridge University Press, 1992).

33. From John M. Clarke and Rudolf Ruedemann, *The Eurypterida of New York: New York State Museum Memoir 14* (Albany: New York State Museum, New York State Education Department, 1912).

34. From Charles D. Walcott, *Cambrian Geology and Paleontology: Middle Cambrian Merostomata* (Washington: Smithsonian Institution Miscellaneous Collections, vol. 57, plate 2, nd). Reproduced courtesy Smithsonian Institution.

35. Neg. No. 124605. Courtesy Department of Library Services, American Museum of Natural History.

36. Neg. No. 124625 (photo by Robert E. Logan). Courtesy Department of Library Services, American Museum of Natural History.

37. Neg. No. 322872. Courtesy Department of Library Services, American Museum of Natural History.

38. Neg. No. 322873. Courtesy Department of Library Services, American Museum of Natural History.

39. Neg. No. 319855 (photo by Robert E. Logan). Courtesy Department of Library Services, American Museum of Natural History.

40. The Natural History Museum, London.

41. From Robert Hooke, *Micrographia* (1665).

42. From Theodor Schwann, *Mikroskopische Untersuchungen über die Uebereinstimmung in der Struktur und dem Wachsthum der Thiere und Pflanzen* (1839).

43. From Oskar Hertwig, *Beiträge zur Kenntniss der Bildung, Befruchtung und Theilung des thierischen Eies* (1876).

44. From A. Schneider, "Untersuchungen über Platyhelminthen," *Oberhessischen Gesellschaft für Natur- und Heilkunden* 14(ca. 1873):69–140.

45, 46. From Walther Flemming, *Zellsubstanz, Kern und Zelltheilung* (Leipzig: Vogel, 1882).

47. From T. Boveri, "Die Befruchtung und Teilung des Eies von Ascaris megalocephala," *Jenaische Zeit.* 22(1888):685–882.

51. From Walter S. Sutton, "On the Morphology of the Chromosome Group in Brachystola magna," *Biol. Bull.* 4(1902):24–39.

53. From T. H. Morgan, *The Physical Basis of Heredity* (Philadelphia: Lippincott, 1919).

56. From H. Henking, "Untersuchungen über die ersten Entwicklungsvorgänge in den Eiern der Insekten," *Zeit. Wiss. Zool.* 51(1891):685–736.

68–70. From John A. Moore, *Heredity and Development* and *Principles of Zoology*. Reproduced courtesy Oxford University Press.

71. From Ernst Haeckel, *The Creation of Man* (New York: G. P. Putnam's Sons, 1905).

78. From W. Vogt, "Gestaltungsanalyse am Amphibienkeim mit örtlicher Vitalfärbung. I. Methodik und Wirklungsweise der örtlichen Vitalfärbung mit Agar als Farbträger," *Roux' Arch. Entwicklungsmechanik* 106(1925):542–610.

79. From W. Vogt. "Gestaltungsanalyse am Amphibienkeim mit örtlicher Vitalfärbung. II. Teil: Gastrulation und Mesodermbildung bei Urodelen und Anuren," *Roux' Arch. Entwicklungsmechanik* 120(1925):385–706.

84. From W. Roux, "Beiträge zur Entwicklungsmechanik des Embryo . . . ," *Virchows Arch.* 114(1888):113–153.

86. (Top) From E. B. Wilson, *The Cell in Development and Inheritance* (New York: Macmillan, 1900).(Bottom) From E. B. Wilson, "The Cell-Lineage of Nereis: A Contribution to the Cytogeny of the Annelid Body," *Jour. Morph.* 6(1892):361–480.

88, 89. From E. B. Wilson, "Experimental Studies on Germinal Localization. I. The Germ-Regions in the Egg of Dentalium," *Jour. Exp. Zool.* 1(1904):1–72.

90–92. From Sven Hörstadius, "The Mechanics of Sea Urchin Development, Studied by Operative Methods," *Biol. Rev.* 14(1939):139–179.

94. From Johannes Holtfreter, "Organisierungsstufen nach regionaler Kombination von Entomesoderm mit Ektoderm," *Biol. Zentralblatt* 53(1933):404–431.

96. From Hans Spemann and Hilda Mangold, "Über Induktion von Embryonalanlagen durch Implantation artfremder Organisatoren," *Roux' Arch. Entwicklungsmechanik* 100(1924):599–638.

97. From Johannes Holtfreter, "Der Einfluss von Wirtsalter und verschiedenen Organbezirken auf die Differenzierung von angelagertem Gastrulaektoderm," *Roux' Arch. Entwicklungsmechanik* 53(1933):404–431.

98. From Johannes Holtfreter, "Regionale Induktionen in xenoplastisch Zusammengesetzen Explantaten," *Roux' Arch. Entwicklungsmechanik* 134(1936):466–549.

Index

Staining (cells), 261–264, 266
Steno, N., 107–108
Stensiö, E., 209–210
Stern, C., 343–345
Stevens, N., 324
Strasburger, E., 275
Sturtevant, A. H., 336, 345–348
Subspecies, 160
Sumerians, 11
Sutton, W. S., 304–314
Symbiosis, in cells, 195
Synapsis, 278
Systematics, see Classification

Tatum, E. L., 361–367
Taxonomy, see Classification
Terrestrial life, 203, 213–215,
Tertiary, 219–221
Thales, 30–31
Theophrastus, 43–45
Theory, see Scientific methods
Theory of the Earth, 120–121
Tiamat, 23–24
Tongue stones, 107, 109
Tools, 15, 19–20
Topsell, E., 71–73
Totemism, 14–15
Transforming principle, 368–370
Triassic, 216
Tschermak, E., 300
Tunicates, 183, 416
Tylor, E. B., 13–14

Uniformitarianism, 119–122
Urbanization, 11

Van Beneden, E., 278
Variation, see Diversity
Venus figurines, 19
Vertebrates, 209–230
Vesalius, A., 79–99, 173
Virchow, R., 261
Vogt, W., 434–437
Von Baer, K., 400, 404–407

Walcott, C., 206
Warrawoona Group, 185–186
Watson, J., 371–378
Weissmann, A., 275, 277–279
Weldon, W. F. R., 300–301
Whewell, W., 247
Whitman, C., 457–458
Whittaker, D. M., 495–496
Whittington, H., 200
Wightman, W., 31
Wilson, E. B., 282–284, 302–303, 310–311,
      324–327, 444–445, 455–456, 458–460, 462–
      467, 469, 495
Wöhler, F., 123
Wolff, C. F., 402
Writing, 18, 21–22

Zeiss, C., 263